Progress in Environmental Assessment Policy, and Management Theory and Practice

Progress in Environmental Assessment Policy, and Management Theory and Practice

Thomas B Fischer

University of Liverpool, UK

ICP

Imperial College Press

Published by

Imperial College Press
57 Shelton Street
Covent Garden
London WC2H 9HE

Distributed by

World Scientific Publishing Co. Pte. Ltd.
5 Toh Tuck Link, Singapore 596224
USA office: 27 Warren Street, Suite 401-402, Hackensack, NJ 07601
UK office: 57 Shelton Street, Covent Garden, London WC2H 9HE

Library of Congress Cataloging-in-Publication Data
Names: Fischer, Thomas B., 1965–
Title: Progress in environmental assessment policy, and management theory and practice / Thomas B Fischer, University of Liverpool, UK.
Description: New Jersey : Imperial College Press, 2015.
Identifiers: LCCN 2015026761 | ISBN 9781783268375 (hardcover : alk. paper) | ISBN 1783268379 (hardcover : alk. paper)
Subjects: LCSH: Environmental impact analysis. | Environmental policy. | Environmental management.
Classification: LCC TD194.6 .F574 2015 | DDC 333.71/4--dc23
LC record available at http://lccn.loc.gov/2015026761

British Library Cataloguing-in-Publication Data
A catalogue record for this book is available from the British Library.

In-house Editor: Philly Lim

Typeset by Stallion Press
Email: enquiries@stallionpress.com

CONTENTS

INTRODUCTION

Environmental Assessment (EA, including both environmental impact assessment [EIA] of projects and strategic environmental assessment [SEA] of policies, plans and programmes) is an internationally applied — *ex-ante* — decision support instrument aiming at reducing, and, if possible, avoiding the negative environmental impacts of actions before they occur and enhancing positive environmental and other outcomes. It was first legally required in the USA based on the National Environmental Policy Act of 1969. Literally all countries now have related experiences, either based on legal requirements, or based on provisions by development banks and organisations, including the World Bank, the Asian Development Bank, the Inter-American Development Bank and others (Fischer and Nadeem, 2014).

The *Journal of Environmental Assessment Policy and Management* (*JEAPM*) is one of the three leading international English language journals for Strategic Environmental Assessment (SEA) (Fischer and Onyango, 2012) and EIA (Fischer and Noble, 2015), next to the journals *EIA Review* and *Impact Assessment and Project Appraisal* (*IAPA*). Now in its 16th year of existence (with a total of 341 published documents appearing on the citation database 'scopus' until 2014), it is the youngest of these three, with *EIA Review* having been published for over 41 years (with 1,346 published documents on 'scopus') and *IAPA* for over 33 years (until 1997 as two journals; *Project Appraisal* and *Impact Assessment*; with 738 published documents on 'scopus').

This publication presents a selection of papers from a total of 22 issues of *JEAPM* over a period of five and a half years, from mid-2009 (the year I took over as the editor-in-chief) to 2014. Fourteen of the 145 papers (next to 16 book reviews and 22 editorials) that have been published since then have been picked, mainly based on their key relevance for the EA topic. Seven of the papers are taken from the seven special issues published since mid-2009, that included 'SEA in China', 'SEA as a tool to contribute to high level policy objectives', 'Spatial data and Geographic Information Systems (GIS) as support tools for EA',

'25 years of the European EIA Directive', 'EA in the context of renewable energy deployment', 'SEA in Latin America' and 'Disaster and risk management: the role of EA'.

Overall, since mid-2009 the journal has seen a good mix of papers, focusing on different world regions and topics. About 40 papers focused on practices in Europe, 30 on North America, 18 on Asia, 15 on South America, 10 on Africa and 7 on Australia and New Zealand. One paper also focused on Antarctica. Nearly 50 papers focused on SEA-related aspects and 25 on aspects revolving around EIA. Another 50 papers dealt with other issues and topics, including wider environmental management as well as other tools, such as life-cycle assessment (LCA), cost-benefit analysis (CBA), health impact assessment (HIA) and risk assessment.

Next to the seven papers that represent the topics of the special issues, the other seven papers included in this publication have various foci. These range from the concept of 'policy windows with regards to SEA', from 'EIA theory', 'the role of change agents in SEA', the 'communication of causality in EIA', the 'preparedness for offshore hydrocarbon energy development in the absence of SEA', 'knowledge management and EA', to 'environmental assessment and management related higher education'.

Summary of Contributions

The first contribution is by Kin-Che Lam, Yongqin David Chen and Jing Wu (all Chinese University of Hong Kong) and is taken from the first special issue of any international English language journal to deal specifically with SEA in China (edited jointly by Thomas B. Fischer and Xu He; Nankai University) in 2009. The authors discuss opportunities, issues and challenges of SEA in China. They do so by looking at progress, emerging issues and problems, based on the Chinese and wider international literature, as well as on case studies and other sources. As an important conclusion, the authors suggest that for the time being a plan-EIA approach may be the most appropriate form of SEA in China. Overall, the authors underline the important role SEA may play in "fostering a sustainable and harmonious society".

Next, in their 2010 publication, the Canadian authors Denis Kirchhoff and Dan McCarthy (both University of Waterloo, Ontario), Debbe D. Crandall (Save the Oak Ridges Moraine Coalition), Laura McDowell (Regional Municipality of York, Ontario) and Graham Whitelaw (Queen's University, Ontario) explore the concept of a policy window as a driver of governmental agenda setting. In this context, they explore SEA-type applications at the municipal level in the regional municipality of York, Ontario. Looking at the development of problem, political and policy streams, they describe how a window of opportunity for implementation of SEA-type practices emerge, leading to policy change in the York region. Taking the example of a trunk sewer project in 2002/2003, they suggest that of particular importance for a policy window to open are a "focusing event" and a "resulting crisis", moving citizens and later the municipality to change.

In his 2010 paper, Joe Weston (at the time of publishing, Oxford Brookes University, UK) presents the EA community with some thought provoking — and controversial — suggestions, reflecting on the development of EIA theory. Most importantly, he suggests that "all the theories [have failed] to adequately explain or even justify EIA". Personally, whilst I do agree with some of his observations on EIA theory, I wouldn't fully sign up to the author's overall conclusions that "EIA fails in both, its procedural role of influencing environmental decision making and its substantive role of changing social values" and that "it is perhaps time that we stop searching for theories to defend it and start campaigning for something radically different".

Richard P. Eales and William R. Sheate (both Collingwood Environmental Planning, UK) next discuss whether SEA and Sustainability Appraisal tools in the UK have helped to deliver more sustainable development in the 2011 special issue on 'SEA as a tool to contribute to high level policy objectives' (guest-edited by Elsa João and Anna McLauchlan; both University of Strathclyde, UK). The authors look at three national policy level assessments in the UK that took place in 2008 and 2009. As one poor area of performance they identify an apparent "blind faith in mitigation", where little evidence is provided about whether mitigation measures will be delivered or what will be done to check that they will be successful. They argue that the current performance by the UK Government in implementing the SEA Directive for national level strategic actions is far from exemplary, claiming that the reasons include the trust in mitigation, and also a poor consideration of alternatives and the weak conception of sustainability adopted. The authors believe that the main problem with SEA implementation is the common perception that having to undertake an assessment and comply with the SEA Directive is a hurdle, rather than a useful mechanism for helping to deliver better and more sustainable policy making.

In another paper from 2011, Lone Kørnøv, Ivar Lyhne, Sanne Vammen Larsen and Anne M. Hansen (all Aalborg University, Denmark) report on how change agents in Strategic Environmental Assessment (SEA) research and practice can steer action in support of sustainability. They base their suggestions on three cases, where PhD researchers were working on their projects in close cooperation with an organisation outside the university. These include the company Energinet.dk on national energy infrastructure development in Denmark; the Greenlandic Self Government on planning and assessing an aluminium smelter in Greenland; and the Danish consultancy Rambøll on river basin management plan SEA preparation. The authors conclude that a researcher, who has high autonomy and interdependence, can function as a change agent for more environmentally sustainable decisions.

The next paper from 2012 is by the guest editor of another special issue on 'Spatial Data and Geographic Information Systems (GIS) as Support Tools for Environmental Assessment', Ainhoa González Del Campo (Trinity College Dublin, Ireland). In her contribution, she provides a review of critical considerations for effective spatial data management and GIS implementation in SEA/EIA, discussing current issues affecting spatial dataset management and use. These include: availability, accessibility, scale, completeness and metadata. Similarly, commonly applied GIS methods for impact assessment and public consultation are described and existing constraints to their application examined. Although many of the issues outlined in this paper are apparent and could be anticipated in applied-GIS, further insights are obtained from their contextualisation to environmental assessment research and practice. Subsequently, González Del Campo formulates recommendations for optimising the contribution of spatial data to environmental planning, and establishing future research and practice needs to enhance GIS use in SEA and EIA.

Anastassios Perdicoúlis (University of Trásos-Montes e Alto Douro, Portugal) and John Glasson (Oxford Brookes University, UK) next look at the explanation of causal relationships in EIA in their 2012 paper 'How clearly causality is communicated in EIA'. Based on an evaluation of 10 US and UK environmental statements, using text samples, they find that associated information was either uncertain or entirely absent. They argue that if EIA is indeed a diagnostic tool, this situation needs to be addressed urgently. They go on to suggest that diagrammatic presentation of causality would be particularly useful for clear communication. They conclude that "authors, reviewers, and even decision makers may need to re-consider their commitment to causal thinking, in a community-wide call for exchanging ideas from the various points of view".

In a 2012 special issue dedicated to the 25th anniversary of the implementation of the EIA Directive (85/337/EEC) in the EU (edited by Thomas B. Fischer; University of Liverpool, UK — UL; Hens Runhaar; University of Utrecht, the Netherlands — UU), Jos Arts (University of Groningen, the Netherlands), Urmila Jha-Thakur (UL), Frank Van Laerhoven (UU), Peter P. J. Driessen (UU) and Vincent Onyango (UL), reflect on the extent to which EIA is effectively addressed in terms of contributing to environmental awareness and environmental protection in the Netherlands and the UK. The authors suggest that although in the UK, EIA seems to contribute to a slightly greater extent to the environmental awareness of proponents and competent authorities than in the Netherlands, the overall impact on the environmental performance of projects is very similar. They conclude that EIA still performs well as an instrument for raising environmental awareness, but not for environmental optimisation.

In 2013, Courtney Fidler and Bram F. Noble (both University of Saskatchewan, Canada) reflect on Canada's preparedness for offshore Arctic hydrocarbon energy development in the absence of Regional Strategic Environmental Assessment (R-SEA). Whilst the authors identify a range of implementation challenges regarding, for example, issues of governance and the nature and scope of the alternatives to be considered in assessment, they conclude that, "R-SEA offers a much needed framework to begin addressing stakeholder concerns about future offshore development in the region."

The contribution by Kedar Uttam, Berit Balfors, Charlotta Faith-Ell and Ulla Mörtberg (all KTH Royal Institute of Technology, Stockholm, Sweden) is part of a further special issue on 'Environmental Assessment in the context of renewable energy deployment' in 2013, focusing on the energy efficiency and green procurement in the building sector. Green procurement involves services, products and also energy that meet environmental requirements. Their paper focuses on the interlinkage between environmental impact and life cycle-based assessments. Interview results with municipal and construction sector stakeholders allowed the authors to identify manifold perspectives on how best to further explore this EA-LCA interlinkage and partnerships to strengthen integration.

Also in 2013, Luis E. Sánchez (University of Sao Paulo, Brazil) and Pierre André (École Polytechnique de Montréal, Canada) report on a knowledge management (KM) survey in Quebec's Environmental Assessment Department and Public Hearings Office. The authors argue that, "EIA is a knowledge intensive activity that would benefit from a highly structured approach to KM." The key aim of their paper is to identify the key elements of KM in the two institutions mentioned above. They find that initiatives for managing knowledge had been adopted with a main emphasis on compiling previous findings, opinions and recommendations from enquiries and reports. The authors conclude that whilst KM was successful in retention and transfer of internal knowledge users, there "is as yet an underexplored learning potential in the follow-up activities after project approval".

Thomas B. Fischer and Urmila Jha-Thakur (both University of Liverpool, UK) go on to describe the baseline, trends, challenges and opportunities of environmental assessment and management (EAM) related master level degree programmes in the EU in 2013. Their findings are in parts based on the findings of the European Erasmus Mundus T_{wo}EA-M project (www.twoeam-eu.net). The authors argue that due to the absence of a clear disciplinary home, understanding of EAM-related higher education degree programmes has remained poor. Furthermore, they stress that teaching of EAM is influenced by how it is practised and therefore differs from country to country. Twenty-three of the then 27 EU member states were found to offer related programmes in 106 institutions with significant differences between them in terms of, for example, disciplinary home and focus, length of study and aspects covered, tuition and geographical focus.

In another special issue on 'environmental assessment in Latin America' (edited by Marcelo Montaño and Thomas B Fischer), Marcelo Montaño, Priscila Oppermann, Anne Caroline Malvestio, and Marcelo Pereira Souza (all University of Sao Paulo, Brazil) reflect on the current state of the SEA system in Brazil. They do so by contrasting it with systems from developed and developing countries, including England, Portugal and Spain, on the one hand, and Angola, Mexico, Mozambique and South Africa on the other. They establish a number of important differences between these eight countries, spanning from practical SEA experiences, over legal requirements, guidelines and an overall supportive culture towards SEA to the role of academia. Based on their evaluation, Brazil is currently in between countries with well developed and with developing systems. Whilst there have been some positive experiences, there are also a range of remaining challenges for the effective application of the instrument in the country.

The last contribution to a special issue on 'disaster and risk management: the role of environmental assessment' is provided by Steve Swain (Environment Agency of England and Wales) and Riki Therivel (Levett-Therivel, UK), describing potential environmental impacts of civil emergency plans, discussing implications of their exemption from SEA. In this context, for a number of emergency plans, the authors identify and categorise mitigation measures, determining whether these could cause significant environmental impacts and whether SEA would be useful in potentially helping to avoid them. The authors conclude that there are indeed a range of possible impacts. These "could affect the marine environment, surface and ground water quality, localised habitat, historical or cultural features". Furthermore, they could also "have resource use and wider carbon and energy use implications". Whilst, in principle, SEA could help address those impacts, the authors suggest that further study is needed in order to be able to provide for some more definite recommendations.

References

Fischer, TB and B Noble (2015). Impact Assessment Research — achievements, gaps and future directions. *Journal of Environmental Assessment Policy and Management*, 17(1), 1501001.

Fischer, TB and O Nadeem (2014). Environmental Impact Assessment Course Curriculum for Higher Education Institutions in Pakistan, IUCN, Pakistan. http://cmsdata.iucn.org/downloads/niap_eia_curriculum_for_hei.pdf.

Fischer, TB and V Onyango (2012). SEA related research projects and journal articles: an overview of the past 20 years. *Impact Assessment and Project Appraisal*, 30(4), 253–263.

CHAPTER 1

Strategic Environmental Assessment in China: Opportunities, Issues and Challenges

Kin-Che Lam*, Yongqin David Chen† and Jing Wu‡

Centre of Strategic Environmental Assessment for China
The Chinese University of Hong Kong
Shatin, N.T., Hong Kong, China
**kinchelam@cuhk.edu.hk*
†ydavidchen@cuhk.edu.hk
‡wujing@cuhk.edu.hk

Based on literature reviews, case studies and other information available on the web, this paper examines the evolution and status of strategic environmental assessment (SEA) in China by reviewing the progress that has been made; discussing the issues and problems encountered and exploring the best way forward. It is suggested that the Chinese environmental impact assessment (EIA) system has evolved over the years to cope with the rapidity and scale of development in China and the current emphasis on Plan EIA is probably the most appropriate form of SEA given the political reality and complicated institution setup. Given China's environmental and resource problems, the paper affirms the potential role of SEA in fostering a sustainable and harmonious society and the need to mainstream sustainability considerations in the formulation of national plans and strategies. To circumvent political resistance from line agencies, the paper suggests that SEA can be applied, in various names and forms, to national and provincial socio-economic plans.

Keywords: Strategic environmental assessment; stakeholder analysis; national socio-economic plan.

Introduction

The year 2003 was a significant milestone in environmental assessment in China. The EIA Law which came in force in that year gave SEA a formal status and much impetus. Much has happened in the first five years since the implementation of the

Law: pilot studies were initiated, regulations promulgated, workshops and conferences held and many studies are underway. It is high time to review the issues encountered, experience learned and grounds gained. It is against this backdrop that this paper aims, based on published academic works and other information available on the web, to review the progress that has been made, discuss the issues and problems encountered and explore the best way forward, for a country which is probably more in need of SEA than others.

The experience of China in implementing SEA has attracted considerable attention, both within and outside China, in the last few years probably because the success or otherwise of the Chinese experience has significant implications for other developing countries. Over these years, quite a number of articles both in English and Chinese have been published. In addition to the conferences and workshop organised by academic and professional bodies, the Ministry of Environmental Protection (MEP and its predecessor State Environmental Protection Administration (SEPA)) has also organised two conferences involving international SEA experts in 2005 and 2007 respectively. In January 2008, MEP also organised a high level forum (SEPA, 2008a) involving key academics and practitioners in the field to discuss the future of SEA in China.

Based on the discussion in these published articles and conferences, we first discuss the necessity and opportunities of SEA due to the rapid socio-economic development in China, then review the development and status of SEA before and after the milestone enactment of the EIA Law in 2003. Furthermore, this paper offers a critical analysis of the current SEA system and practices. Finally, a summary of the opportunities and challenges of using SEA as an important tool for securing sustainability in China's development will enable us to make a general proposal for the way forward.

Opportunities for SEA

If the fundamental goal of SEA is to assure sustainable development, there is probably no other country in the world which is in greater need for SEA than China. The pace of development, scale of some projects, disparity between the coastal and inland areas, and the potential ramifications of a single policy on the society, call for top-level strategic decisions to be made. This is exactly where SEA can make a contribution to put China's development on the sustainable path.

During the 30-year period between 1978 and 2007, the GDP of China increased 15-fold. In the last decade, the annual GDP growth has for most of the time been more than 8%, largely as a result of the rapidly growing export and foreign investment. At the same time, China has paid a significant environmental price for the growth. A report of the Standing Committee of the National People's Congress (NPC, 2006) revealed that one-third of China's water bodies are seriously polluted, one-third of its territory suffers from acid rain, and the percentage of cities meeting primary and secondary air quality objectives are only 4.3% and 58.1% respectively. The green accounting undertaken by SEPA put the environmental costs, largely in terms of health and loss of productivity, at 3.05% of GDP for the year 2004 (SEPA and NBSC, 2006). Probably not reflected in the above figures is that economic growth in the past has depended heavily on the consumption of China's natural resources at efficiencies which are well below those of developed countries. Many of the commodities are so resource-intensive that China is in effect engaging in the "virtual transfer" of resources out of the country through foreign trade, an act which can be seen as "banking" on the future (Oleson, 2008).

Whether China's resource base can sustain her recent pace of growth has attracted considerable concern, particularly if the societal development is to attain the target level of "xiao kang" (moderately rich) by the year 2020. If such a level is to be achieved, the per unit GDP burden on the environment has to be reduced nine-fold taking into account the increase in resource use and population growth (Qian, 2007). The significant leap forward in resource use efficiency called for can hardly be achieved through technological advancement alone. It must be supplemented by significant policy changes and restructuring of the Chinese economy that enhance resource use efficiency.

Confounding the problem is the scarcity of some key resources, notably water, energy and agricultural land, which has significantly hampered development in many parts of the country. China is already reliant on foreign import of over 50% of her demand for iron and aluminium ore and over 70% of copper to sustain her manufacturing base. The resource scarcity problem is particularly severe in areas where water is critically short and the natural ecosystems fragile, such as in the Great West, and also where development is seen by many as urgently needed to address the regional disparity issue.

The Chinese leaders are not unaware of the constraints imposed by resource scarcity and environmental pollution on sustained growth. There are targets in China's 11th Five-Year Plan (2006–2010) to slash the emission of major pollutants by 10%, save energy by 20% and increase the per unit GDP use of 15 major resources by as much as 25%. Some of the targets, particularly those related to energy, are binding. There are already indications that such targets can hardly be met (NPC, 2008). These are ambitious targets that call for strategies and policy instruments well orchestrated and applied at various levels.

The driving forces for pollution reduction and greater efficiency in resource use have originated within and outside the country. On the international platform, pressure is mounting, through trade negotiations and multilateral agreements, calling on China to reduce greenhouse gas emissions and comply with more stringent product environmental standards. Internally, China has to reduce the gap between the rich and poor and address regional disparity through development. In this context, development aims not only to alleviate poverty but also to achieve social harmony.

The notion of development for a harmonious society has taken a new dimension recently. There has been a call by the Chinese Communist Party and Central Government to subject the pathway of development to scientific discourse. Such a call has also been echoed in the subtle change in the mode of development, from the very early emphasis solely on speed, to the more recent one on speed and efficiency, to the latest on efficiency and speed. The change in the emphasis reflects the concern of the Chinese leaders that the mode of development must be in the long-term interest of the country. The call to subject development to scientific discourse bespeaks the realisation that China's goal of sustainable development can only be achieved by resolving a number of strategic issues. The top ones include how to reduce pollution loading with increasing resource use, how to restructure the economy to take the pressure off a dwindling resource base, how to guide development in different parts of the country with regard to their strategic advantage and resource constraints, how to preserve biodiversity and maintain ecological integrity, and how to designate land use zoning that favours or restricts certain uses.

The above are critical issues which China must resolve, failing which China will pay a high price for water, air and land pollution, further development will be impeded by water shortage and ecological degradation, and economic growth will be significantly constrained by resource scarcity. Resolution of all these problems and difficulties call for strategic decisions informed by the findings of holistic analysis and scientific debate. These are areas of discourse where SEA can help clarify objectives, define goals, explore options and alternatives, and provide pertinent information on the environmental and socio-economic consequences of various scenarios and options.

The need to address such strategic issues in China has provided an opportunity for SEA to contribute in a significant way, a role which has been affirmed by Chinese scholars and SEA professionals. SEA is seen, at least from the practitioners, as a mechanism that translates science into knowledge that can inform policies and strategies.

Development and Status of SEA

Like in many other countries, SEA has evolved and developed from the project-based Environmental Impact Assessment (EIA) system first outlined in the Environmental Protection Law of China (for trial implementation) enacted in 1979. Details of the procedures, scope and responsibilities were subsequently elaborated in various directives and regulations issued at different times (Gu and Sheate, 2005) setting the framework for the EIA process.

The Chinese EIA system reached a new milestone in 1993 with the promulgation of the "NEPA Circular on Strengthening of Environmental Management of Construction Projects" in which the notion of "regional EIA (REIA)" was first introduced. The extension of assessment from construction project to regional development plans and programmes can be seen as a response to rapid development, some of large scale, since the opening of China to the outside world in the early 1980s. Recognising the limited scope of the earlier project-based EIA system, the process was extended to cover new comprehensive developments, urban regeneration programmes and river basin management plans. The development of REIA in China in the 1990s was paralleled by the realisation of EIA professionals in other countries that project-based EIA could hardly grapple with cumulative and second order impacts, small incremental changes with large aggregate impacts, and cross-boundary and cross-media environmental consequences. To address issues arising from large-scale regional developments, new assessment methodologies, such as carrying capacity and total emission control, were developed at that time. While some researchers see REIA as an embryonic form of SEA, Bina (2007) argued that the assessment of cumulative impacts is embodied in the US National Environmental Policy Act of 1969, which is widely acknowledged to be the first legislation in the world introducing EIA. Putting aside the issue of whether or not REIA is a form of SEA, the extension of EIA practices to cover regional development is a manifestation that the principle and practice of EIA can be moved "upstream" to include higher level decision-making and planning, as argued by Wallington *et al.* (2007).

Another significant milestone in the development of EIA as well as SEA in China was marked by the promulgation of the EIA Law in China in 2003. Seizing the opportunity to enact a comprehensive piece of legislation on environmental impact assessment, it was the original intention of SEPA as well as other academics and legislators to incorporate the assessment of plans, programmes and policies in the proposed law. Because of strong resistance from some government agencies, the new EIA Law covers only "plans", but not policies. Whilst this is seen by many as a very significant compromise for the sake of getting the legislation through, the extension of EIA to "plans" is a significant step given the significance and scope of central planning in China.

Under the EIA Law, the application of environmental assessment to "plans" (Plan EIA, hereafter PEIA) can go beyond a district or even a region. Plans that are subject to assessment include land use plans, regional, watershed and coastal development plans, as well as the development plans of ten major sectors (industry, agriculture, animal husbandry, forestry, energy, water conservation, transportation, urban construction, tourism and resource exploitation). It can be seen that whilst the omission of the assessment of policies is a compromise, there is scope in the EIA Law to interpret "plans" as flexibly as possible. Such a liberal interpretation is made possible not only by defining the type of plans to be covered, but also by the preamble (Article 1) which states that the Law "is formulated for the purpose of realising sustainable development strategy, preventing adverse impacts on the environment from implementation of plans and construction projects, and promoting coordinative development of the economy, society and environment" (NPC, 2002). Bao *et al.* (2004) observed that, "The salient feature of SEA is considering and designing preventive and mitigation measures at higher level and wider scope in an early period of decision-making, in contrast to project EIA."

The enactment of the EIA Law was also a response to China's Agenda 21 which explicitly called for the establishment of sustainable development impact assessment system in law making. The Agenda also urged that "when government departments formulate proposals of regulations, policies, plans, programs and projects, the evaluation of their impacts on sustainable development should be implemented." (ACCA, 1994). Such a move was also consonant with the call at the 5th National Environmental Protection Conference held in January 2002 to integrate environmental considerations into the development plan in China.

Following the introduction of the 2003 EIA Law, SEPA was keenly aware of the necessity to spell out in explicit terms the type of "plans" that are under purview of the Law, the procedures for public participation and review, and to prescribe the contents and methodologies of different studies in the form of technical guidelines. Table 1 gives a list of directives, decrees and regulations related to PEIA promulgated since 2003. While the list has addressed the major areas of concern in the implementation of PEIA, it is understood that more technical guidelines are needed to facilitate preparation and compilation of the PEIA studies.

To facilitate and promote the implementation of PEIA, SEPA has launched several batches of pilot studies, covering different types of projects under the purview of the EIA Law. These projects include, for example, the 11th Five-Year

Table 1. List of directives, decrees and regulations of PEIA since 2003.

Name of regulation	Note
Environmental Impact Assessment Law of the People's Republic of China	Law adopted by the National People's Congress in 2002
Regulations for Database of Environmental Impact Assessment Review Experts (Order No. 16 of SEPA)	Issued by SEPA in 2003, established an online database of PEIA experts to help related agencies select experts.
Measures for Reviewing on Environmental Impact Assessment Report of Specific Plan (Order No. 18 of SEPA)	Issued by SEPA in 2003, regulating the context and procedure of reviewing PEIA
List of Plans Requiring Environmental Impact Assessment Report and List of Plans Requiring Environmental Impact Assessment Chapter or Statements (On trial) (File No. 2004-98)	Issued by SEPA in 2004 with a detailed list of Plans subject to EIA Law
Interim Measures for Hearing the Administrative License in Respect of Environmental Protection (Order No. 22 of SEPA)	Issued by SEPA in 2004, regulating the hearing process
Measures for the Disclosure of Environmental Information (for Trial Implementation) (Order No. 35 of SEPA)	Issued by SEPA in 2007
Technical Guidelines for Plan Environmental Impact Assessment (On trial) (HJ/T130-2003)	Guidelines for PEIA, issued by SEPA in 2003
Technical Guidelines for Environmental Impact Assessment of Development Area (HJ/T131-2003)	Guidelines for EIA of development area, issued by SEPA in 2003

Plan of Inner Mongolia Autonomous Region, the 11th Five-Year Plan of Wuhan, development of the railway sector in the 11th Five-Year Plan, industrial development plan of the Three Gorges Area and the Nansha District of Guangdong Province, respectively. There is as yet no official statistics on the number of PEIAs undertaken in the first five years since the implementation of the EIA Law. Some officials put the number at around 500. However, according to Zhu and Ru (2008) very few of these PEIAs were related to plans initiated by non-environmental ministries.

Critique of the Current System and Practices

In the last five years since the implementation of PEIA under the EIA Law, quite a number of articles, in English and Chinese, have been published. SEPA has also organised two conferences, in collaboration with the International Association of Impact Assessment and other Chinese institutions, focusing on SEA in China. In January 2008, SEPA organised a forum for renowned academics and researchers in the field to comment on the SEA/PEIA process (SEPA, 2008a). There was also a special session in the National People's Congress held in October 2008 reviewing the progress in the implementation of the EIA Law since 2003 (Xinhua, 2008). Several books and a number of articles have also been published, mainly in Chinese, on actual SEA/PEIA cases in China. Some are pilot studies earmarked by SEPA (SEPA, 2006) for the purpose of highlighting issues and problems, and identifying areas for further improvement; others are summary descriptions given in workshops and conferences, such as "First China International Forum on Environmental Impact Assessment" held in December 2004 in Boao, Hainan Island (SEPA, 2005) and the international workshop on "Strategic Environmental Impact Assessment in China" held in November 2007 in Beijing (Au *et al.*, 2008; SEPA, 2008b).

It is based on the above data and information that this critique is written. Recognising the difficulty of gathering the large number of case studies and other materials, formally published or available on the internet, there is no claim the analysis reported below is a comprehensive one. Our primary objective is to highlight and contrast the major concerns of different stakeholders involved in the SEA process in China, including those who are with the administration (speeches of principal officials of SEPA and subsidiary institutions); in the legislature (i.e., discussion note of the National People's Congress in October 2008 (NPC, 2008)) and advisory bodies (Focused Group Meeting organised by SEPA in January 2008 (SEPA, 2008a)); practitioners who are working the SEA projects and preparing the reports, and academics who are concerned with the development of SEA in China.

It is worthwhile to mention that academic articles written in the last few years reviewing the progress and issues of implementing SEA in China fall broadly into two categories. The first, exemplified by the works of Bina (2008), Zhu and Ru (2008) and Gu and Sheate (2005), tends to focus more on broader issues of the purpose and objectives of SEA, effectiveness of the current process, particularly in the context of the Chinese bureaucratic and institutional setup and its relationship with environmental governance. The other category, represented by the works of Bao *et al.* (2004), Zhu *et al.* (2005), Xu *et al.* (2003), Wu and Xu (2007) and Sun *et al.* (2005), are more practice-oriented, addressing mainly the operational issues and problems of PEIA under the new EIA Law.

By analysing the contents of the speeches, writings and arguments of the five major stakeholder groups, we have tabulated the relative importance of the key concepts advocated; issues mentioned and proposed solutions made by various parties (Table 2). It should be noted the numbers are indicative rather than quantitative because of the practical difficulty of sampling and analysing the writings of all concerned parties systematically and rigorously.

It can be seen from Table 2 that there is a notable difference in the perceived ultimate purpose (*raison d'être*) of SEA between the principal officers/legislators and some academics. Whilst there is no disagreement that the ultimate aim of SEA is to assure sustainability, the principal officers, legislators and key advisors see

Table 2. Key concepts advocated, issues mentioned and suggestions made by the five groups of stakeholders.

	Principal government officers	Legislators	Advisors	Academics	Practitioners
Ultimate aim (*raison d'être*) of SEA					
SEA for sustainability	2			3	3
As a strategy to influence decision making				1	2
Scientific development	2	1	2		2
Harmonious society	2	1			3
Building a resource efficient & environmentally friendly society	1				2
Target of assessment					
Five-Year Plan		2	1		1
Socio-economic Plan		2			1
Sectoral Plan		1			2
Regional Plan		1			1
Watershed Plan		1			1
SEA as an instrument for					
Strategic positioning			3		1
Industrial location	1		1		1
Industrial policy			1		1
Resource allocation	1				1
Resource efficiency — circular economy	3				2
Integrated economic, social & environmental plan	2		1	1	1
Comprehensive plan — interwoven	1				1
Governance — transparency & public participation	1			2	1

Table 2. (*Continued*)

	Principal government officers	Legislators	Advisors	Academics	Practitioners
Environmental equality — compensation for ecological services, regional disparity	1	3		2	1
Issues					
Resistance & circumvention by line agencies and local authorities	1	2	1	2	1
Limitation & difficulties of PEIA	3	2	2	2	2
Unclear roles & responsibilities				2	1
Inadequate standards & guidelines		1		1	2
Inadequate enforcement & follow up		2	1	2	1
Narrow focus on biophysical aspects (little human, econ & political)				2	3
Timing — late start				2	2
Too solution-oriented, not enough diagnostic				1	3
Too reactive and mitigation-based				1	2
Over-emphasis on procedural technicalities instead of sustainability		1		1	3
Too project-oriented				3	1
Too technocratic-rational				2	1
Little interaction between assessors and planners				1	1
Lack of transparency				3	1
Limited public participation				2	1
Doubtful effectiveness				2	1
Root of problem					
Putting economic first		2		2	1
Leaders wanting to do something at all cost		1			1
Institutional setup, particularly "Tiao Kuai"		1		1	
Lack of dialogue and coordination among agencies				3	1

Table 2. (*Continued*)

	Principal government officers	Legislators	Advisors	Academics	Practitioners
Solution					
Make SEA "strategic"			1	3	1
SEA of macroeconomic strategies	3		1		1
SEA of policies and decisions, laws	1	1	2	1	
SEA of comprehensive plans ahead of assessment		1	1		1
Institution building	1				1
Regulations/guideline on scoping, procedures, funding, methodology, review, approval & post-decision variations		3	1	2	2
Quality assurance — review, certification		3		1	
Public participation		1	1	3	2
EIA follow up		2			
Governance and accountability		1	1	1	
Capacity building	2				2
Methodology development				2	2
Data source, sharing & common data bank		1	1	2	2
Pilot studies	1				2
Post analysis — post hoc analysis		1			1

Note: The numbers indicate the level of attention or concern expressed by various parties (the larger the number, the higher the level).

SEA as a means to achieve a harmonious society through scientific discourse, some academics, notably Bina (2008), argued that SEA should be a strategy to influence top-level government decision making, particularly in the formulation of plans, programmes and policies. Her main thesis is that good information does not necessarily influence decision making and a SEA study will have little effect if it does not influence decision outcomes. To be effective in influencing decision outcomes, she maintained that the findings of SEA should have a place in the interactions between key decision makers. Her observation has significant bearing on how SEA should be conducted and its findings being considered. The difference is also rooted in the priorities of different stakeholders. While academics accord the highest priority to sustainability, the principal officers see SEA as an instrument to solve some of the most difficult problems in building a sustainable and harmonious society, namely what is the strategic advantage of one region/city over another, what functions can be allowed or restricted in a particular area, what is the most appropriate policy for particular industries, how to allocate the limited resources to various areas and sectors, and ways and means to use them efficiently. These are questions of great strategic importance in a vast country with many competing demands and conflicts of interests. Resolution of these questions calls for taking a holistic and integrated view, balancing conflicting claims and ensuring the gains in one area/sector will not create losses in another. The resolution of these issues in fact goes beyond the scope of the current PEIA process.

The legislators on the other hand venture beyond the above strategic debate and ask how SEA can be applied to the formulation of Five-Year Plans for the nation, the next of which is due for 2010. At the same time, they also wish to apply the SEA process to the formulation of socio-economic plans for various provinces, regions and cities. Academics on the other hand are keen to see the SEA process as the catalyst for greater transparency in policy making and more opportunities for public participation, which are key ingredients for better environmental governance and ways to influence top-level decision making.

All actors involved in SEA recognise the limitations of PEIA and are aware of the resistance of some sectoral ministries to undertake EIA. Zhu and Ru (2008) observed that whilst the EIA Law highlights ten sectors of which their plans are subjected to the SEA process, sectoral SEAs only account for a disproportionately small share of all SEAs undertaken. For those sectoral SEAs which have done, some see SEA as a bureaucratic hurdle to overcome (Zhu and Ru, 2008). With such an attitude, one can hardly expect SEA to have any significant influence on the planning process and outcome.

Concerning the implementation of PEIA, many actors are aware of the ambiguity, soft spots and loopholes in the provisions of the EIA Law. There are criticisms that the roles and responsibilities of various parties are not clearly defined, regulations and standards are open to interpretation, and enforcement has been weak and inconsistent (Bina, 2008; Wu and Jiang, 2005; Zhu and Ru, 2008). One critic commented that there is a lack of clarity on important issues such as who, what, when and how. The administration is not unaware of these issues and has strived, as outlined in Table 1, to remedy these deficiencies by supplementary directives and decrees. It was also the wish of MEP to have the whole package of regulations passed by the legislature in 2008.

Some critics opined that whilst it is necessary to have these legal and procedural mechanisms in place, such efforts alone do not add much value to the SEA process (Bina, 2008), because the PEIA system, as it stands now, is too reactive and mitigation oriented. As such, the assessment is often done after the plans are formulated and there is hence little opportunity for the assessors to interact with, and provide timely input, to the planners. Others criticise the scope as being too narrowly focused on the biophysical aspect of the environment, overlooking the social and economic dimension of the society. Bina (2008) and Zhu and Ru (2008) maintained that the current technocratic-rational approach contributes little to the building of a sustainable society that meets the aspirations of the public and allows for the development of the latent ecological and social capabilities of the society. Some believe that unless these issues are addressed, the effectiveness of the SEA process can be significantly undermined (Sun *et al.*, 2005; Zhu and Ru, 2008).

Many explanations have been advanced to account for the limited effectiveness of the current system. Some attribute the difficult situation to the political resistance and passive attitude of line agencies and individual ministries. Others observed that because of the political resistance, the original EIA Bill was significantly compromised in the legislative process. It has also been pointed out the "putting economic development first" ethos is still prevalent in some line agencies and local administrations. The problem is confounded by an institutional structure (known locally as "tiao-kuai") with local environmental authorities both answerable to the central Ministry of Environment and local government (Lieberthal, 2003; Ma, 1998; Zhong, 2003).

As regards the way forward, one can see slight difference in emphasis between various actors (Table 2). Some academics (Bina, 2008; Ahmed and Ernesto, 2008), based on political theory and policy making experiences, promoted treating SEA as a "strategy" to influence plan making. Others, particularly the principal officers, legislators and the advisors, advocated the application of SEA to policies, major legislation, macro economic decisions and government decisions. There are subtle differences between the two; the former focuses on SEA as a process, the latter an end. There is probably little difference in the calls of various actors to hasten institution and capacity building, to encourage public participation, to assure the quality of the SEA findings through a rigorous and impartial review process and to foster greater accountability within the government machinery.

Summary and Conclusion: The Way Forward

The development of SEA in China is at a defining moment and at a crossroads. It is at a defining moment because on the one hand, the future development of China will be significantly constrained by environmental degradation and resource scarcity; and on the other hand, the building of a sustainable and harmonious society is the national policy, affirmed time and again by the Chinese leaders. There are many opportunities, at different levels and in numerous sectors, where SEA can contribute. It is a test of if, and how, SEA can make an influence on China's pathway to sustainability.

We have in this paper reviewed the development of EIA/SEA in China, highlighting how it has responded over the years to the needs of the society by adapting its scope and focus. We have shown how EIA took the form of REIA, with its new assessment tools and methodology, to cope with the rapid and massive development in the 1990s. We have also maintained that the more recent top-level commitment to building a harmonious society through scientific discourse has provided a window for SEA to play a key role in decision-making.

The enactment of the EIA Law in 2003 is a significant milestone in the evolution and maturing of the EIA process in China. Whilst the new PEIA system introduced in the law does not encompass policies, it does allow environmental assessment to be applied to regional and sectoral plans of different kinds. The obvious omission of the assessment of policies is partly made up by interpreting the law liberally to include plans of different types and levels. The limitations of narrow scope and technocratic bias of PEIA are well recognised and the administration has strived hard to iron out deficiencies and clarify ambiguities. Within the bounds of PEIA, the approach and emphasis have admittedly been technocratic and rational. Given the political realities in China, the current PEIA is probably the most appropriate and attainable at the moment. In fact, the pathway which China has embarked on is probably not too different from that of Europe and the USA (Bina, 2008; Lam, 2007).

Notwithstanding the above, there is no room for complacency, particularly in view of the need to put China's development on the sustainable path given the country's many environmental problems and resource constraints. The major thrust of the work of the government in the last few years has been in the setting of procedures, methods, mechanisms and subsidiary regulations. We maintain that the attention given to these legalistic and procedural mechanisms are justified because of the complicated institutional setup of China. Setting the norm and defining the perimeters are after all needed to assure quality, without which the right information cannot be provided to the right people at the right time.

We are at the same time mindful that the most fundamental question for all SEA practitioners remains to be answered, i.e., how SEA can influence decision making that underpins China's sustainability. To answer this question, there are two important considerations. The first is how to overcome the political resistance of line agencies and sectoral ministries; and the other is at what level of decision making should SEA be applied.

The first consideration is a matter of ownership. Currently SEA is seen as an administrative hurdle, often imposed by MEP on other ministries and different levels of government. If the proponents envisage SEA as a hurdle rather than a process to add value, they will only look for ways to resist and circumvent the system. The promulgation of directives, guidance notes and regulations will only compel them to undertake the assessment, but not induce them to use the findings. The only way for SEA to make an influence on those plans is make the proponents envisage, and own, the benefits of SEA. This can be achieved only if the local authorities or line agencies believe SEA is the right tool they can use to execute their tasks effectively and achieve the given targets efficiently.

The second consideration is at what levels of decision making SEA should be applied. As afore-mentioned, there is a call, both from within and outside the administration, to apply SEA to policies in China, which was an earlier intent aborted in the making of the EIA Law. There is probably no dispute on the need to apply SEA to policies, but in doing so, one must not underestimate the difficulty. One must be prepared to face the hard and uncomfortable reality that widening of the scope of assessment to encompass policies, through the amendment of the EIA Law, will provoke the same, if not more, resistance as it did when the EIA bill was before the legislature a few years ago. No matter how hard it may be, the move of SEA "upstream" to the assessment of policies is inevitable if SEA aims to have an influence on policy making. After all, the assessment of plans is, by nature, re-active, and can hence exert little influence on decision making.

It is because of the difficult situation that the development of SEA in China is at a crossroads. We believe the best way forward, at this particular juncture, is to promote the application of SEA, to those high level policies and strategies that have significant bearing on the nation's sustainability, which is outside the statutory framework of the current EIA Law. The obvious candidate for application is the Five-Year Plan which China has to update every five years. SEA can also be applied to the socio-economic development plans for various provinces, cities and municipalities. In fact, these are being undertaken already, not always in the form of SEA but in the name of environmental planning. There are three distinct advantages of adopting such an approach. Firstly, there is much less political resistance, as the study is seen as an internal matter rather than something imposed from outside. Secondly, it would be much easier for the proponents and assessors to interact; and for the proponent to own the process and the benefits arisen therein. Thirdly, the formulation of Five-Year Plans and local socio-economic plans are under the scrutiny of the State Council and the National Development and Reform Commission (NDRC) which have an authority higher than that of MEP. By transferring the task to the State Council and NDRC, it helps take the pressure off MEP.

This is not to say that MEP does not need to perfect the current PEIA system. There is still much to be done on the PEIA system, on the regulations and guidance notes, and on measures to encourage public participation and promote transparency. But building up the capacity and a good governance institution, the role of MEP can gradually change from being a regulator to a facilitator and allow the public to play a more significant role in assuring quality of the SEA process and accountability of the government.

By implementing SEA outside the current statutory PEIA framework, it is hoped that the senior administration can appreciate how SEA can contribute to the formulation of sustainable policies for the nation and various line agencies and ministries will gradually "buy in", envisage and own the benefits.

Acknowledgement

The authors gratefully acknowledge the funding support of Focused Investment Scheme of The Chinese University of Hong Kong (Project 1902041) and a grant from the Research Grants Council of the Hong Kong SAR Goverment (CUHK 445809).

References

ACCA (1994). China's Agenda 21. Retrieved from http://www.acca21.org.cn/english/index.html.

Ahmed, K and ST Ernesto (2008). Using strategic environmental assessments to design and implement public policy. In *Strategic Environmental Assessments for Policies: An Instrument for Good Governance*, K Ahmed and ST Ernesto (eds.). Washington, DC: World Bank Publications, pp. 181–207.

Au, WK, KC Lam, T Zhu and M Partidario (eds.) (2008). *International Experience on Strategic Environmental Assessment*. Beijing, China.

Bao, CK, YS Lu and JC Shang (2004). Framework and operational procedure for implementing strategic environmental assessment in China. *Environmental Impact Assessment Review*, 24(1), 27–46.

Bina, O (2007). A critical review of the dominant lines of argumentation on the need for strategic environmental assessment. *Environmental Impact Assessment Review*, 27, 585–606.

Bina, O (2008). Context and systems: thinking more broadly about effectiveness in strategic environmental assessment in China. *Environmental Management*, 42(4), 717–733.

Gu, L and WR Sheate (2005). Institutional challenges for EIA implementation in China: a case study of development versus environmental protection. *Environmental Management*, 36(1), 125–142.

Lam, KC (2007). UNEP's approach to integrated assessment of trade-related policies: Evolution and recent progress: Comments. *Regional Development Dialogue*, 27, 45–47.

Lieberthal, K (ed.) (2003). *Governing China: From Revolution to Reform*, 2nd edn. New York: W W Norton & Co Inc.

Ma, LH (1998). Discussing tiao-kuai relations within the management of government. *Political Science Research*, 4, 68–74.

NPC (2002). Environmental impact assessment law of the People's Republic of China. Adopted at the 30th meeting of the Standing Committee of the Ninth National People's Congress on October 28, 2002.

NPC (2006). Review of the Implementation of Some Legislations of Environmental Protection. Retrieved from http://www.npc.gov.cn/npc/oldarchives/cwh/common/showrecord.jsp@pdmc=fywj&hyid=0210023&lmid=310.htm.

NPC (2008). Reviewing the Progress in the Implementation of the EIA Law. Retrieved from http://www.npc.gov.cn/npc/xinwen/jdgz/2008-10/27/content_1455370.htm.

Oleson, K (2008). Integrating the environment into economics in China — A focus on exports. In *International Experience on Strategic Environmental Assessment*, Au, WK, KC Lam, T Zhu and M Partidario (eds.). Beijing, pp. 113–120.

Qian, Y (2007). Circular economy and new industrialization. *The Tsinghua People*, 22(4), 41–42. (In Chinese).

SEPA (2005). *Proc. of the 1st China International Forum on Environmental Impact Assessment*, SEPA (eds.). Bo'ao, China.

SEPA (ed.) (2006). *Comments on Strategic Environmental Assessment Cases*. Beijing: China Environmental Science Press.

SEPA and NBSC (2006). China Green National Accounting Study Report 2004. Retrieved from http://www.sepa.gov.cn/plan/gongwen/200609/P020060908545859361774.pdf.

SEPA (2008a). Integrate Strategic Environmental Assessment into Decision Making Forum. Retrieved from http://www.mep.gov.cn/hjyw08/200801/t20080129_117779.htm.

SEPA (2008b). *Proc. of the 2nd China International Forum on Environmental Impact Assessment-SEA in China*, SEPA (eds.). Beijing, China.

Sun, YJ, XG Chen, CR Mo, YH Sang, XC Peng and R Fan (2005). Opportunities and challenges of planning environmental impact assessment. *Environmental Protection*, 334(9), 42–45. (In Chinese).

Wallington, T, O Bina and W Thissen (2007). Theorising strategic environmental assessment: fresh perspectives and future challenges. *Environmental Impact Assessment Review*, 27(7), 569–584.

Wu, J and H Jiang (2005). Discussion on the implementation of plan EIA in China. *Environmental Protection*, 330(4), 81–97. (In Chinese).

Wu, J and H Xu (2007). SEA on the 11th five-year plan for national economic and social development of Wuhan City in China. *Management of Environmental Quality: An International Journal*, 18(3), 340–352.

Xinhua (2008). Reviewing the Progress in the Implementation of the EIA Law. Retrieved 11/07, 2008, from http://news.xinhuanet.com/newscenter/2008-10/27/content_10260662.htm.

Xu, H, T Zhu and SG Dai (2003). Strategic environmental assessment of wastewater reuse policy: a case study from Tianjin in China. *Journal of Environmental Assessment and Management*, 5(4), 503–521.

Zhong, Y (2003). *Local Government and Politics in China: Challenges from Below*. New York: ME Sharpe.

Zhu, D and J Ru (2008). Strategic environmental assessment in China: motivations, politics, and effectiveness. *Journal of Environmental Management*, 88(4), 615–626.

Zhu, T, H Xu and J Wu (eds.) (2005). *Strategic Environmental Assessment*. Tianjin: Nankai University Press. (In Chinese).

CHAPTER 2

A Policy Window Opens: Strategic Environmental Assessment in York Region, Ontario, Canada

Denis Kirchhoff*, Dan McCarthy†, Debbe D. Crandall‡,
Laura McDowell§ and Graham Whitelaw¶

**Department of Geography and Environmental Management*
University of Waterloo
200 University Avenue West, Waterloo
Ontario N2L3G1, Canada
dkirchho@uwaterloo.ca

†Social Innovation Generation
Department of Environment and Resources Studies
University of Waterloo, Canada

‡Save the Oak Ridges Moraine Coalition
Canada

§Engineering Water and Wastewater Branch
Transportation and Works Department
Regional Municipality of York, Canada

¶School of Urban and Regional Planning
School of Environmental Studies
Queen's University, Canada

Government agenda setting has been a focus of research in the field of policy sciences for over two decades. The concept of a policy window is explored as a driver of governmental agenda setting. The Regional Municipality of York, Ontario, Canada was chosen as a case study for exploring the application of strategic environmental assessment at the municipal level through a policy window lens. Problem, policy and political streams converged to provide the necessary conditions for improved environmental assessment and

*Corresponding author.

infrastructure planning in York Region. A focusing event and the resulting crisis motivated stakeholders to identify and act on the problem. An SEA-type approach was initiated as one key response. A variety of activities were initiated by York Region including the development of a Sustainability Strategy, synchronisation of master planning, wider consideration of alternatives at the master plan level and improved public consultation. Conclusions are drawn and several recommendations are presented and discussed.

Keywords: Strategic environmental assessment; policy windows; growth management; Regional Municipality of York; sustainability; master plan.

Introduction

Why do specific matters become priorities for governments? Why do decision makers pay attention to particular policy, program or planning issues rather than others? At any given time, there are a number of issues competing for decision makers' attention, yet, only a few of those can actually gain serious attention at the same time. Government agenda setting has been a focus of research in the field of policy sciences for over two decades. A provocative research contribution in the area is presented by Kingdon (1984; updated 1995) using the concept of policy windows to describe how individual matters become governmental priorities. Policy windows are temporally discrete opportunities for innovative policy change that result from abrupt and uncommon phenomena, often referred to as focusing events (Kingdon, 1995).

Within the strategic environmental assessment (SEA) literature, few have attempted to understand how SEA adoption becomes a priority (see for example Fischer, 2004). Most SEA scholars have focused their research on practical and process issues, methodological and governance and decision making (Abaza *et al.*, 2004; Brown and Thérivel, 2000; Caratti *et al.*, 2004; Cashmore, 2004; Dalal-Clayton and Sadler, 2005; Fischer, 2007; Hildén *et al.*, 2004; Jones *et al.*, 2005; Retief, 2007; Schmidt *et al.*, 2005; Stinchcombe and Gibson, 2001; Thérivel, 2004). The Regional Municipality of York, Ontario, Canada was chosen as a case study for exploring SEA application at the municipal level. During the research process the concept of a policy window emerged as highly relevant in York Region's adoption of an SEA-type approach to reframe and restructure its planning and decision-making processes in particular, with reference to infrastructure planning and development. We use the term "SEA-type" to describe approaches that do not meet formal specifications or definitions of SEA, but have some of their characteristics or elements. Changes in York Region's approach to planning and decision making, as well as the story behind them will be highlighted throughout this article through a policy window lens.

This research was designed and implemented in collaboration with municipal infrastructure and planning staff from York Region. The research included multi-stakeholder workshops, participant observation, document analysis and key-informant interviews. The paper is structured as follows: background information on policy windows, Strategic Environmental Assessment (SEA) and the York region case study; research methods; findings; and, implications and recommendations.

Policy windows

Policy windows are transitory opportunities during which the likelihood of adopting new policy or legislative proposals is greater than usual (Solecki and Michaels, 1994). In this context, the policy windows concept is useful in describing and analysing what happens in the agenda setting process in public bureaucracies. As indicated by Kingdon (1995, p. 166), policy windows "open infrequently, and do not stay open long, but despite their rarity, the major changes in public policy result from the appearance of these opportunities".

Kingdon (1995) describes how three related policy components or "streams" (problem, political and policy) converge to move an issue onto the public policy agenda and toward potential governmental action. The problem stream refers to how decision makers learn about new issues and define and understand their evolution. The solution stream (policy stream) includes the set of potential solutions to the problem. The third stream refers to the political climate and public opinion with reference to the issue (Michaels *et al.*, 2006). Kingdon labels the term "coupling" to describe when these three major streams converge, and finds that these independent streams "flow" through the system all at once. They become coupled or emerge onto the policy agenda when a window of opportunity is opened (Thurber, 2003). Kingdon asserts that an issue is most likely to achieve public agenda status when public problems, policy alternatives, and political opportunities intersect. Furthermore, these windows present themselves when there is a major incident or accident that requires action. The idea of a focusing event is used to describe a sudden, uncommon event (such as a disaster or a crisis) that leads to policy change (Birkland, 1997).

As will be discussed subsequently, this study identified a focusing event associated with one of the projects derived from a wastewater collection system project in York Region. The crisis caused by this event, generated significant public opposition, and served as the focusing event that led to the adoption of an SEA-type process for infrastructure planning and decision making in York Region.

Strategic environmental assessment

Since the formal introduction of environmental assessment (EA) in North America through the USA National Environmental Protection Act (NEPA) in 1969 and later the Canadian Environmental Assessment and Review Process (EARP) in 1973, EA has become more efficient, effective and fair (Gibson, 2000). Since then, there has been growing interest in the environmental implications of initiatives at scales beyond the project level, vis a vis the policy, plan and program levels. In recent years, SEA received considerable attention as a means of integrating environmental considerations into the development, approval and implementation of policy, plans and programs, as well as other strategic undertakings.

Primary motivations for the development of SEA include: counteracting limitations of project EA; and, promoting sustainability principles and practice (Bina, 2003; Clark, 2000; Dalal-Clayton and Sadler, 2005; Fischer, 1999; Petts, 1999; Sadler and Verheem, 1996; Thérivel and Partidário, 1996). While project-level assessment processes have generally led to more environmentally-informed, transparent and participatory decision-making, project-level EA has, in general, not been able to deal effectively with broader concerns (e.g., cumulative effects, broad objectives and alternatives, underlying policy conflicts and longer-term options). Project assessments are usually reactionary, rushed, narrowly mandated and come too late in decision making to be generally effective vehicles for examining strategic concerns and options (Alshuwaikhat, 2005; Gibson, 2007; Partidário, 2000). Moreover, if sustainability is to be the guiding set of principles for future development, then a strategic approach in development processes is required. A strategic approach implies: linking long-term vision to medium-term targets and short-term action; "horizontal" linkages across sectors, so that there is a coordinated approach to development; "vertical" spatial linkages, so that local, regional and national policy and governance are mutually supportive; and genuine partnership between government, business, and community and voluntary organisations since the problems are too complex to be resolved by any group acting alone (Dalal-Clayton and Bass, 2002).

SEA emerged as a promising means of dealing directly with strategic issues in a way that has the advantages of project EA processes (e.g., firm obligations, integration of environmental concerns in planning and decision making, more transparent and open process, etc.), but also has the necessary scope and mandate to influence strategic level decisions (Benevides *et al.*, 2008). In addition, SEA is intended to occur at a stage in the process and a scale that can provide guidance to subsequent, lower-tier strategic undertakings as well as overall project planning. The main purpose of SEA is to improve strategic planning and decision making, that is, "to make it more integrated, farsighted, open, efficient, credible and defensible, and most importantly, more likely to bring consistent delivery of lasting benefits from strategic initiatives" (Benevides *et al.*, 2008).

Increasing awareness of, and concern about, some of these effects has encouraged greater use of SEA, both to ensure better attention to environmental and sustainability considerations in the development of policy, plans and programs and to address sustainability issues through new strategic initiatives at the sectoral or regional scale. SEA may be applied at the supra-national, national, and sub-national levels of governance in arenas such as legislation, economic policy (budgeting, privatisation, subsidies, and taxation), land use planning, infrastructure development planning, resource management, and social policy. Examples of SEA in practice include (although not always under the SEA label) applications to the transportation, land use planning, energy, and water and wastewater sectors (e.g., Carter and Howe, 2006; Fischer, 2002; Stockholm Environmental Institute, 2001; Tang, 2008).

After more than 20 years of experience with SEA worldwide, perceived benefits include: wider consideration of effects and alternatives; improving the evidence base for strategic decisions; facilitating consultation with stakeholders and addressing their concerns; and streamlining other processes such as environmental assessment of individual development projects by establishing an appropriate context for the project level undertaking (i.e., tiering) (Alshuwaikhat, 2005; Bina, 2003; Dalal-Clayton and Sadler, 2005; Dusik *et al.*, 2003; Fischer, 1999, 2003, 2006; Jones *et al.*, 2005; Nooteboom, 2000; Partidário, 2005; Sadler and Verheem, 1996; Stinchcombe and Gibson, 2001; Thérivel, 2004; Vicente and Partidário, 2006; Thérivel, 2004, p. 18) expresses the potential benefits of SEA as being able to, "help achieve clearer, more environment-friendly and more publicly acceptable strategic actions that are approved more quickly." While there are a growing number of jurisdictions that have SEA-type processes or apply some elements of SEA, few countries have established formal arrangements or legislative requirements for SEA (Dalal-Clayton and Sadler, 2005).

York Region and background information

The York Region approach to implementing an SEA-type decision-making process involves integrated planning, rather than just a contribution of information to planning and decision-making processes. Recently, there has been some attempt to introduce elements of SEA good practice[1] into their overall growth management

[1] Elements of SEA such as the ones described in IAIA (2002). Strategic environmental assessment performance criteria. Retrieved 20 April 2005, from http://www.iaia.org/publicdocuments/special-publications/sp1.pdf.

planning, including infrastructure planning processes, such as the review and update of both the Transportation and Water and Wastewater Master Plans (York Region, 2007e,d), and the publication of a Sustainability Strategy (York Region, 2007d), making explicit York Region's commitment to sustainability.

Replacing the former York County in 1971, the Regional Municipality of York is an upper-tier municipality in south central Ontario, Canada. It covers 1,776 square kilometres from Lake Simcoe in the north to the northern boundary of the City of Toronto in the south. It borders Simcoe County and Peel Region in the west and Durham Region in the east (see Fig. 1). York Region is one of six Regional Governments in Ontario and is among the fastest growing regional municipality in Canada, with its population expected to reach 1.5 million residents by 2031 (Ministry of Public Infrastructure Renewal, 2006). The entire Region is part of the Greater Toronto Area (GTA), the Greater Golden Horseshoe and over 69% of York Region is designated part of Ontario's Greenbelt, which includes the Oak Ridges Moraine (York Region, 2008a).

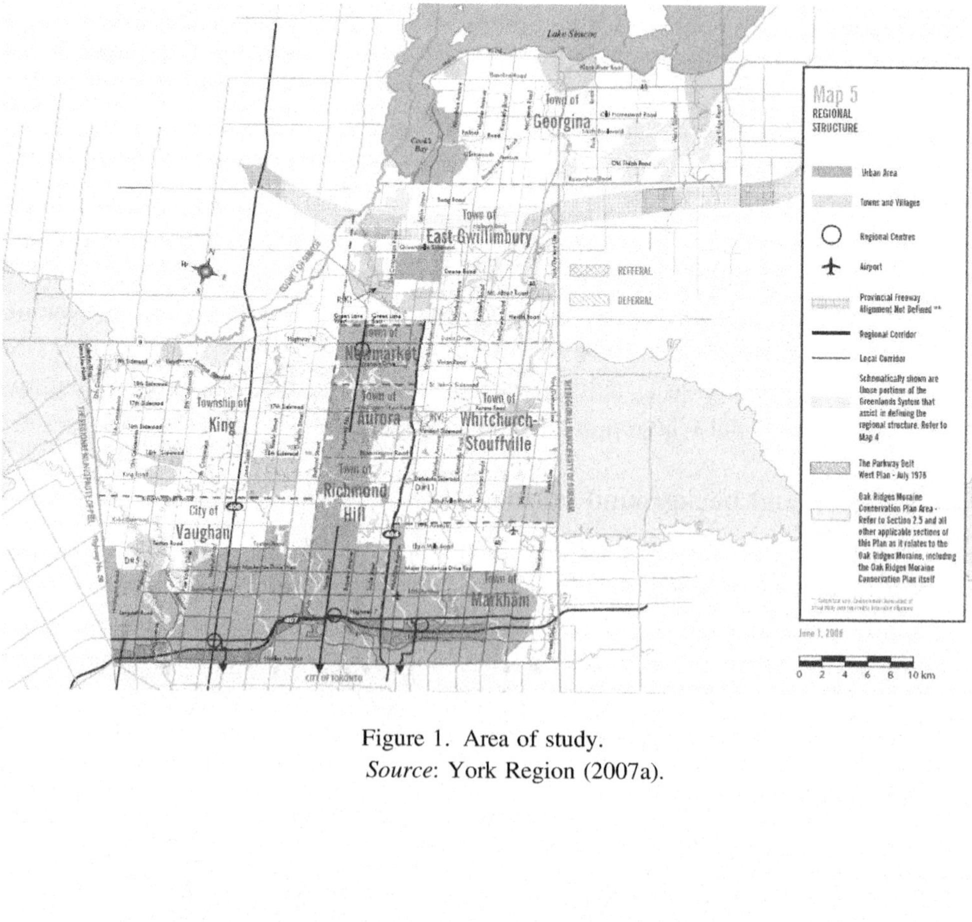

Figure 1. Area of study.
Source: York Region (2007a).

York Region provides an upper-tier governance structure to nine lower-tier municipalities, including: Town of Aurora, Town of East Gwillimbury, Town of Georgina, Township of King, Town of Markham, Town of Newmarket, Town of Richmond Hill, City of Vaughan and Town of Whitchurch-Stouffville. York Regional Council consists of 20 elected representatives from each of the constituent towns and cities in the region, including each of the nine mayors and 11 regional councillors (York Region, 2008a).

York Region's landscape includes farmlands, wetlands and kettle lakes, forests and the Oak Ridges Moraine, a unique protected landform that is an immense glacial moraine that runs east to west, north of, and parallel to, Lake Ontario. One of the moraine's most important functions is as a water recharge/discharge area — sustaining the health of the many watersheds, which originate in the moraine and directly providing drinking water to over 250,000 people (STORM coalition, 2008). The moraine is described as southern Ontario's rain barrel — its permeable sands and gravels absorb and collect precipitation, which slowly recharge the deep aquifers below the ground (see Fig. 2). In York Region, the Oak Ridges Moraine is the headwaters for many rivers and streams, including: the Schomberg; Black and Holland Rivers, which flow north to Lake Simcoe, and the Humber; and Don and Rouge Rivers, which flow south towards Lake Ontario. The natural

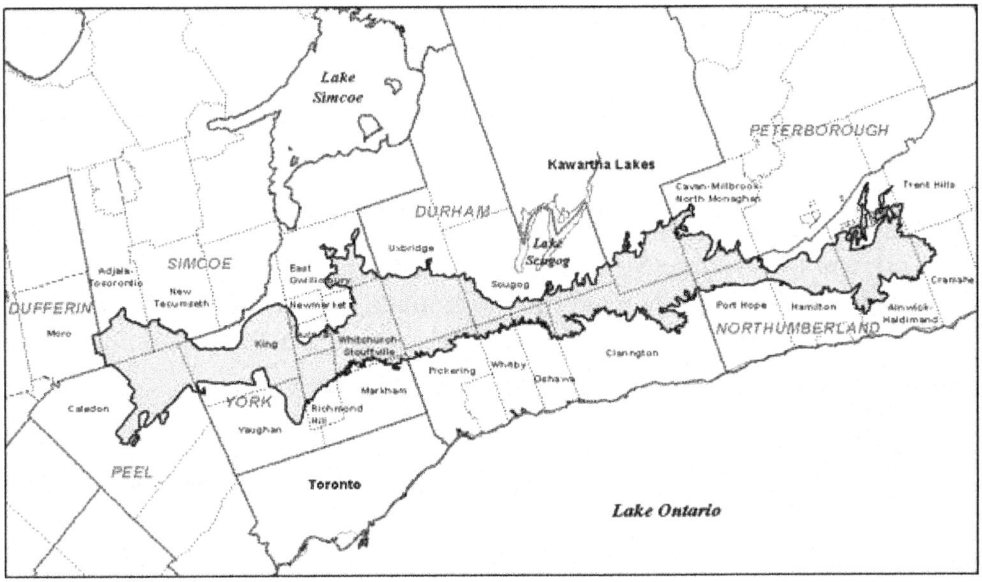

Figure 2. The Oak Ridges Moraine Area.
Source: MMAH (2008).

environment is characterised by many forested areas, wetlands and kettle lakes, which are part of a provincially-recognised planning area (MMAH, 2008; York Region, 2007a).

York Region continues to experience significant development pressure. This rapid growth has made the Region one of the most prosperous areas in Canada, and has resulted in the conversion of approximately 160 square kilometres (62 square miles) of countryside to urban uses since 1971 (York Region, 2009). The Region grew from 169,000 people in 1971 to 759,000 people by 2001, and is expected to reach 1.5 million by the year 2031 (Ministry of Public Infrastructure Renewal, 2006). Approximately 350,000 employees work in the Region's 21,000 businesses. Employment is forecast to double by the year 2031. Most of the growth in the Region has occurred within the southern municipalities of Vaughan, Richmond Hill and Markham. For example, Markham grew from a population of approximately 37,500 in 1971 to a population of 273,805 in 2006. The City of Vaughan experienced the highest growth rate of 8.2% over the 35 year period (York Region, 2008f).

Growth led to enormous pressure for new and updated infrastructure, especially in terms of transportation, water and wastewater. York Region has one of the largest infrastructure budgets of all Canadian municipalities, with $400 million spent in 2005 for water and wastewater and road infrastructure, and $3 billion to be spent over the next 10 years (York Region, 2008b). It is in this context that the Region of York started a growth management initiative in 2006 entitled *Planning for Tomorrow* (York Region, 2008c). This involved the development of a Sustainability Strategy and the update of Infrastructure Master Plans, which culminated in an update to the Regional Official Plan in December 2009 (York Region, 2009).

In 1994, York Region's first Regional Official Plan was published and in 1997, the York Durham Sewage System (YDSS) master plan[2] (later updated in 2002) was completed. The YDSS master plan identifies current conditions and future alternatives necessary to provide the infrastructure to meet population projections

[2] The idea for the YDSS — a centralised sewage treatment system — dates back to 1965 when the province of Ontario decided that no additional sewage treatment plants could be built on the Don and Rouge Rivers and Duffins Creek. The original YDSS servicing scheme then was started by the Province of Ontario to meet environmental goals emerging at the time. The conceptualisation, design and construction of this project occurred over a 25-year period and was completed in the early 1980s. For more information, see York Region (2007b). YDSS — Needs and Justification. Retreived 25 July 2008, from http://www.york.ca/NR/rdonlyres/faanyvi3l7vt4t2xnajvwwiswbcyurw25i3x76jeglbw3kbd4evztahep7buueiag4thkrhmhj4hyhg2cpqa4rvegd/Need+and+Justification+Report+2007.pdf).

in the Region (York Region, 2008e). The "preferred alternative" was to double the capacity by twinning the existing YDSS. The YDSS is a complex wastewater collection system and it converges at the jointly operated (York Region and Durham Region) Duffins Creek Water Pollution Control Plant (WPCP) located in Pickering, Ontario, Canada and discharges treated wastewater into Lake Ontario.

The 1997 YDSS master plan identified several projects that would need an environmental assessment, including the 16th avenue trunk sewer phase I project, which was 6.7 km in length, with a pipe diameter of 2600 mm. Numerous problems emerged associated with the construction of this project. Tunnelling was the method of construction used, which required "de-watering" to lower the water level in the immediate vicinity. In order to safely construct the shafts and the tunnels of this sewer to the required depth, the groundwater level was lowered by pumping in the deep Thorncliffe aquifer to below the depth of the shaft and tunnel (York Region, 2006). Some of this water was returned to area streams and some was discharged into the YDSS (York Region, 2007c). Due to unanticipated sub-surface conditions, more de-watering than originally anticipated was necessary (York Region, 2006). This impacted private wells outside what was originally defined as the zone of influence during the EA process. These events, coupled with inaction by the EA project team to address concerns early on, led to a major media campaign by local residents, environmental movement organisations and politicians criticising the efficacy of the original EA. This was a focusing event, which we describe in greater detail in the *Findings* section below.

Research Methods

Methodological approach

Our methodological approach incorporated participatory action research and the use of a case study. "Participatory action research" (PAR) is a generic term that can refer to several related streams of collaborative forms of research, including PAR (Whyte, 1991), Action Science (Argyris and Schon, 1974, 1978; Schon, 1983) and Co-operative Inquiry (Reason, 1988, 1994; Reason and Heron, 1995), among others. For the purposes of this research, the focus is primarily on Whyte's PAR.

> In participatory action research, some of the people in the organisation or community under study participate actively with the professional researcher throughout the research process from the initial design to the final presentation of results and discussion of their action implications (Whyte, 1991, p. 20).

Our project was conceived, developed and implemented collaboratively with representatives from York Region and Save The Oak Ridges Moraine Coalition and with researchers from the University of Waterloo and Queen's University.

Case studies involve the exploration of a single entity or phenomenon bounded by an event and process. As a research strategy, case studies are used in many situations to "improve our knowledge of individual, group, organisational, social, political and related phenomena" (Yin, 2003 p. 1). The need for case studies arises out of the desire to understand complex social phenomena; a case study allows investigators to retain the holistic and meaningful characteristics of real-life events, such as organisational and managerial processes (Yin, 2003).

Multiple research methods

For the purposes of this research, data was obtained from and verified using the following:

(i) semi-structured interviews;
(ii) analysis of relevant government policy documents; and
(iii) participant observation/workshops.

Semi-structured interviews of key informants were the main source of data about the emerging SEA-type approach in York Region. Twenty-one people were interviewed (see Table 1). Semi-structured interviewing is an interaction between an interviewer and respondent in which the interviewer has a general plan of inquiry but not a specific set of questions that must be asked in particular words and in a particular order (Babbie, 2004). In the semi-structured interview, interviewers "develop, adapt, and generate questions and follow-up probes appropriate to the given situation and the central purpose of the investigation" (Berg, 1995, p. 32). Employing semi-structured interviews allows the researcher to probe a consistent set of issues and hear a variety of perspectives (Berg, 1995; Hughes, 2002).

In addition, each respondent was asked to identify the key documents for understanding the issues and events associated with the history and evolution of York Region's EA and land-use planning processes. All documents were systematically reviewed as a means of triangulation and verification of the qualitative data from the interview, literature review and participant observation. Each policy document was reviewed in terms of its purpose, content and significance for the research. The analysis of policy documents was used to verify details regarding key events, key individuals, groups, organisations and agencies involved in the

Table 1. Interviews conducted.

Interviewee number	Date	Affiliation
1	November 5, 2007	York Region staff
2	November 6, 2007	York Region staff
3	November 14, 2007	York Region staff
4	November 14, 2007	City of Vaughan staff
5	November 14, 2007	York Region staff
6	November 16, 2007	Former Ministry of Natural Resources (MNR) staff
7	November 22, 2007	Environmental NGO representive
8	November 27, 2007	York Region staff
9	November 27, 2007	York Region staff
10	November 28, 2007	York Region staff
11	November 28, 2007	Former Toronto and Region Conservation Authority (TRCA) staff
12	November 29, 2007	TRCA staff
13	November 30, 2007	Former Ministry of the Environment (MOE) staff
14	December 21, 2008	York Region staff
15	March 10, 2008	Consultant
16	June 9, 2008	Markham resident
17	June 20, 2008	Markham councillor
18	July 18, 2008	York Region staff
19	July 18, 2008	York Region staff
20	August 15, 2008	TRCA staff
21	September 11, 2008	Environmental NGO representive

history and evolution of York Region's EA and related master planning and land-use planning processes.

Participant observation refers to "research that involves social interaction between researcher and informants in the milieu of the latter, during which data are systematically and unobtrusively collected" (Taylor *et al.*, 1984, p. 15). This interpretation of participatory observation "assumes that knowledge develops from experience, particularly the experience of social-political action" (Newman, 2000, p. 24). Throughout the research, we engaged participants in various forums and observed participants to complement the in-depth interviews and policy document review. The research team was invited to attend various York Region internal and public meetings and project-related workshops involving York Region staff (eight meetings and two research team-led workshops).

Findings — Evidence of a Policy Window

The following section presents a summary of our analysis from interviews, observations and reports based on the three separate policy window "streams" (problem, political and policy) that converged to move York Region infrastructure planning onto the public policy agenda, i.e., the adoption of an SEA-type approach to planning and decision making in York Region.

The focusing event and the problem stream

One influence on policy change is "the inexorable march of problems pressing in on the system" (Kingdon, 1995 p. 16), and a crisis or a focusing event that brings one issue to the forefront and defines the problem stream. In the York Region case, many issues associated with growth management were pressing on decision makers. Problem recognition, in this case, emerged as a result of a focusing event in the form of a crisis generated during the construction phase of the 16th avenue trunk sewer project in 2002/2003 (one of the 1997 YDSS Master Plan projects). A number of interviewees suggested that a crisis is often needed to promote change (Interviews #3, 10, 14, 15).

More extensive dewatering than predicted through the environmental assessment was required for this project, which caused many residential wells to run dry. Residents in the affected area rely on wells and initially were unable to secure appropriate mitigation. In contacting the contractor, first responses included "you didn't look after your well, it's your problem" to even "the environmental assessment study says this is the zone of influence, and your well is outside, so it must be something else" (Interviews # 15, 16 and 17). Given that this was a design/build contract, the responsibility for mitigation measures was left with the contractor rather than the Region of York (Interviews #8, 15). Since the contractors' profits would be impacted when mitigation measures were required, this led to feelings of a potential conflict of interest by some residents (Interviews #15, 16, 17 and 21). This conflict was the basis for growing public opposition during the construction phase. One York Region staff explained, "we did not address these problems upfront so relationships with the public, the major stakeholders and the area municipalities got strained. We stonewalled and did not admit the problem for a while" (Interview #10).

Another component of the problem stream was associated with the level of dewatering necessary for this project resulting in changes to base flow of nearby streams negatively impacting fish habitat (Interview #21). Robinson Creek was affected in the Rouge system as the recipient creek of pumped water from the project. This led to flow augmentation and erosion problems including changes to fish habitat (i.e., Robinson creek is known for being a great trout stream in the Rouge). Therefore, impacts on nearby streams were also one of the reasons for public and key stakeholders' outcry.

Besides these more visible issues as an outcome of the project, which certainly focused the attention of the public, was a general public perception of a lack of transparency in the planning/EA process that further heightened the confrontational relationship (Interviews #6, 7 and 16). A related concern raised during interviews was the perception that contractors cannot be seen as independent experts, because they were working for the Region and the Region's interest/goals (Interviews #16 and 21). This perception amplified the general lack of trust in the whole process by affected residents.

The corporate culture in the Region of York, before this focusing event, encouraged EAs to be completed at minimal costs and on short timelines (Interviews #2, 3 and 14). This project was illustrative of this culture and resulted in the problems noted above. The focusing event forced the Region to recognise the need for additional background studies especially in terms of hydrogeology and increased public participation throughout the process. The resulting higher costs and extended timelines of subsequent EAs were acknowledged by many interviewees to be a necessary condition for better overall project-level assessment and decision making (Interviews #3, 5, 14, 15). This view especially resonated with a number of interviewees based on the cost of delays, negative publicity associated with the project, social impacts to citizens of York Region and ecological impacts (Interviews #3, 10, 18 and 20).

A final element of the problem stream was associated with the water and wastewater master planning process. This process foreclosed viable project-level alternatives because the master plan favoured large-scale engineering solutions leading to the YDSS project. This foreclosed conservation and smaller-scale infrastructure alternatives at the project level. Interviewees noted that the only viable alternatives, as a result of the water and wastewater master plan, were to minor routing alternatives to the "big pipe" (Interviews #2, 5 and 11).

Political stream

The main issue discussed above (i.e., dewatering impacting wells and streams) was the catalyst for public opposition to the project. Additional issues cited by interviewees related to the 16th avenue trunk sewer project that contributed to greater public opposition and pressure to change included poor communication with affected residents (Interviews #6, 7, 15, 16) and lengthy response times by the Region to address construction-related problems (Interviews #15, 17, 20).

After requests to the Ministry of the Environment (MOE) to have the environmental assessment bumped-up to an individual EA[3] (all bump-up requests were denied), the response from the Minister in 2004 through a letter to York Region required York Region to address over 40 conditions for current and future YDSS-related EAs. Among the conditions included more monitoring of the aquifer and making the information available to the public on a website. Interestingly two YDSS-related projects (i.e., Southeast collector project and the Upper Leslie street truck project) were voluntarily bumped-up by York Region to Individual EAs (a more extensive process requiring MOE involvement).

Overall, York Region received extremely negative press and strong public opposition to the 16th Avenue process. As a result, the Region, together with other agencies, developed a $30 million Environmental Management Plan and Well Mitigation Plan to manage impacts resulting from the future phases of the project. The conditions laid out in the Minister's letter were raised by some interviewees as beneficial and crucial in determining the direction and resolution of the problems associated with the identified focusing event (Interviews #2, 15 and 21). A few York Region staff however, argued that the Minister's letter was not a top down imposition of completely new conditions, but rather a catalyst that further accelerated the pace of a few initiatives already underway in York Region, such as the development of the York Region Sustainability Strategy (Interviews # 10 and 18). Regardless, MOE's involvement and the Minister's letter definitely played an important role in the evolution of York Region's approach to EA and infrastructure planning.

Furthermore, Regional Council was identified as having played a critical role in relation to improving the Region's planning and EA approach. As stated by interviewees # 1, 3 and 18, Regional Council was "open to a new approach" especially because they considered that there was "no other choice but to do things differently" due to public concerns and municipal credibility. The focusing event described above resulted in Regional Council's openness to a new, more strategic approach to EA and infrastructure planning. The Minister's letter and regional council's interest in changing York Region's reputation provided the political will to open the policy window.

Policy stream

An SEA-type approach to planning and decision making represented a key policy option, or solution, to the problem identified in the problem stream. York Region

[3] An Individual EA is a more strict process as compared to the municipal class EA and requires the involvement of the MOE in the approval procedure. It also takes more time and has greater public participation.

embarked on a number of longer-term visioning initiatives designed to inform both infrastructure master planning and related project-level EAs including the creation of the "Towards Sustainability Advisory Group", which drafted a strategic document called "York Region's Sustainability Strategy" (York Region, 2007d). This document is intended to inform all York Region-related activities on the policy, plan and program levels through project-level implementation. Public consultation processes were recognised as being critical for transparency and informed decision making. As a result, York Region increased public engagement and education efforts, and interviewees noted that there has been "a lot of investment in terms of how they consult with the public and the manner in which they do the environmental assessment work" (Interview #10). The Region allocated substantial resources in support of these initiatives, including substantial in-kind contributions to our SEA-focused research.

Integration and tiering is another aspect that was slowly put in place, as Master Plans increase their relationship and guidance to project-EAs. This kind of guidance appears to be present in a recent Individual Environmental Assessment Sewage Servicing project that is taking into consideration the sustainability principles and criteria developed for the Water and Wastewater Master Plan (York Region, 2010; Interviews #14 and 15). Since this is a very recent initiative, it is still early to say whether this will get to the ground or not, but what seems to be happening is a "filtering down" of the York Region Sustainability Strategy principles going through and guiding Master Plans and further influencing the consideration and assessment of alternatives at the project level (usually referred to as tiering in the SEA literature).

The Region has also synchronised transportation with water and wastewater master planning to ensure increased cross-departmental or horizontal coordination and knowledge sharing (Interviews #1, 3, 8, 10, 18). In addition, the Region has committed to collaborative planning processes with the intention of creating genuine partnerships with private sector corporations and community and voluntary organisations (Interviews # 3, 10, 14, 18, 19). This may be interpreted as a first step in recognising that complex issues associated with EA and infrastructure planning require discourse and integration of multiple perspectives.

Discussion

The discussion around the focusing event and the three streams highlight the opening of a window of opportunity (i.e., policy window) that led to policy change in York Region. The main argument is that a focusing event and the resulting crisis moved citizens, and later the municipality, to identify and act on the perceived problem. One key response was the approach initiated by York Region, which had remarkable similarities to what is called in the literature an "SEA-type" approach (i.e., an approach to planning and decision making that does not carry the SEA label, nor does it follow formal SEA specifications, but has to some degree essential SEA elements in place, including an approach that is: applied early and proactively, participative, adaptive, sustainability-centred, tiered and integrated).

Furthermore, the analysis revealed plenty of potential for SEA, as a proposed solution, to improve strategic decision making in York Region. Accordingly, two main expected benefits include:

- better informed, more credible and more broadly beneficial strategic initiatives; and
- clearer and more timely guidance for subsequent undertakings.

In this perspective, Kingdon's (1995) policy window concept shows promise as a useful instrument for identifying windows of opportunities for SEA adoption.

Conclusions and Recommendations

This paper provides a first attempt to better understand how SEA adoption becomes a government priority by looking at this issue through a policy window lens. As such, this paper described the opening of a policy window in York Region as a result of a crisis generated during the construction phase of a sewer trunk project in 2002/2003. Analysis revealed that opportunities for change, articulated as a policy window, appeared after problems were recognised during the 16th Avenue project-level undertaking (i.e. the focusing event). Problem, policy and political streams converged, providing the necessary conditions for York Region to embark on initiatives focused on improving environmental assessment and infrastructure planning in the Region.

It is probably reasonable to say that it will take years for many of the outcomes of the SEA-type approach in York Region to become evident. However, procedural and some structural changes have occurred including the development of a sustainability strategy (York Region, 2007d), the synchronisation of master planning (York Region, 2007e, 2008d), proper consideration of a full suite of alternatives at the master plan level (Interviews #3, 8 and 14), improved relationship — tiering — between different levels of decision making (Interviews # 1, 8, and 10) and improved public consultation (York Region, 2007d). The resilience of these changes is unknown at the time of publication.

The policy window approach proved useful as an evaluation instrument, clearly identifying the problem stream (i.e., environmental assessment crisis), a viable solution (i.e., a more strategic, collaborative and integrated approach to planning and environmental assessment) and the political will to enact the solution (in the form of a letter from the Minister of Environment and a willing municipal council) and how all three streams converged, or "coupled", to provide a viable opening for policy change in York Region.

Based on the insights gleaned from our analysis the following three recommendations are presented:

(1) *Taking advantage of policy windows*: Planning, EA practitioners and other stakeholders should become familiar with the concept and requirements of the opening of a policy window, in the form of "coupling" the problem, policy and political streams to better take advantage of such opportunities to make innovative policy change.
(2) *The policy stream and SEA*: SEA holds significant potential to inform the solution stream (or policy stream) associated with policy windows that may open within the domain of environmental assessment and infrastructure planning. EA practitioner and stakeholders should apply SEA approaches and tools to counteract the limitations of project-level environmental assessment and complement the overall assessment system in place.
(3) *Future research*: Follow-up research should be carried out to: further examine the utility of using the policy window approach for research on SEA; and examine the long-term implications and resilience of the policy solutions developed in the York Region case.

Acknowledgements

The authors wish to thank the Canadian Environmental Assessment Agency's Research and Development Programme and support staff for providing the funding and resource support that allowed this project to be completed. We thank the Regional Municipality of York for generously supporting this project with staff time, resources and access to meetings and their consultants. We also thank two anonymous referees for their useful suggestions and comments.

References

Abaza, H, R Bisset and B Sadler (2004). Environmental impact assessment and strategic environmental assessment: towards an integrated approach. Retrieved 8 Dec 2004, from http://www.unep.ch/etu/publications/textONUBr.pdf

Alshuwaikhat, HM (2005). Strategic environmental assessment can help solve environmental impact assessment failures in developing countries. *Environmental Impact Assessment Review*, 25(4), 307–317.

Argyris, C and DA Schon (1974). *Theory in Practice: Increasing Professional Effectiveness*. San Francisco: Jossey-Bass.

Argyris, C and DA Schon (1978). *Organizational Learning: A Theory of Action Perspective*. Reading, Mass: Addison-Wesley Publishing Company.

Babbie, ER. (2004). *The Practice of Social Research*, 10th ed. Belmont, CA: Wadsworth Thomson Learning.

Benevides, H, D Kirchhoff, RB Gibson and M Doelle (2008). Law and policy options for strategic environmental assessment in Canada. Ottawa, Canada: Report prepared for the Canadian Environmental Assessment Agency.

Berg, BL (1995). *Qualitative Research Methods for the Social Sciences*, 2nd ed. Toronto: Allyn and Bacon.

Bina, O (2003). Re-conceptualising strategic environmental assessment: theoretical overview and case study from Chile. Unpublished PhD Thesis, University of Cambridge, Department of Geography.

Birkland, TA (1997). *After Disaster: Agenda Setting, Public Policy, and Focusing Events*. Washington, DC.: Georgetown University Press.

Brown, AL and R Thérivel (2000). Principles to guide the development of strategic environmental assessment methodology. *Impact Assessment and Project Appraisal*, 18(3), 183–189.

Caratti, P, H Dalkmann and R Jiliberto H (eds.) (2004). *Analysing Strategic Environmental Assessment: Towards Better Decision-Making*. Cheltenham, UK; Northampton, MA: Edward Elgar.

Carter, J and J Howe (2006). The water framework directive and the strategic environmental assessment directive: exploring the linkages. *Environmental Impact Assessment Review*, 26, 287–300.

Cashmore, M (2004). The role of science in environmental impact assessment: process and procedure versus purpose in the development of theory. *Environmental Impact Assessment Review*, 24(5), 403–426.

Clark, BD (2000). Making EIA count in decision making. In *Perspectives on Strategic Environmental Assessment*, MR Partidário and R Clark (eds.). Boca Raton, Fla.; London: Lewis Publishers.

Dalal-Clayton, DB and B Sadler (2005). *Strategic Environmental Assessment: A Sourcebook and Reference Guide to International Experience*. London, UK: Earthscan.

Dalal-Clayton, DB and S Bass (2002). *Sustainable Development Strategies: A Resource Book*. Retrieved 26 May 2005, from http://www.nssd.net/working/resource/indexa.htm#contents

Dusik, J, TB Fischer and B Sadler (2003). *Benefits of Strategic Environmental Assessment*. Retrieved 20 Sept 2005, from http://europeandcis.undp.org/files/uploads/Peter/SEAengl.pdf

Fischer, TB (1999). Benefits arising from SEA application – a comparative review of North West England, Noord-Holland, and Brandenburg-Berlin. *Environmental Impact Assessment Review*, 19(2), 143–173.

Fischer, TB (2002). *Strategic Environmental Assessment in Transport and Land Use Planning*. London; Sterling, VA: Earthscan Publications Ltd.

Fischer, TB (2003). Strategic environmental assessment in post-modern times. *Environmental Impact Assessment Review*, 23, 155–170.

Fischer, TB (2004). Transport policy making and SEA in Liverpool, Amsterdam and Berlin — 1997 and 2002. *Environmental Impact Assessment Review*, 24(3), 319–336.

Fischer, TB (2006). Conference Report: IAIA Conference on "International Experience and Perspectives in SEA", 26–30 September 2005, Prague, Czech Republic–Conclusions Stream C: linkages between SEA and other assessment or planning tools. *Journal of Environmental Assessment Policy and Management*, 8(4), 495–504.

Fischer, TB (2007). *The Theory and Practice of Strategic Environmental Assessment: Towards a More Systematic Approach*. London, Sterling; VA: Earthscan.

Gibson, RB (2000). Specification of sustainability-based environmental assessment decision criteria and implications for determining "significance" in environmental assessment. Retrieved 13 Mar 2009, from http://www.ceaa-acee.gc.ca/default.asp?lang=En&n=086E7767-1

Gibson, RB (2007). *Lecture — Strategic Environmental Assessment*. University of Waterloo, Ontario, Canada: University of Waterloo.

Hildén, M, E Furman and M Kaljonen (2004). Views on planning and expectations of SEA: the case of transport planning. *Environmental Impact Assessment Review*, 24(5), 519–536.

Hughes, M. (2002). Interviewing. In *Research Methods for Postgraduates*, T Greenfield (ed.), 2nd ed. London: Oxford University Press Inc.

IAIA. (2002). *Strategic Environmental Assessment Performance Criteria*. Retrieved 20 Apr 2005, from http://www.iaia.org/publicdocuments/special-publications/sp1.pdf

Jones, CE, M Baker, J Carter, S Jay, M Short and C Wood (eds.) (2005). *Strategic Environmental Assessment and Land Use Planning: An International Evaluation*. London; Sterling, VA: Earthscan.

Kingdon, JW (1984/1995). *Agendas, Alternatives, and Public Policies*, 2nd ed. New York: Longman.

Michaels, S, NP Goucher and D McCarthy (2006). Policy windows, policy change and organizational learning. *Environmental Management*, 38, 983–992.

Ministry of Public Infrastructure Renewal (2006). *Places to Grow — Growth Plan for the Greater Golden Horseshoe*. Retrieved 21 Jul 2008, from http://www.placestogrow.ca/images/pdfs/FPLAN-ENG-WEB-ALL.pdf

MMAH (2008). *Full map of the Oak Ridges Moraine Area — Land Use Designation Map (Ontario Regulation 140/02)*. Retrieved 5 Aug 2008, from http://www.mah.gov.on.ca/Page1744.aspx

Newman, WL (2000). *Social Research Methods: Qualitative and Quantitative Approaches*. 4th ed. Boston, USA: Allyn and Bacon.

Nooteboom, S (2000). Environmental assessments of strategic decisions and project decisions: interactions and benefits. *Impact Assessment and Project Appraisal*, 18(2), 151–160.

Partidário, MR (2000). Elements of an SEA framework — improving the added-value of SEA. *Environmental Impact Assessment Review*, 20(6), 647–663.

Partidário, MR (2005). The contribution of strategic impact assessment to planning evaluation. In *Beyond Benefit Cost Analysis: Accounting for Non-Market Values in Planning Evaluation*, D Miller and D Patassini (eds.), pp. 151–162 Aldershot; Burlington, VT: Ashgate.

Petts, J (ed.) (1999). *Handbook of Environmental Impact Assessment*. Oxford; Malden, MA: Blackwell Science.

Reason, P (1988). *Human Inquiry in Action: Developments in New Paradigm Research*. London: Sage.

Reason, P (1994). Three approaches to participative inquiry. In *Handbook of Qualitative Research*, NK Denzin and YS Lincoln (eds.). Thousand Oaks: Sage.

Reason, P and J Heron (1995). Co-operative inquiry. In *Rethinking Methods in Psychology*, R Harre, J Smith and L Van Langenhove (eds.), pp. 122–142. London: Sage.

Retief, F (2007). Effectiveness of Strategic Environmental Assessment (SEA) in South Africa. *Journal of Environmental Assessment Policy and Management*, 9(1), 83–101.

Sadler, B and RAA Verheem (1996). *Strategic Environmental Assessment — Status, Challenges and Future Directions:* Ministry of Housing, Spatial Planning and the Environment of the Netherlands.

Schmidt, M, EM João and E Albrecht (eds.) (2005). *Implementing Strategic Environmental Assessment*. Berlin: Springer.

Schon, DA (1983). *The Reflective Practitioner: How Professionals Think in Action*. New York: Basic Books.

Solecki, W and S Michaels (1994). Looking through the postdisaster policy window. *Environmental Management*, 18(4), 587–595.

Stinchcombe, K and RB Gibson (2001). Strategic environmental assessment as a means of pursuing sustainability: ten advantages and ten challenges. *Journal of Environmental Assessment Policy and Management*, 3(3), 343–372.

Stockholm Environmental Institute. (2001). *Strategic Environmental Assessment: Methodology and Application in the Energy Sector*. Retrieved 10 Jun 2005, from www.sei.se/policy/sea-e.html

STORM coalition (2008). *The Moraine*. Retrieved 5 Aug 2008, from http://www.storm-coalition.org/pages/moraine.html

Tang, Z. (2008). Integrating the principles of strategic environmental assessment into local comprehensive land use planning. *Journal of Environmental Assessment Policy and Management*, 10(2), 143–171.

Taylor, SJ, J Steven and R Bogdan (1984). *Introduction to Qualitative Research Methods.* New York: John Wiley and Sons.

Thérivel, R (2004). *Strategic Environmental Assessment in Action.* London; Sterling, VA: Earthscan.

Thérivel, R and MR Partidário (eds.) (1996). *The Practice of Strategic Environmental Assessment.* London: Earthscan Publications.

Thurber, JA. (2003). Foreword. In *Agendas, Alternatives and Public Policies*, JW Kingdon (ed.). New York: Longman.

Vicente, G and MR Partidário (2006). SEA: enhancing communication for environmental decisions. *Environmental Impact Assessment Review*, 26, 696–706.

Whyte, WF (1991). *Participatory Action Research.* Newbury Park, California.: Sage Publications.

Yin, RK (2003). *Case Study Research: Design and Methods.* 3rd ed. Thousand Oaks, Calif: Sage Publications.

York Region (2006). *York Durham Sewage System — 16th Avenue, Phase I.* Retrieved 29 Jul 2008, from http://www.york.ca/NR/rdonlyres/o2nra4u5qip3z5dm647lbymbbc43b7ppok3x-lyzlhysrydl6zrtplsk3jqlstez tjzygnxxxyvvhqgun643oqkqdff/A6+–16th+Ave-phase+one+-+March+2006.pdf

York Region (2007a). *The Region of York Official Plan.* Retrieved 5 Aug 2008, from http://www.york.ca/Departments/Planning+and+Development/Long+Range+Planning/ROP.htm

York Region (2007b). *YDSS – Needs and Justification.* Retrieved 25 Jul 2008, from http://www.york.ca/NR/rdonlyres/faanyvi3l7vt4t2xnajvwwiswbcyurw25i3x76jeglbw3kbd4-evztahep7buueiag4thkrhmhj4hyhg2cpqa4rvegd/Need+and+Justification+Report+2007.pdf

York Region (2007c). *York Durham Sewage System Overview.* Retrieved 25 Jul 2008, from http://www.york.ca/NR/rdonlyres/uha34eyjy33qooqab25c6mxh7ralshrt3iggjwbvqphxj-qygsoyrqhf7vesm5gjp7lm3orqtehwixsflrvdpowxf6g/A9+–+YDSS+Overview+-+February+ 2007.pdf

York Region (2007d). *York Region Sustainability Strategy — Towards a Sustainable Region.* Retrieved 19 Dec 2008, from http://www.york.ca/NR/rdonlyres/ddvav3nrw2657f4dlj-c4m3kbxa4v35qzqdtw62f3zce5cvwur4wbcymuvn oobehvgdkpzgsjrzeohstbduin2jlsmd/Final+Sustainability+document.pdf

York Region (2007e). *York Region Transportation Sustainability – Principles, Objectives and Example Performance Indicators.* Retrieved 19 Dec 2008, from http://www.york.ca/NR/rdonlyres/xsm4evyds5hvf734obmhbq5wlgfpb54jic45wmfcu4baeowdcxkl5l63dcw7oco2hgfkyqp533x65prp7vpzgmlcbb/%28IP%29+TMP+-+Sustainability+Principles+%28Nov+02- 07%29.pdf

York Region (2008a). *Council Members.* Retrieved 29 Jul 2008, from http://www.york.ca/regional+government/council/default+council.htm

York Region (2008b). *Infrastructure Planning Projects and Initiatives.* Retrieved 22 Jul 2008, from http://www.york.ca/Services/Regional+Planning/Infrastructure/default_InfrastructurePlanning.htm

York Region (2008c). *Planning for Tomorrow — York Region Growth Management.* Retrieved 19 Dec 2008, from http://www.york.ca/Departments/Planning+and+Development/Growth+Management/default+Growth+M anagement.htm

York Region. (2008d). *Water and Wastewater Sustainability Strategy.* Retrieved 19 Dec 2008, from http://www.waterwastewatermasterplan.ca/york/files/Water%20and%20Wastewater%20Sustainability%2 0Strategy%20FINAL.pdf

York Region (2008e). *York Durham Sewage System Information Centre.* Retrieved 25 Jul 2008, from http://www.york.ca/York%2c+Durham+Sewer+System+Front+Page.htm

York Region (2008f). *York Region Facts — Population Growth in Municipalities.* Retrieved 22 Jul 2008, from http://www.york.ca/About+Us/York+Region+Facts/Population+GrowthinMunicipalities.htm

York Region (2009). *York Region Official Plan.* Retrieved 1 Apr 2010, from http://www.york.ca/Departments/Planning+and+Development/Growth+Management/default+Growth+M anagement.htm

York Region (2010). *UYS Solutions Information.* Retrieved 7 Apr 2010, from http://www.uyssolutions.ca/en/information/ProjectDetails.asp

CHAPTER 3

EIA Theories: All Chinese Whispers and No Critical Theory

Joe Weston

Department of Planning, Oxford Brookes University
Oxford, United Kingdom
wjweston@brookes.ac.uk

There have been a large number of attempts to develop a theory or theories of Environmental Impact Assessment in order to justify its use in environmental decision-making. A review of academic articles demonstrates that these theories are largely drawn from planning theories. Planning theories are themselves a development of sociological theories of decision-making and from one particular strand of sociological theory. In this review of the theories of EIA it is argued that an understanding of wider sociological theory is necessary to fully understand both planning and EIA.

Keywords: Environmental Impact Assessment; planning; theory; critical theory.

Since 1997 and Lawrence's (1997) call for theory building in Environmental Impact Assessment (EIA) there has been a great deal of discussion in the academic literature on the ideas that seek to explain and justify the use of EIA in environmental decision-making. Using electronic Library databases a key word search was made for EIA articles dealing with the theory of EIA. While developments in EIA, such as Strategic Environmental Assessment (SEA), have also been subject to a theoretical examination in academic journals, this research has been limited to EIA as it is the original source from which those developments have sprung and hence the area of environmental assessment where most theorising has taken place. The search revealed that between 1997 and June 2010 31 such articles were published in the journals: *Environmental Assessment and Review*; *Impact*

Assessment and Project Appraisal; *Journal of Environmental Assessment Policy and Management*; *Journal of Environmental Management*; and *Journal of Environmental Planning and Management*. These journals were chosen for the search as these deal more directly with EIA than other journals and it is acknowledged that EIA theory is discussed in a very wide array of other academic journals on planning, the environment and related fields. However the search was limited to the above articles in order for the search to be manageable within time and resource constraints.

The theories of the role of EIA in environmental decision-making discussed in these articles have not developed in a vacuum and have grown out of other theoretical frameworks. The references and sources cited in the 31 articles identified above were reviewed to determine the main source of the theories being discussed. Planning theory appears to be the major source for writers on EIA theory within the reviewed articles. Within the 31 articles prominent planning theorists are cited 86 times with Allmendinger (2002), Faludi (1973), Flyvbjerg (1998), Forester (1989), Healey (1997) and Tewdwr-Jones (1996) accounting for almost 72% of these references.[1] The dominance of planning theory in the discussion of EIA theory in these articles does not suggest that planners have been the sole source of theories into EIA, and Bartlett and Kurian (1999) provide a useful typology of EIA theories. However, as suggested by Glasson *et al.* (2005, pp. 13–14) and Lawrence (2000) planning theory and EIA theory have very strong ties.

From planning theory two major strands of EIA theory have developed: rationalist procedural theories that emphasise the role of EIA in the project or plan decision-making process; and substantive theories that see EIA as part of a wider social learning and communicative process in the project of sustainable development. That these planning theories have been developed from sociological decision-making theories appears to be less well recognised. Of the 31 articles reviewed 20 included references to major sociological figures who have made contributions in the development of the decision-making theories that underpin much of the theorising on EIA. However, in total only 37 direct references were made to these core writers. Weber (1864–1920), a founding father of sociology and the source of the theories on rationality in decision-making was cited only twice. Simon (1957, 1976), whose work is the basis for much of the rational procedural theories of EIA is mentioned only 3 times and critiques of the pure rational model and developers of alternative models of decision-making, Lindblom (1959), Dror (1964) and Etzioni (1967), are collectively only cited

[1] The references referred to here are examples rather than the full list of works referred to in the text of the articles.

11 times between them. Where these theorists are referred to in the text it is usually only in passing with citations such as "see for example Lindblom and Etzioni" or in footnotes that steer the reader to a discussion of the theories. Only Persson (2006) of the reviewed articles provides a detailed discussion of Weber and his notions of rationalism. Habermas' (1984) theory of communicative rationality lies at the heart of the so-called substantive theory of EIA and is cited only nine times. Again the majority of the citings in the reviewed texts are cursory, with only Elling (2009) providing an in-depth consideration of Habermas' work.

While the articles reviewed for this paper draw on sociological and planning theory they do so only partially. In the development of EIA theories there appears to have been an eagerness to embrace the sociological perspectives of Weber and yet no real examination of the sociological critiques of Weberian theories that come from the Marxist school of sociology. This reluctance to understand a critical theory examination of planning theories by writers on EIA weakens the whole basis of EIA theory building. EIA theories are a development of planning theory and planning theory has been largely undermined by a critical theory analysis.

There is now a widespread acceptance that EIA has its theoretical roots in the 1960s dominant view in planning, and environmental management generally, that rationalism and rational procedures and methods lead to more rational (better) decisions (see Barrow, 1997, p. 6; Caldwell, 1988; Weston, 2004). For Lynton Caldwell (1913–2006), the main author of the USA's 1969 NEPA (Eccleston, 2006), rationalist planning theory was one of the core principles on which EIA was built (Caldwell, 1988). Rationalist planning theory has its roots in the sociological traditions of Max Weber and Weber's work is largely based on a critique of Marx and he is often portrayed as being in a constant battle with Marx (Giddens, 1986, p. 13; Löwith, 1982). Weber rejected Marx's historical materialism and dismissed the structure — superstructure model as naïve (Forte, 2008). For Weber it was the development of particular beliefs (Calvinism) and its links with rational economic ethics that were the causal determining factors in the development of capitalism (*ibid.*). While for Marx the key to understanding modern society was an analysis of capitalism and the social structure and superstructure it created, for Weber the key was the inevitable and unstoppable development of rationalism (Weber, 1979; Forte, 2008).

Weber believed that the roots of capitalism and the modern world lay in an increasing move towards the rationalisation of society, which was characterised by the bureaucratisation of organisations (Weber, 1979; Abrams, 1982, p. 4). This rationalisation of society had, for Weber, been occurring throughout history and was irreversible (Lee and Newby, 1985, p. 177). Abrams (1982) argues that for Weber rationalisation would lead to a society where faith is replaced by specialist scientific knowledge and objective experts in science and technology would become the dominant source of intellectual knowledge (Abrams, 1982, pp. 83–86). The distinctive feature of capitalism was, for Weber, the development of routinised procedures, that he termed instrumental rationalism, that were based upon impersonal, 'rational legal' rules (Weber, 1947, pp. 324–333, 1997; Giddens, 1986, p. 80). Giddens (2006) describes Weber's notion of rationality as the "*organisation of social and economic life according to principles of efficiency on the basis of technological knowledge*" (p. 114). Rules become the key to ensuring uniformity and, together with rational decision-making, the character, opinions and values of the individual decision-maker become irrelevant (Gore *et al.*, 1992, p. 4). Under the rational legal model created by Weber a bureaucratic organisation must be based upon a division of labour cast within a hierarchy of decision-making (Weber, 1997). Within that hierarchy there had to be experts and a bureaucracy would need to maximise the use of expert knowledge in order to be efficient (see Hill, 2005, pp. 199–202). In this way, it is argued, the instrumental rationality of decision-making procedures subordinates subjective judgement to the structures and procedures of decision-making (Adams and Balfour, 1998, p. 166).

Following Weber, Herbert Simon (1957, 1976) applied the concept of rationalism to an understanding of decision-making in organisations and he is seen as one of the major contributors to the development of decision-making theory (Hill, 2005, p. 146). The basic assumption within Simon's ideal model of rational decision-making is that decision-makers act purely in the interests of the organisation and, following Weber, their training and professionalism allow them to put all other considerations to one side. Decisions are arrived at following a:

- systematic consideration of all possible alternatives;
- an assessment of all possible solutions; and
- an analysis of all available information which is objectively assessed as it is always quantifiable.

Simon's rational model is an "ideal type" that is put forward as a measure against which real decisions should be made (Hill, 2005, p. 172). Critics of the pure rationalist model developed theories that rejected the notion of value-free, objective decision-makers. Lindblom's (1980) incrementalism argued that big rational decisions are never made and decision-makers only take small steps away from an existing and politically preferred position. Etzioni's (1967) "mixed scanning" and Dror's (1964) 'optimal' decision-making theories suggest that decision-makers make use of rational tools as part of essentially political decisions (see Weston, 2000). All these theories have their roots in the notion that a rational decision-making process, free of values, will lead to 'better' decisions. Because of its links with, and roots in, notions of rational scientific process, rational approaches to decision-making have dominated many areas, including environmental, of administrative decision-making. Habermas, for example, took the view that instrumental rationality has colonised the whole of economic and administrative systems (Dillar and Ruchala, 2005). Indeed, rationalism became part of the social paradigm of modernity and Gamble (1981) argued that much of the construction of modern society was based upon the lending of authority to rational knowledge and the very project of modernity was to give value to "*only those things counted as knowledge which could be shown to be rational...*" (p. 107). Where such a paradigm exists it sets the agenda for the legitimising of decision-making and as such decision-making processes like planning and procedures such as EIA must, at the very least, aspire to be rational.

For Taylor (1998), planning itself can be seen as part of the project of rationalist modernity as the very notion that cities would be improved by the application of rational thought and action is surely based upon a belief in rationality as a force for good (p. 164). Hall (1982) describes post-1960s planning as being based upon a desire for "*systematic value free consideration of a range of alternatives to achieve a policy objective*" (p. 274). Faludi (1983), a major figure in the development of planning thought, says that "*decisions should be taken by evaluating the consequences of the alternatives that offer themselves to the decision maker. To be more precise, all consequences of all alternatives need to be considered*." This approach to planning theory remains a major theme, Leknes (2001) for example, believes that planning decisions should be based upon the consideration of problems, solutions, and alternatives by "experts" within a value free bureaucratic structure. For Rydin (2003) the notion that planning is a rational process using expert knowledge provides planning decisions with their legitimacy (p. 3). With its emphasis on systematic procedures, a consideration of alternatives by objective experts, it is not difficult to see why rationalist planning theory and EIA are so inextricably linked. Glasson *et al.* (2005, p. 13) point to the 'climate of a rational approach to decision-making' that prevailed in the 1960s when EIA first emerged. Indeed the language of rationalism dominates almost any discussion of what EIA is, what it does and how it should be performed (see Bartlett and Kurian, 1999; Weston, 2000).

Marxist critical theory is largely dismissive of urban planning and planning theory generally because the basis of Marxist theory is its historical deterministic social constructionism that explains rationalism and the predominantly rationalistic views of planning theorists, as part of the ideological superstructure of capitalism. Central to an understanding of a Marxist explanation of planning is the role of the state in capitalism (Allmendinger, 2009, p. 84). The state is seen as part of the superstructure of capitalism and there to protect the interests of capital and facilitate the accumulation of capital. A Marxist theory of planning explains it as a mechanism for legitimising decision-making by the state through the claim of value-free rationalism and scientific method. For Paden (2003) this allows the values of the market to be part of the planning process without being contested as those values themselves are characterised as rational and "natural". The rational models of planning theory are seen in critical theory as being based upon values of economic efficiency and notions of what represented the public good and public interest. Campbell and Marshall (2002) argue that the whole legitimisation of planning intervention in the market is predicated on the notion of the public interest. Planning theorists have long debated what constitutes the public interest in planning and there is recognition that the concept is problematic for planners (Campbell and Marshall, 2002). Healey (1997), for example, accepts that there can be a plurality of what would constitute the public interest and from a collaborative planning perspective argues that the public interest can only be discovered 'discursively' (p. 297). A critical theory approach to notions of what constitute the "public good" or 'public interest' is to ask who defines what the public good or public interest is. The "public interest" is determined by the agents of the state and in the UK, the public interest for planning decisions is defined by the government through national policy statements. The "public interest", and the values that underpin it, will therefore coincide with what is in the interests of capitalism at any particular point in time. From a critical theory perspective the public interest is an ideological social construct that will support, or at least not question, capitalism (Paden, 2003).

Furthermore, Paden (2003) argues that Marxism questions the whole legitimacy of the rationalist theories on which planning is justified. The use of rationalism as the legitimising force behind planning and EIA, permits social problems to be explained as manifestations of bureaucratic policy failure, rather than the direct logic of capitalism (Allmendinger, 2009, p. 98). The failure of planning to deal with the market forces that shape the urban environment is not that urban planners have failed, but that planning is part of the superstructure of capitalist social relationships and was put in place to help not hinder the market (Taylor, 1998, pp. 105–107). Indeed, the planning process in the UK acts to increase, maintain and protect property values by creating an artificial form of scarcity. To grant planning permission increases the value of land while to refuse planning permission maintains or protects the value of neighbouring land (Allmendinger *et al.*, 2003).

Since the 1960s the rationalist paradigm became increasingly under attack by Marxist critical theorists and postmodernists alike. Rationalism is seen both as part of the ideological superstructure of capitalism that justifies and reinforces capitalist relationships or alternatively, for the post-modernists, a denier of values and culture. In critical theory rationalism is seen as being value based and inextricably linked to capitalism and that the use of terms such as "rational decision" are inherently decisions that support capitalist economic relationships (Agger, 1991; Weston, 2006). Indeed Weber believed that rationalisation was the engine for capitalism (Giddens, 2006, p. 113). The Frankfurt School of sociology, Marcuse, Horkheimer and Adorno, saw positivist science and rationality, developed from the Enlightenment onwards, as the dominant ideology within capitalism that teaches people to accept the world as it is as natural and in this way rationalism perpetuates capitalism as a given system that cannot be changed (Agger, 1991). Rational science is seen as justifying capitalism in that as rules or laws of nature existed so do rules or laws of economics. Based upon his social constructionist perspective, Marcuse argued that science and technology were not free of the power base of capitalism and that any claims to be neutral, objective and rational are simply an attempt to hide the true nature and purpose of science in a capitalist system (Feenberg, 2008).

This congruence between rationalism and capitalism means that, for Smith and May (1997), rationalism is inherently value-based: something that is said to be "rational" is widely seen as being good, correct and efficient as it will represent this 'natural' condition of capitalist relationships and values. Furthermore, critics of rationalist, objective science point to the distinctive and selective nature of scientific knowledge and the discounting of other forms of knowledge as 'irrational', mystic or ideologically driven, while much of science is itself ideologically driven and supports the prevailing social, economic and political paradigm of capitalism. There have, of course, been very many critical assessments of the links between capitalism and scientific method (see, Agger, 1991; Pepper, 1984, Chapter 5; Huxham and Sumner, 2000, p. 71; Yearley, 1988 and 1991). For Walker (2001) science becomes 'socially constructed' as scientific questions are created, shaped and manipulated by the very many, often conflicting, interests that participate in the development of scientific investigation (p. 280). The use of scientific method in EIA, a cornerstone of the legitimacy of its use, provides a good example of the value-laden nature of science. Values and interests underpin the "science" of EIA just as they do in other aspects of the process such as screening and scoping (see Weston, 2000). Some of these values and interests may come from the training of the assessors and the rationalist culture that lies behind their discipline. For despite the many criticisms of rationalism and the scientific method they remain ingrained in our culture and belief systems and are difficult to dislodge from our understanding of how to assess change (Owens *et al.*, 2004;

Weston, 2006). Other values used to determine the type of science used in EIA prediction will come from the developer who commissions the investigations. In most EIA systems, and certainly in the UK, it is the project proponent that is responsible for producing the Environmental Statement (ES) and for Morgan (1998, p. 180) it is the project proponent who therefore steers, on the basis of their own values, the whole agenda. The project proponent, mainly, but not exclusively, a private sector developer is seeking, in a capitalist society, to make profit from their proposals. Their main focus is therefore likely to be on their ultimate aim of securing planning permission, rather than on the environmental consequences of their proposals. The determination of both definition and method in EIA are therefore intrinsically linked to the values and interests of those who establish the framework of the investigation and Wilkins (2003) and Morgan (1998, p. 180) argue that the choice of impact prediction methodology, the assumptions used in predictions and the conclusions drawn are themselves based upon value judgements and not "science". Kontic (2000) argues that personal value systems and beliefs are inherent in the process of evaluation and interpretation of the results of assessments. Furthermore, Kontic's (2000) research into the use of "experts" found that the credibility of expert opinion largely depends less on the "scientific correctness" of their work and more on the interests or perspective of those that review it. So even if the science in EIA could be "good science" the values and interests involved in the production and review of EIA documents will undermine that science. The environment is socially created and subjectively experienced and valued, and, as a result, the conversion of environmental impacts into quantitative data can never tell the full story.

For Rorty (1991) what is really meant by a claim of rationality is that the subject of that claim is in accordance with the current standards of a particular system (pp. 23–26) and for Flyvbjerg (1998) the very concept of rationality is defined by the powerful and then used by them to reinforce that power. Feenberg (2008) argues that rationalism provides a silencing effect in that the claim to being rational legitimises capitalist forms of production and power relationships. Within this context, critical theory sees rationalism as part of the ideological superstructure of capitalism, providing justification and legitimacy to decisions that are deemed to be rational and value-free but ultimately serve the interests of maintaining the capitalist system of power relationships. For critical theory, decision-making procedures such as land use planning and decision-making tools, such as EIA, are little more than a smokescreen behind which decisions are legitimised. Dillar and Ruchala (2005) argue that Weber's conception of instrumental rationality, the procedural basis of much environmental decision-making and particularly in the use of decision-making tools such as EIA, is means-rather than ends-based. That is, rational decision-making procedures provide legitimacy for decision-making without any requirement to appraise the legitimacy of the outcome. By claiming a decision is based upon a rational, objective scientific consideration of its impacts and its alternatives, an essentially political value judgement can be made to appear logical, inevitable and "right" (Boehmer-Christiansen, 1994).

Critical theory's attack on rationalism and rationalist planning theory resulted in what Taylor (1998) sees as an acknowledgement, in the 1980s, by planning theorists that planning as an activity could not be separated from the social, economic and political context in which it takes place (p. 107). The critical theory attack on rationalism is also seen as driving planning theorists to look for new models (*ibid.*). Healey (1997) also sees planning theory as slowly beginning to "*escape*" from the "*trap*" of rational theory. But rather than accept a Marxist critique of planning, mainstream planning theorists took the view that the project of modernity, rather than their theories, had failed and that in a post-modernist world new theories needed to be developed. The post-modernist planners accepted the political nature of planning choices and the inherent conflict of values that exist in planning practice. This led to a new batch of communicative theories based upon Habermas' concept of communicative rationality (Paden, 2003). The sociological theory of Habermas is based upon a rejection, on the one hand, of Marxism as an explanation of history and social change and yet, on the other hand, an acceptance of much of Marx's views on the nature of capitalism (Giddens, 2006, p. 118). For example, while Habermas embraces much of the Weberian notions of social action and rationalism, he rejects the neo-Weberian post-modernist view that society has moved beyond capitalism as being little more than neo-conservative and irrational (Cremer, 2002). Habermas is probably best known for his development of the theory of 'communicative rationality' where society should be ruled by reason and where communicative rationality is the use of language to further the process of modernism and reason (Elling, 2008, p. 75). For Habermas (1984) the project of rationalism is one of emancipation and is driven forward by communication where rational statements and rational knowledge are challenged and developed. Such views are, of course, reliant on society being pluralistic and democratic, where change can be achieved by debate and discussion and the marshalling of evidence (Flyvbjerg, 2000).

The "communicative" and "collaborative" planning theories developed from Habermas' work are usefully reviewed by Fainstein (2000) and Richardson (2005) and what is apparent in those reviews is that lying behind notions of power within communicative theory remains a belief in varying forms of Weberian sociology and rationalism. The concept of power, for example, is not well defined in planning theory, but it does follow pluralist models such as Dahrendorf's (1959) notions of competing interests and Foucault's notion that power is linked to knowledge (Giddens, 2006, p. 22 and p. 846). Allmendinger (2002), for example, tries to combine a post-modernist pessimism of rational science as the basis for the future while at the same time embracing aspects of a Habermasian belief that rational communicative planning will bring change. In his examination of EIA Elling (2008) states that, "*Knowledge can lead to action, and that knowledge can lead to refrain from action. But also that communication in all directions across society can prevent knowledge monopolies, both in the sense of the distribution of knowledge and in the sense of what type of knowledge is attributed*" (p. 262). Communicative planning theory sees the role of the planner as either an advocate, objectively and professionally pursuing the interests of his/her client or a conduit through which the different interests are communicated and negotiated. The planner is seen as a counsellor, facilitating change through non-coercive negotiation (Huxley and Yiftachel, 2000; Fainstein, 2000; Taylor, 1998, p. 162). Healey and Gilroy (1990) plead for planning officers to actively seek out ways in which stakeholders can be involved in decisions. In communicative planning theory, public involvement in decision-making processes is seen as a mechanism for conflict resolution. As Friedmann (1998) puts it "*it seeks to forge a limited consensus through negotiated settlements among contesting parties*". In this view, if experts provide the right information to those affected by decisions their fears over development and change will be allayed.

Just as the planning theorists took a "communicative turn" so EIA theorists have followed. The failure of EIA to conform to the rationalist model is well documented (Beattie, 1995; Cahmore *et al.*, 2004; Weston, 2000) and led to the development of a new approach to the justification of its use in environmental decision-making. Holder (2004) identifies a new substantive theory of EIA in what she calls 'cultural theory' where EIA is seen as part of a broader programme to change attitudes towards the environment and the impact of development (pp. 19–29). This cultural theory of EIA is very close to Elling's (2008) characterisation of EIA as an ecological modernisation process (p. 229) and a reflexive communicative process that allows citizens to share in the "insights" and "experiences" of the expert systems (p. 261). Here he sees EIA as part of Habermas' communicative rationality programme where EIA experts communicate information to stakeholders through the EIA process. This communicative theory of EIA sees the purpose of EIA as not limited to its procedural role of informing decisions but part of a far more long-term substantive project to increase environmental awareness and contribute to the achievement of sustainable development. In communicative EIA theory it is seen as a mechanism for changing values and building consensus to bring a long-term transformation of the priorities of decision makers (Jay *et al.*, 2007). In this view environmental assessment acts as a mechanism for social learning through its "public participation, transparency" and "promotion of discourse" (Wilkins, 2003). This model is based on a particular understanding of society that believes that in communicating information, consensus can be built that will bring change in a democratic pluralist society. It is a perspective that is based on sociological understanding that Flyvbjerg (2000), considers to be "utopian" and "sociologically naïve". Indeed, critics of Habermas highlight the almost complete disregard within his work of power relationships within society. For example, Flyvbjerg (2000) describes Habermas as a "top-down moralist" concerned with the procedural rationality of communicative rationality and yet there is the "bottom-up" belief in a process where the objectives and outcomes of debate are defined by the participants. He goes on to argue that the main thrust of Habermas' communicative rationality is that the only form of power in society is (or will be) the ability to put forward better arguments (*ibid.*). Yet even Castells (2007 and 2009), the arch proponent of mass communication as an empowering and democratising force, accepts that the control of communication is increasingly being taken over by large corporate interests (2009, p. 414).

Habermas, communicative rationality, and the planning and EIA theories that have developed from it, are also "sociologically naïve" in that there is an apparent lack of understanding of competing values and interests in environmental decision-making (see Bartlett and Kurian, 1999; Friedmann, 1998; Wilkins, 2003). There is what appears to be a view that the outcome of decisions can be rational without any real sense that there can be a plurality of concepts of what would constitute a rational decision (McCarthy, 1994). The Nolan Commission Report on UK planning and local government made clear that planning decisions result in "*winners and losers*" (Committee on Standards in Public Life, Third Report, 1997, p. 69). Planning is, almost by definition, an adversarial process that requires decisions to be made over whose interests and values should prevail in any given debate over the future use of land. The outcome of such decisions can be seen as completely rational to the winners and irrational to the losers, and very rarely will there be any agreement that the "right" decision has been made. The communicative rationality model of EIA theory also fails to recognise that there can be irresolvable conflicts of interests and values within the environment itself. It is not the clear-cut choice between economic interests and environmental interests that seem to dominate the debate, for there are often competing interests over what part or what element of the environment should prevail. In some instances this can be a debate between local impacts and global environmental concerns, as in the case of windfarm developments (see for example Warren *et al.*, 2005; Toke, 2005). The substantive purpose of EIA, the furthering of the project of sustainable development, also falls foul of the plurality of interests and values in the environment. As a socially constructed concept, there can never be a true consensus on what would constitute sustainable development and hence how EIA can contribute to its realisation (Cashmore, 2004; Cashmore *et al.*, 2004). The concept of sustainable development itself has inherent conflicts and tensions between the different interests that make up the different strands within it (Campbell, 1996). Furthermore, there is nothing in the European Directive or the UK's EIA Regulations that raise the importance of environmental considerations above anything else (Alder, 1993; Krämer, 2007, p. 136; Weston, 1997). While environmental considerations, such as climate change, have, arguably, become more important in planning decision-making since the introduction of EIA in the UK in 1988, this has been a result of policy change rather than anything flowing from the use of EIA. As EIA does not have a specifically stated purpose of preventing environmental harm or creating environmental benefits, as Jay *et al.* (2007) acknowledge, the likelihood of EIA making any meaningful contribution to the project of sustainable development "*seems remote*". Despite attempts to develop environmental assessment procedures so that they focus on sustainability, such as the UK's sustainability appraisals of development plans, because SEA and other assessment frameworks take place within the same legal and policy context and constraints as EIA, the likelihood of their success seems equally remote.

Wilkins' (2003) notion that EIA acts as a mechanism for social learning through its "public participation, transparency" and "promotion of discourse" is not supported by the examination of EIA in practice. While public consultation and participation is a mandatory part of most EIA systems, its influence on the assessments and outcomes is marginal (Benson, 2003; Wood, 1997, p. 225). The adversarial nature of planning decision-making and the perceived costs and delays that public involvement can bring are all seen as reasons why public involvement is largely relegated to little more than is required by regulation (Benson, 2003; Glasson *et al.*, 2005, p. 170; Weston, 2004). In the context of the plurality of environmental values and the conflicts that emerge between these values and interests in planning decision-making, the drive for some form of consensus building role for EIA would seem, to use Owens *et al.* (2004) phrase, "*heroic*". While the substantive theorists see the whole process of EIA as a social learning tool the only real tangible outcome of the EIA process that the majority of the public have access to is the ES. As a social learning tool an ES is particularly weak, as the only people who read them are likely to be those who already have a particular perspective on the project in hand and whose values are unlikely to be changed by anything contained in an ES produced by the proponents of a project. A further major problem for this social learning model of EIA is the risk society phenomenon identified by Beck (1992, 1995). In a risk society we no longer trust science, technology or the claims made of them by "experts" and the assessments and statements produced by the EIA process are not trusted by the public or the decision makers to which they are aimed. Evidence suggests that the few people who do read an ES do not believe or trust its content and are highly sceptical that consultancies working for a project proponent can ever provide unbiased information (Weston, 2003, 2004). Planning and EIA are largely a top down state-led activity, and in the eyes of a critical theorist, takes place as part of the superstructure of capitalism. In that context, no amount of public involvement, consultation and communication will change the essential nature of the social system that creates the divisions and problems that communicative rationality seeks to solve (Paden, 2003).

Within the texts on EIA theory that were examined for this paper there is one convincing argument that emerges and that is, when measured against practice, all the theories fail to adequately explain or even justify EIA (see in particular Cashmore, 2004; Cashmore *et al.*, 2004; Pische and Cashmore, 2006; Weston, 2000). One explanation for this failure is that even though most developments in EIA theory have been as a result of a critique of the pure rationalist model they remain, as with the planning theories from which they are drawn, firmly in the rationalist mould. This continuing demand to ground EIA within notions of rationalism has been an attempt to justify its use in environmental decision-making against a background of its failure to deliver its procedural role. Borrowing from planning theory, EIA proponents still try to justify its use under one banner of rationalism or another. However, it is not enough to borrow from planning theory to develop a theory of EIA without also subjecting those theories to the same scrutiny that has so undermined planning theory. Particularly as Taylor (1998) argues that all the variants of planning theory that have developed since 1945 have done so as a result of the failures of the previous typologies of planning theory to deal with the criticisms levelled at them by critical theorists (*ibid.*). Even Allmendinger (2002) accepts that a Marxist critique has made a telling contribution to the debate over planning theory but seemingly dismisses it as normative "*in that the interpretation of evidence to support it is value-driven*" (p. 91). Evidence perhaps of a failure on Allmendinger's part to recognise that the rational models of planning theory are themselves value-based and that in environmental decision-making it is impossible to separate values from decision-making processes.

Marxist analysis has not been particularly fashionable in the last two or three decades and is often today seen as simplistic. However, and as Forte (2008) argues, while Marx's historical materialism, structure — superstructure model, may be considered today as rather unsophisticated, its use as a starting point for sociological analysis, particularly in terms of the role of the state and state action within society, remains compelling. We only need to go as far as looking at how the state bailed out failing banks in the 2008–2009 crisis and the subsequent payment of that bail-out by the poor, the disabled and the disenfranchised to understand the state's role in capitalism. As a state-led process of development management, both planning and EIA are there to serve the interests of capitalism and they do that by trying to provide a rationalist justification for the outcome of environmental decision-making.

That EIA fails in both its procedural role of influencing environmental decision-making and its substantive role of changing social values is evidence that the world has moved on from a belief in all things rational, whereas EIA remains stuck in the past. The risk society factors that have undermined trust in politicians, science and experts, the social construction of environmental values and critical theory's telling critique of the theories that seek to justify it have all eroded any meaningful role that EIA may once have had. It is perhaps time that we stop searching for theories to defend it and start to campaign for something radically different.

References

Abrams, P (1982). *Historical Sociology*. London: Open Books.

Adams, GB and DL Balfour (1998). *Unmasking Administrative Evil*. Thousand Oaks CA: Sage Publications.

Agger, B (1991). Critical theory, poststructualism, postmodernism: their sociological relevance. *Annual Review of Sociology*, 17, 105–131.

Alder, J (1993). Environmental impact assessment — the inadequacies of English Law. *Journal of Environmental Law*, 5(2), 203–220.

Allmendinger, P (2002). *Planning Theory*. Basingstoke: Palgrave.

Allmendinger, P, M Tewdwr-Jones and J Morphet (2003). Public scrutiny, standards and the planning system: assessing professional values within a modernised local government, *Public Administration*, 8(4), 761–780.

Allmendinger, P (2009). *Planning Theory*, 2nd edn. Basingstoke: Palgrave Macmillan.

Barrow, C (1997). *Environmental and Social Impact Assessment*. London: Arnold.

Bartlett, R and A Kurian (1999). The theory of environmental assessment: implicit models in policy making. *Policy and Politics*, 27(4), 415–433.

Beattie, RB (1995). Everything you already know about EIA (but don't often admit). *Environmental Impact Assessment Review*, 15(2), 109–114.

Beck, U (1992). *Risk Society: Towards a New Modernity*. London: Sage.

Beck, U (1995). *Ecological Politics in an Age of Risk*. Cambridge: Polity.

Benson, JF (2003). What is the alternative? Impact assessment tools and sustainable planning. *Impact Assessment and Project Appraisal*, 21(4), 261–266.

Boehmer-Christiansen, S (1994). Politics and environmental management. *Journal of Environmental Planning and Management*, 37(1), 69–85.

Caldwell, L (1988). Environmental impact analysis: origins, evolution and future directions. *Review of Policy Research*, 8(1), 75–83.

Campbell, H and R Marshall (2002). Utilitarianism's bad breath? A re-evaluation of the public interest justification for planning. *Planning Theory*, 1(2), 163–187.

Campbell, S (1996). Green cities, growing cities, just cities? Urban planning and the contradictions of sustainable development. *Journal of American Planning Association*, 62, 296–313.

Cashmore, M (2004). The role of science in environmental impact assessment: process and procedure versus purpose in the development of theory. *Environmental Impact Assessment Review*, 24, 403–426.

Cashmore, M, R Gwilliam, R Morgan, D Cob and A Bond (2004). The interminable issue of effectiveness: substantive purpose, outcomes and research challenges in the advancement of environmental impact assessment theory. *Impact Assessment and Project Appraisal*, 22(4), 295–310.

Castells, M (2007). Communication, power and counter-power in the network society. *International Journal of Communication*, 1, 238–266.

Castells, M (2009). *Communication Power*. Oxford: Oxford University Press.

Committee on Standards in Public Life (1997). Third Report: Standards of conduct in local government in England, Scotland and Wales: Volume 1: Report CM3702. London: Stationery Office.

Cremer, D (2002). Marx and the postmodernist debates: an agenda for critical theory by Lorraine Y Landdry, book review. *Journal of Political Ecology: Case Studies in History and Society*, 9(2), http://jpe.library.arizona.edu/volume_9/903cremer.html (accessed 19/07/10).

Dahrendorf, R (1959). *Class and Class Conflict in Industrial Society*. Stanford: Stanford University Press.

Dillar, J, F and L Ruchala (2005). The rules are no game: from instrumental rationality to administrative evil. *Accounting, Auditing and Accountability Journal*, 18(5), 608–630.

Dror, Y (1964). Muddling through — 'science' or inertia'. *Public Administration Review*, 24, 153–157.

Eccleston, CH (2006). Perspective: Lynton Caldwell, a founding father of America's "Environmental Magna Carta". *Environmental Practice*, 8(4), 206–207.

Elling, B (2008). *Rationality and the Environment: Decision-making in Environmental Politics and Assessment*. London: Earthscan.

Elling, B (2009). Rationality and effectiveness: does EIA/SEA treat them as synonyms. *Impact Assessment and Project Appraisal*, 27(2), 121–131.

Etzioni, A (1967). Mixed-scanning: a "third" approach to decision-making. *Public Administration Review*, 27(5), 385–392.

Fainstein, SS (2000). New directions in planning theory. *Urban Affairs Review*, 35, 451–478.

Faludi, A (1973). *Planning Theory*. Oxford: Pergamon Press.

Faludi, A (1983). Critical rationalism and planning methodology. *Urban Studies*, 20, 265–278.

Feenberg, A (2008). From critical theory of technology to the rational critique of rationality. *Social Epistemology*, 22(1), 5–28.

Flyvbjerg, B (1998). *Rationality and Power: Democracy in Practice*. London: University of Chicago Press.

Flyvbjerg, B (2000). Ideal theory, real rationality: Habermas versus Foucault and Nietzsche. *Political Studies Association's 50th Annual Conference, The Challenges for Democracy in the 21st Century*, London School of Economics and Political Science, 10–13 April.

Forester, J (1989). *Planning in the Face of Planning*. Berkeley CA: University of California Press.

Forte, JM (2008). Religion and capitalism: Weber, Marx and materialist controversy. *Philosophy and Social Criticism*, 34(4), 427–448.

Friedmann, J (1998). Planning theory revisited. *European Planning Studies*, 6(3), 245–253.

Gamble, A (1981). *An Introduction to Modern Social and Political Thought*. London: Macmillan.

Giddens, A (1986). *Sociology: A Brief Critical Introduction*. London: MacMillan Press.

Giddens, A (2006). *Sociology*. 5th edn. Cambridge: Polity Press.

Glasson, J, R Therivel and A Chadwick (2005). *Introduction to Environmental Impact Assessment*, 3rd edn. London: UCL Press.

Gore, C, K Murray and B Richardson (1992). *Strategic Decision-Making*. London: Cassell.

Habermas, J (1984). *The Theory of Communicative Action, Volume One: Reason and the Rationalisation of Society*. Boston: Beacon Press.

Hall, P (1982). *Urban and Regional Planning*. 2nd edn. London: George Allen and Unwin.

Healey, P and R Gilroy (1990). Towards a people sensitive planning. *Planning Practice and Research*, 5(2), 21–29.

Healey, P (1997). *Collaborative Planning: Shaping Places in Fragmented Societies*. Basingstoke: Macmillan.

Hill, M (2005). *The Policy Process*. 4th edn. London: PearsonLongman.

Holder, J (2004). *Environmental Assessment: The Regulations of Decision-Making*. Oxford: Oxford University Press.

Huxley, M and O Yiftachel (2000). New paradigm or old myopia? Unsettling the communicative turn in planning theory. *Journal of Planning Education and Research*, 19, 333–342.

Huxham, M and D Sumner (2000). *Science and Environmental Decision-Making*. Harlow: Prentice Hall.

Jay, S, C Jones, P Slinn and C Wood (2007). Environmental impact assessment: retrospective and prospect. *Environmental Impact Assessment Review*, 27, 287–300.

Kontic, B (2000). Why are some experts more credible than others? *Environmental Impact Assessment Review*, 20, 427–434.

Krämer, L (2007). The development of environmental assessment at the level of the European Union. In Holder, J and D McGillivray (eds.). *Taking Stock of Environmental Assessment*. London: Routledge-Cavendish.

Lawrence, D (1997). The need for EIA theory-building. *Environmental Impact Assessment Review*, 17(2), 79–107.

Lawrence, D (2000). Planning theories and environmental impact assessment. *Environmental Impact Assessment Review*, 20, 607–625.

Lee, D and H Newby (1985). *The Problem of Sociology*. London: Hutchinson.

Leknes, E (2001). The role of EIA in the decision-making process. *Environmental Impact Assessment Review*, 21, 309–344.

Lindblom, CE (1959). The science of muddling through. *Public Administration Review*, 19(1), 79–88.

Lindblom, CE (1980). *The Policy Making Process*. 2nd edn. Englewood Cliffs, NJ: Prentice Hall.

Löwith, K (1982). *Max Weber and Karl Marx*. London: George Allen and Unwin.

McCarthy (1994). We don't have a theory of rationality, and we never will — a philosophically pragmatic point of view. In Katz M (ed.). *Philosophy of Education*. Urbana, Ill., Philosophy of Education Society. http://www.ed.uiuc.edu/EPS/PES-Yearbook/94_docs/MCCARTHYHTM (accessed 19/07/10).

Morgan, RK (1998). *Environmental Impact Assessment: A Methodological Perspective*. London: Kluwer.

Owens, S, T Rayner and O Bina (2004). New agendas for appraisal: reflections on theory, practice and research. *Environment and Planning*, 36, 1943–1959.

Paden, R, (2003). Marxism, utopianism, and modern urban planning. *Utopian Studies*, 14(1), 82–111.

Pepper, D (1984). *The Roots of Modern Environmentalism*. Beckenham: Croom Helm.

Persson, J (2006). Theoretical reflections on connection between environmental assessment methods and conflict. *Environmental Impact Assessment Review*, 26, 605–613.

Pische, F and M Cashmore (2006). Decision-oriented environmental assessment: an empirical study of its theory and methods. *Environmental Impact Assessment Review*, 26, 643–662.

Richardson, T (2005). Environmental assessment and planning theory: four short stories about power, multiple rationality and ethics. *Environmental Impact Assessment Review*, 25, 341–365.

Rorty, R (1991). *Objectivity, Relativism and Truth*. Cambridge: Cambridge University Press.

Rydin, Y (2003). *Conflict, Consensus, and Rationality in Environmental Planning: An Institutional Discourse Approach.* Oxford: Oxford University Press.

Smith, G and D May (1997). The artificial debate between rationalist and incrementalist models of decision-making. In Hill, M (ed.). *The Policy Process: A Reader.* 2nd edn. London: Prentice Hall Wheatsheaf.

Simon, H (1957). *Administrative Behaviour.* 2nd edn. New York: Macmillan.

Simon, H (1976). *Administrative Behaviour.* 3rd edn. London: Free Press.

Taylor N (1998). *Urban Planning Theory Since 1945.* London: Sage.

Tewdwr-Jones, M (1996). Reflective planning theorising and professional protectionism. *Town Planning Review*, 67, 235–243.

Toke, D (2005). Explaining wind power planning outcomes: some findings from a study in England and Wales. *Energy Policy*, 33, 1527–1539.

Walker, KJ (2001). Uncertainty, epistemic communities and public policy. In Handmer, J, T Norton and S Dovers (eds.). *Ecology, Uncertainty and Policy*, Harlow: Prentice Hall.

Warren CR, C Lumsden, S O'Dowd and RV Birnie (2005). Green on green': public perceptions of wind power in Scotland and Ireland. *Journal of Environmental Planning and Management*, 48(6), 853–875.

Weber, M (1947). *The Theory of Social and Economic Organisation.* New York: Oxford University Press.

Weber, M (1979). The essentials of bureaucratic organisation. In Worsley, P (ed.). *Modern Sociology.* 2nd edn. London: Penguin Books.

Weber, M (1997). Rational legal authority. In Hill, M (ed.). *The Policy Process: A Reader.* 2nd edn. London: Prentice Hall Harvester Wheatsheaf.

Weston, J (1997). Introduction: EIA in the UK. In Weston, J (ed.). *Planning and EIA in Practice.* London: Longmans.

Weston, J (2000). EIA decision-making theory and screening and scoping in UK practice. *Journal of Planning and Environmental Management*, 43(2), 185–204.

Weston, J (2003). Is there a future for EIA? *Impact Assessment and Project Appraisal*, 21(4), 278–280.

Weston, J (2004). EIA in a risk society. *Journal of Planning and Environmental Management*, 47(2), 313–325.

Weston, J (2006). Common sense in environmental impact assessment — it cannot be avoided. *Impact Assessment and Project Appraisal*, 24(1), 12–13.

Wilkins, H (2003). The need for subjectivity in EIA: discourse as a tool for sustainable development, *Environmental Impact Assessment Review*, 23, 401–414.

Wood, C (1995). *Environmental Impact Assessment: A Comparative Review.* Harlow: Longman.

Yearley, S (1988). *Science, Technology, and Social Change.* London: Unwin Hyman.

Yearley, S (1991). *The Green Case: A Sociology of Environmental Issues, Arguments and Politics.* London: HarperCollins Academic.

CHAPTER 4

Effectiveness of Policy Level Environmental and Sustainability Assessment: Challenges and Lessons from Recent Practice

Richard P. Eales* and William R. Sheate†

Collingwood Environmental Planning Limited
1E The Chandlery, 50 Westminster Bridge Road
London, SE1 7QY, United Kingdom
**r.eales@cep.co.uk*
†w.sheate@cep.co.uk

This paper explores the challenges and lessons from recent practice and experience of applying tools such as Strategic Environmental Assessment (SEA) and Sustainability Appraisal at the policy level in the UK and Europe. It investigates whether or not these tools have been effective and helped to deliver more sustainable development at the high level of national policy development. The analysis is illustrated by case examples from the UK, such as Eco-towns and Energy Planning. The paper concludes that the current performance by the UK Government in implementing the SEA Directive for national level strategic actions is far from exemplary. At the root of the problem is the poor consideration and evaluation of reasonable alternatives, the fundamentally weak conception of sustainability adopted and the apparent perception that having to undertake an assessment and comply with the SEA Directive is a hurdle, rather than a useful mechanism for helping to deliver better and more sustainable evidence-based policy making.

Keywords: Strategic environmental assessment; sustainability appraisal; national policy; decision-making; effectiveness; alternatives; institutional inertia.

*Richard Eales is founder and Managing Director of Collingwood Environmental Planning Limited, London, UK.

†William Sheate is Technical Director, Collingwood Environmental Planning Limited and Reader in Environmental Assessment, Imperial College London, Centre for Environmental Policy. Corresponding author.

Introduction

In the European Union, the Directive 2001/42/EC on the assessment of the effects of certain plans and programmes on the environment (OJ, 2001) (known as the SEA Directive), which requires certain public strategic actions to undergo an environmental assessment before they are adopted, includes "promoting sustainable development" in its key objective:

> "to provide for a high level of protection of the environment and to contribute to the integration of environmental considerations into the preparation and adoption of plans and programmes with a view to promoting sustainable development…" (SEA Directive, Article 1).

However, there is considerable debate about how Strategic Environmental Assessment (SEA) contributes to sustainability, as will be discussed in this paper, see for example Partidário (2000), Kørnøv and Thissen (2000), Sheate *et al.* (2003), Owens *et al.* (2004), Gibson *et al.* (2005), Gibson (2006), Dalal-Clayton and Sadler (2005), Sadler (2005), Pope and Grace (2006), Morrison-Saunders and Fischer (2006), Wallington *et al.* (2007). This paper explores the challenges and lessons from recent practice and experience of applying environmental and sustainability assessment tools at the policy level in Europe and the UK. In particular the authors consider whether SEA and Sustainability Appraisal have been "effective" at delivering more sustainable development at the high level of national policy development. The paper starts by analysing the role of SEA in delivering more sustainable development. The paper then establishes the meaning of "effectiveness" with respect to SEA and Sustainability Appraisal. This is followed by discussion of experience and lessons from Europe and the UK in applying SEA and other similar forms of strategic assessment. Three case studies were selected to demonstrate how effective, or otherwise, some of the assessments have been at the national level in the UK:

- The 2008 Eco-towns Planning Policy Statement and Programme;
- The 2009 Energy and Ports National Policy Statements (NPS); and
- The 2008 Scottish National Planning Framework (NPF2).

The paper concludes with a discussion of the particular challenges of strategic assessment at the national level and suggestions as to how practice could be improved.

Strategic Assessment and its Potential Role in Delivering More Sustainable Development

Common conceptions of sustainable development talk of balancing environmental, social and economic factors, accepting trade-offs between these factors in the process — this is often referred to as "weak" sustainability (Neumayer, 2003). Other conceptions recognise that ultimately all economic and social activity is dependent on the natural environment, its resources and ecosystem services that it provides — sometimes referred to as "strong" sustainability. The most commonly used definition of sustainable development is that from the Brundtland report (WCED, 1987), but usually only the first part of the definition is quoted, forgetting the important second part:

> "Sustainable Development is development that meets the needs of the present without compromising the ability of future generations to meet their own needs. It contains within it two key concepts:
>
> - the concept of "needs", in particular the essential needs of the world's poor, to which overriding priority should be given; and
> - the idea of limitations imposed by the state of technology and social organisation on the environment's ability to meet present and future needs." (WCED, 1987, p. 43)

The Brundtland report therefore recognises the strong social and environmental imperatives underlying sustainable development (as does the SEA Directive in Article 1) and that this is not therefore simply a matter of *balancing* environmental, social and economic factors, which is otherwise effectively business-as-usual. That does not mean that economic factors are not important since they contribute to human wellbeing, and the SEA Directive, in requiring assessment of population, human health, material assets and cultural heritage for example (Annex I (f)), already defines the environment broadly. But the purpose of SEA is very much to ensure that environmental considerations are integrated into strategic decision-making, recognising that traditionally this has not been done sufficiently, and in so doing this will promote sustainable development (Sadler, 2005).

Many strategic actions will have a strong economic rationale to their promotion so, in the absence of SEA, that is likely to be the determining factor. This dual purpose of SEA — having an *"advocative" role*, where its primary purpose is to raise the profile of the environment, and an *integrative role* where environment, social and economic considerations are combined in a more "objective" way — is widely recognised (Kørnøv and Thissen, 2000). In an integrative role, SEA requires decision-making to recognise a *strong* conception of sustainability, since otherwise the economic benefits are likely to be double counted both in the rationale for the strategic action and in the SEA. This is particularly pertinent in the context of the application of Sustainability Appraisal of spatial plans in the UK, as well as "Appraisals of Sustainability" which is an alternative term used for a largely indistinguishable technique applied recently by the UK government to the assessments of National Policy Statements (NPS), which seek to assess environmental, social and economic factors together.

The SEA Directive follows the same broad principles as the Environmental Impact Assessment (EIA) Directive in the European Union (Council Directive 85/337/EEC on the assessment of the effects of certain public and private projects on the environment, as amended by Directives 97/11/EC and 2003/35/EC), not necessarily changing the decision on whether the strategic action should go ahead, but ensuring that it is more informed so significant adverse effects can be avoided and positive benefits enhanced. This reflects the more traditional rationalist perspective on decision-making inherited from EIA.

In reality most strategic decision-making is recognised as being characterised by "bounded rationality" because of the more iterative, political and dynamic nature of such decision-making and greater uncertainty at strategic levels (Nilsson and Dalkman, 2001). Consequently SEA can also bring a range of other benefits to strategic decision-making, not just informing the decision, but improving the whole planning process and strategic thinking. One of the key ways SEA can do this is, by providing a framework where more strategic participation of the public and stakeholders can take place. The stages of SEA provide excellent opportunities for the inclusion of participation, in order to inform better the development of options (e.g. at scoping stage), the assessment of options and the drafting of the strategic action itself. SEA can also therefore provide opportunities for social and institutional learning by those undertaking and using SEA (Wallington *et al.*, 2007; Runhaar, 2009; Hertin *et al.*, 2009; Fischer, 2010) and exchange of knowledge among stakeholders and participants (Sheate and Partidário, 2010). Effective SEA, therefore, can present a much wider range of potential and more long-term benefits than simply informing the decision-making process.

Defining "Effectiveness" with Respect to Environmental and Sustainability Assessment

Despite the fact that evaluating the effectiveness of tools like SEA is notoriously difficult, there have been plenty of attempts at evaluating or promoting effectiveness, internationally (e.g. Sadler and Verheem, 1996; IAIA, 2002; Retief, 2007) and particularly in the EU (e.g. Fergusson *et al.*, 2001; Sheate *et al.*, 2001, 2003; Fischer and Gazzola, 2006; Gazzola, 2008; Cashmore *et al.*, 2009). Evaluating effectiveness is difficult given the problems in determining the role SEA might have played among multiple factors influencing decision-making. Therefore context is recognised by many of these authors as important. What works in one jurisdiction may not transfer readily to another. Cashmore *et al.* (2010) recognise that there is a plurality of views of what effectiveness in environmental assessment might mean, depending on the perspective taken. But a simple understanding of effectiveness may be that for a tool to be considered effective it must at least achieve its own purposes (which may still be open to different interpretations). In the case of SEA this will be to ensure that environmental considerations influence the decision-making process and, given the underlying "sustainable development spirit" of the SEA Directive, this may entail a change in the mindset of the actors involved (Fischer, 2005). This influence may occur at various stages throughout the planning process: early on in influencing the options considered, and at later stages to inform mitigation and monitoring.

Arguably it is at the earliest stages where SEA can be most effective in influencing the overall direction and objectives of the strategic action under consideration. A truly strategic consideration of alternatives requires the assessment of alternative options (Noble, 2000, 2002), i.e. alternatives for meeting the objectives set, such as alternative modes of transport for a transport strategy. This is in contrast to a consideration of option alternatives, for example, alternative locations or routes that might occur in an EIA of a road scheme where the option of constructing a new a road has already been decided.

In the EU, the SEA Directive requires the evaluation of "reasonable alternatives" (Article 5 (1)), which includes (through the reference in Article 5 (1) to Annex I), the "measures envisaged to prevent, reduce and as fully as possible offset any significant adverse effects on the environment of implementing the plan or programme" (para (g), Annex I) and "reasons for selecting the alternatives dealt with" (para (h), Annex I). That sequence of *prevent, reduce or offset,* reflects an important principle of the SEA Directive, that it seeks to avoid impacts from occurring rather than merely their mitigation. This is reflected in the preamble to the Directive (Recital 1, invoking the precautionary principle, and Recital 5 that SEA procedures "should contribute to more sustainable and effective solutions"). This creates an important lever in ensuring that alternatives are properly considered and a potential focus for legal challenge if they are not. See for example recent UK case law on SEA such as the *Seaport* case in Northern Ireland (Weatherup, 2007), the East of England Plan case (Mitting, 2009) and the Review of the Eco-towns Sustainability Appraisal Reports (Sheate, 2008).

Three critical issues for sustainable development (and a focus for the Earth Summit in Rio de Janeiro in 1992), are also recognised in the SEA Directive:

- climate change (included as "climatic factors" in para (f) Annex I of the SEA Directive);

- biodiversity loss (also reflected in para (f) of Annex 1, and para (d) in relation to Natura 2000 sites, and reference to the Habitats Directive 92/43/EC in Article 3. Recital 3 of the preamble to the Directive further emphasises the requirement of the Convention on Biological Diversity for Parties to integrate "the conservation and sustainable use of biological diversity into relevant sectoral or cross-sectoral plans and programmes"; and
- public participation, which should be "early and effective" (Article 6 (2) of the SEA Directive) and in accordance with the Aarhus Convention 1998 (UNECE, 1998).

Given their spatial and temporal scales, these issues assume a particular importance at the very strategic policy level (João, 2007). SEA, if it is to be effective, therefore needs to ensure that these critical issues, among others, are properly considered in the assessment process alongside the primary purposes of the Directive.

Experience and Lessons from Europe

The SEA Directive came into force in the European Union in July 2004. During 2009 the European Commission published two reports on the application and effectiveness of the SEA Directive. The first was a report by COWI and Milieu (COWI and Milieu, 2009) based on a survey of Member States on how they had transposed the Directive and their experience of implementation and the relationship with other key environmental Directives, such as the EIA and Habitats Directives. The second report, drawing on the first and other reports, was the Commission's formal report to the Council and the European Parliament (CEC, 2009). Both reports recognised the wide variation in experiences of implementing the SEA Directive, but concluded that most Member States believed that SEA had a positive effect on the organisation and structure of the planning process, and that the formal requirements for consultation with environmental authorities and the public had led to increased transparency.

These reports also identified that in many cases, SEA changed the content of the strategic action. Some experiences showed that, at the highest level of national strategic actions, a significant number of the SEA findings had a strong influence on the substance of the strategic actions, including in the selection of the alternatives or by incorporating important suggestions from the SEA. Examples quoted (COWI and Milieu, 2009) included a National Programme for Ports Development in Bulgaria where two of the proposed terminals were rejected because of conflicts with protected areas for biodiversity; a regional waste management plan in Romania where the best environmental alternative was selected because of the SEA analysis; in Hungary a water management programme where the SEA helped to prevent a very expensive project; and in Ireland the importance of alternatives having to be considered by law had resulted in a number of strategic actions being changed in favour of an alternative action because of the SEA, whereas previously that would not have happened.

The Commission concluded: "Overall [...] the SEA Directive contributes to the systematic and structured consideration of environmental concerns in planning processes and better integration of environmental considerations upstream. In addition, by means of its requirements (environmental report, consultation and information of the authorities and public concerned etc.) it ensures better and harmonized planning procedures, and contributes to transparent and participatory decision-making processes." (CEC, 2009, p. 11)

The Commission reports draw heavily upon the views and reporting of Member States themselves, so caution is needed in the interpretation of the findings. In addition, the focus was on the *implementation* of the Directive and SEA procedures, rather than on the *quality* of assessments, changes to the strategic actions and wider benefits. The experience of applying the SEA Directive is still in its infancy and this means that there is not yet a large body of academic literature evaluating the experience of SEA under the SEA Directive. There is some, for example Theophilou *et al.* (2009), Therivel *et al.* (2009), Fischer (2010), Weiland (2010), but it can only be a snapshot at this stage.

The Commission reports highlight that biodiversity and climate change are both areas which have been relatively poorly addressed to date. The former is often too focused on species and/or habitats and not yet much on ecosystem services and functions, and the latter needs to address the issues of uncertainty, particularly how climate change might impact on different groups in society (Weiland, 2010). Indeed the Commission is now developing guidance, in cooperation with the Member States, on the better integration of climate change and biodiversity issues in SEA.[1]

In the UK recent studies by the Sustainable Development Research Network (SDRN, 2008), Department of Communities and Local Government (CLG, 2010), Phillips and Sheate (2010) and Scottish Government (2010) have identified some strengths and weaknesses in SEA implementation in the UK. All found similar weaknesses, including a need for:

- strengthening the evidence base;
- better consideration of alternatives;
- better focusing of the assessment through scoping; and
- better integration with the process of developing the strategic action.

[1] The authors are members of the team undertaking the production of this guidance in 2011.

In addition, the quality of SEA was often compromised by misplaced effort and a lack of capacity and skills in SEA.

Therefore, in the UK overall, it can be concluded that "could do better" sums up much of the experience to date. This is both in terms of how SEA and Sustainability Appraisal are undertaken (i.e. in terms of the techniques and methodologies used) and how they are integrated with or influence the strategic action making processes. In Scotland, where SEA is more widely applied than in England following the Environmental Assessment (Scotland) Act 2005, the early years saw a tendency to try to minimise workload through scoping out of some key environmental issues and a rigid adherence to detailed guidance and templates under the (false) impression that was the only way to do SEA (Phillips and Sheate, 2010). In England, research focused on Sustainability Appraisal by the Department of Communities and Local Government (CLG, 2010), suggested that practitioners should be bolder at scoping out issues, and that SEA should follow an approach that concentrates more on evaluating strategic actions against baseline information rather than "aspirational objectives". An important lesson from these studies should be that over-prescription through guidance can suffocate innovation and can work against delivering what might be most effective in any particular circumstance (Phillips and Sheate, 2010).

Much of the reporting to date, therefore, has focused on the problems of implementation rather than the positive benefits that strategic assessments such as SEA can bring to the plan-making process and to stakeholders. Some other European experience (e.g. Partidário *et al.*, 2009; Sheate and Partidário, 2010) suggests there are real benefits that can accrue which may be less tangible than just changes to actual strategic actions, for example in the relationships created with stakeholders, the opportunities created for better long-term thinking about plans and strategies, and knowledge exchange. However, reaping the benefits requires time, sufficient resources and recognition by plan makers that SEA can provide beneficial support to strategic action-making processes.

The more strategic the assessment, the more flexibility is needed to make the techniques and methodologies adopted fit for purpose. What is required is greater capacity and knowledge of SEA within responsible authorities, so that those undertaking SEA and Sustainability Appraisal, or commissioning it, understand better how and why they are doing it. The European experience (CEC, 2009; COWI and Milieu, 2009) shows that some Member States see SEA as having had a very positive effect on national level strategic actions, for example through the proper consideration of alternatives. The question arises therefore: is the same true for national level plans in the UK?

Current Challenges in the UK: Lessons from Strategic Case Examples

Several case examples within the UK, in the energy and spatial planning arenas in particular, were evaluated with regards to the effectiveness of the assessments of strategic actions at the national level. The following cases and their assessments were considered:

- The 2008 Eco-towns Planning Policy Statement and Programme;
- The 2009 Energy and Ports National Policy Statements; and
- The 2008 Scottish National Planning Framework.

In the UK, national strategic actions, such as planning processes in England and Wales (National Policy Statements) and in Scotland (National Planning Framework), present new challenges for SEA/Sustainability Appraisal. The UK Government has generally accepted that the SEA Directive applies in these cases, i.e. that they are "plans or programmes" for the purposes of applying the SEA Directive and therefore that they set the framework for projects likely to be subject to EIA. Therefore these case studies were selected because they were:

- national in geographical coverage;
- high-level "policy"-like strategic actions;
- prepared by national governments; and
- required to comply with the SEA Directive (including those where SEA is incorporated into Sustainability Appraisals or Appraisals of Sustainability).

Consequently the case studies represented relatively new levels of decision-making for SEA application in the UK and therefore opportunities to learn from these initial experiences.

The review of the case examples draws on work undertaken by Collingwood Environmental Planning Ltd (CEP) as part of a study commissioned by the Royal Society for the Protection of Birds (RSPB) and the World Wide Fund for Nature (WWF), as well as other studies undertaken by CEP. The RSPB/WWF study sought to review the current state of play of SEA in the UK and particularly to review some of the strategic assessments being undertaken at the national level for their effectiveness in delivering the requirements and objectives of the SEA Directive. All the cases were reviewed against a broad set of review criteria which drew on those developed by the Institute of Environmental Management and Assessment (IEMA, 2004), but particularly focusing on those aspects required by the SEA Directive. In addition, to consider effectiveness and not just the quality of the reports, additional criteria were applied, seeking to ask whether:

- a strong or weak approach to sustainability was adopted;
- an advocative or integrative approach to SEA was used;
- the assessment sought to further the primary purpose of the SEA Directive (Article 1);
- the attitude to reasonable alternatives was proactive or reluctant; and
- the assessment sought to further the principles of early and effective participation of Art. 6 (2) of the SEA Directive, and the Aarhus Convention.

While much of the experience of SEA and Sustainability Appraisal in the UK has been through local authority spatial planning, experience of applying SEA/Sustainability Appraisal at national level and/or to high-level strategy is more limited, although it is increasing. In Scotland, SEA is now being applied to high-level strategies developed by Government authorities and agencies and the Scottish Government itself (e.g. the Scottish Forestry Strategy, Deer Commission Strategy for Wild Deer, National Planning Framework 2). The devolved administrations also undertook SEA of their respective Rural Development Plans/Programmes under the European Commission's Rural Development Council Regulation No. 1698/2005 in 2006–2008, which were very strategic documents relating to funding programmes, e.g. for agri-environment schemes. In England, Regional Spatial Strategies have been undergoing Sustainability Appraisal for several years since they were introduced under the Planning and Compulsory Purchase Act 2004.[2] Case law in the UK, such as *Seaport* in Northern Ireland (Weatherup, 2007), has raised the bar in terms of the need for substantial compliance with the requirements of Article 5 and Annex 1 of the Directive relating to the contents of the environmental report.

Assessments of the Eco-towns Planning Policy Statement and Programme

The UK Government's Department for Communities and Local Government published a draft Eco-towns Planning Policy Statement and Programme in 2008

[2] A number of Regional Spatial Strategies have been faced with re-visiting issues around the consideration of reasonable alternatives following the judicial review of the *East of England Plan* (Mitting, 2009), which found the last minute addition of extra housing figures without proper evaluation of alternatives was not in compliance with the SEA Directive (note that the coalition government in the UK is proposing to abolish this regional level of spatial planning).

(CLG, 2008a, b) and, as part of its development, decided to undertake a Sustainability Appraisal incorporating the requirements of the SEA Directive. The Eco-towns concept was intended to assist in meeting the twin challenges of providing additional housing, and mitigating and adapting to climate change. The Planning Policy Statement aimed to promote the development of these new settlements which "will have sustainability standards significantly above equivalent levels of development in existing towns and cities" (CLG, 2008a).

In 2008 CEP undertook a review of the Sustainability Appraisals prepared of the Planning Policy Statement and Programme for the BARD Campaign (a local group campaigning against the Middle Quinton proposed eco-town, see Sheate, 2008 and Sheate and Partidário, 2010). This review identified many problems with what the authors considered to be exceptionally poor Sustainability Appraisal reports for both the Planning Policy Statement and the Programme. A number of the failings could have offered potential grounds for legal challenge with respect to non-compliance with the SEA Directive. These included:

- failure to consider and evaluate reasonable alternatives — only a "have a draft Planning Policy Statement" or "do not have a draft Planning Policy Statement" options were considered;
- the inadequate consideration of the evolution of the baseline without the plan or programme;
- failure to produce an acceptable non-technical summary;
- failure to provide the data necessary to undertake the assessment of likely significant effects;
- failure to assess all likely significant effects;
- poor methodology which fails to utilise current knowledge and methods of assessment;
- failure to undertake a compliant assessment of cumulative and synergistic effects; and
- failure to provide adequate proposals on monitoring.

The SEA Directive is quite explicit about tiering ("set the framework for future development consent of projects", Article 3 (4)). In the Eco-towns case while the Planning Policy Statement set the framework for the consent of subsequent projects subject to EIA, those projects have also effectively set the framework for the Planning Policy Statement, i.e. the programme and projects constrained the very nature of the "reasonable alternatives" available to the Sustainability Appraisal of the draft Planning Policy Statement. In other words, the types of alternatives available at the strategic action level were not considered "reasonable" in relation to the objectives of the draft Planning Policy Statement, because that might open up options other than projects (Eco-towns), such as urban extensions or smaller scale developments. In the authors' view this potentially represented a circumvention of the purposes of the SEA Directive, in a similar way to well-established European Court of Justice case law on EIA, which has addressed attempts to circumvent the purposes of the EIA Directive (Stookes, 2011).

At the time there appeared to be potential scope for a formal complaint to the European Commission over failure to transpose the SEA Directive properly with respect to this type of policy statement and related programmes. The concern was that the Eco-towns draft Planning Policy Statement and Programme Sustainability Appraisals may set a precedent for what could be repeated in the National Policy Statements and related programmes (and indeed was, see below) — essentially the same potential problems of "tiering" existed. In the end, the Middle Quinton eco-town proposal was dropped from the draft Planning Policy Statement/Programme, leaving no incentive for the BARD Campaign to pursue the complaint or to proceed with a potentially costly judicial review at that stage.

Another area of poor performance for the Eco-towns Sustainability Appraisals was the apparent "blind faith" in mitigation, where little evidence was provided as to how the mitigation measures would be delivered. At such a strategic level there are no means of guaranteeing or securing either that mitigation measures will be delivered or that they will be successful. This over-reliance on mitigation appears to pervade the Sustainability Appraisal reports, including consideration of cumulative effects and habitats.

Assessments of the National Policy Statements

In 2009 the UK Government published a number of draft National Policy Statements (NPSs) which provided an important opportunity to reflect on how SEA was being implemented in the UK, particularly with respect to strategic actions at the national level. An important question is whether the desire to speed up the decision-making process, one of the motivations for the NPSs, may compromise the environmental assessments or Appraisals of Sustainability being undertaken and particularly whether they may fail to comply with the requirements of the SEA Directive. If this was the case, then the speeding up of decision-making may actually not happen, even though it is the intention of the NPS. If NPSs and Nationally Significant Infrastructure Projects end up being challenged in the courts, there can be no assumption that the planning process itself and delivery of infrastructure will necessarily be any quicker in the long run.

The term *Appraisals of Sustainability* refers solely to those assessments undertaken for NPSs, though in all other respects they are the same as Sustainability

Appraisals undertaken for spatial plans in the UK and need to be screened for applicability of the SEA Directive. In this case all the Appraisal of Sustainability reports reviewed sought to comply with the SEA Directive. The most important issues identified, particularly where the assessments raise potential concerns about compliance with the Directive, are discussed below in more detail.

Seven NPS Appraisal of Sustainability reports that were published in 2009 were reviewed. One was an overarching NPS appraisal for the energy sector, with four technical annexes.[3] A separate NPS for new nuclear power stations was prepared with a separate appraisal. All of these were produced by the UK's Department of Energy and Climate Change (DECC) with the appraisals undertaken by a number of consultants. One Ports NPS and its appraisal were produced by the Department for Transport (DfT), supported by consultants (DECC, 2009a, b; DfT, 2009).

The approaches adopted for the appraisals varied across the sectors and NPSs, and contained some inconsistencies. The appraisals were, however, generally consistent in that they were limited in terms of their consideration of alternatives, based on the objectives set for the NPSs. Central to the quality of assessment, if an objective-led approach is adopted, is the selection of assessment objectives against which the strategic actions are assessed. In the case of the draft NPS Appraisals of Sustainability these varied considerably. The Overarching Energy appraisal drew on an appropriate range of appraisal objectives that were suitably aspirational (which is essential if an objectives-led approach is taken — the assessment is meant to be against a *desired* future environment). However, the Nuclear NPS appraisal adopted an unusual set of objectives to assess the draft NPS which were almost entirely "unaspirational" — they sought only to avoid potential adverse impacts, and *not* to protect or enhance the environment. Cumulative effects were also generally poorly addressed across the appraisals, often superficial where they were considered or not at all in the case of the energy technology-specific appraisals. Overall, it was concluded that the quality of the Appraisals of Sustainability reports was variable, often poor, and in some cases with critical failings.

The most problematic areas identified stemmed from the overall approach adopted for the assessments and the context in which they were being used. The speeding up of the planning process does little to facilitate better Appraisal of Sustainability or SEA, since such assessments take time to undertake properly and to engage effectively with stakeholders and the public. This can result in the SEA Directive appearing as a hurdle rather than being a tool that can bring real benefits for better strategic action planning.

[3] 1. Fossil fuel electricity infrastructure, 2. Renewables (onshore and offshore wind, energy from waste), 3. Gas and oil infrastructure, 4. Electricity networks.

How sustainability appears to be conceived by the UK Government in the Appraisals of Sustainability also potentially militates against an effective assessment of the likely significant environmental effects of the draft NPSs. Sustainability in the Appraisals of Sustainability is generally conceived of as "*weak sustainability*" where environmental, economic and social factors are balanced and therefore capable of being traded-off against each other (Eales and Sheate, 2010). This runs counter to the tenets of the Bruntland definition of sustainable development which recognises the imperatives of environmental limits and social needs (see above). In many cases, especially true in the context of the energy sector NPSs, there also appears to be an assumption that low carbon can be equated with sustainability when the two may not be compatible.

The authors found the way the draft NPS objectives were defined was highly variable. There were large differences in the purposes ascribed to NPSs when under the Planning Act 2008 there is a clear intention that they are to establish criteria for decision-making by the Infrastructure Planning Commission. Only the draft Nuclear NPS made this explicit as its primary objective, others proffered a range of additional and multiple objectives at the same time as the UK Government's stated intention that the NPSs were only to further *clarify* policy and not to *develop* policy.

How the draft NPS objectives were framed was such that the UK Government then sought to justify a seemingly deliberately narrow interpretation when it came to defining the nature of the alternatives studied (e.g. "have an NPS" or "do not have an NPS"). Therefore it is highly questionable whether the alternatives studied in the appraisals for the draft Overarching Energy NPS and Ports NPS can be considered "reasonable" (in the context of the SEA Directive). This is exacerbated by the fact that the alternatives studied for the draft Nuclear NPS were constrained because the policy decision to promote nuclear power had not been subject to any environmental assessment of the role nuclear should play in the wider energy mix. The alternatives taken into account were often those considered reasonable for Government rather than reasonable from the point of view of being possible and plausible alternatives to the objectives of the plan in question, as required by the SEA Directive.

This *de minimis* approach to the implementation of SEA through the NPS appraisals extends further to the often poor quality of the assessments actually undertaken. In the Ports NPS appraisal, for example, the baseline information provided about the current state of the environment was considered by the authors to be inadequate. The SEA Directive is clear that sufficient information on the state of the environment, and an assessment of the likely evolution of the environment without the strategic action, are required. Good baseline information at the appropriate level of detail for the strategic action concerned is essential to be able to understand what the impacts might be of the proposed strategic action. Again in the case of the Ports NPS appraisal, as well as the Overarching Energy NPS appraisal, how the assessment against the appraisal objectives was undertaken did not assess the real impact of an NPS on the environment. The very existence of an NPS, for example, has the potential to encourage larger, more centralised approaches to energy generation because the Nationally Significant Infrastructure Projects process applies only above a certain threshold and could be perceived as a faster route to approval. Consequently, it is questionable whether many of the Appraisals of Sustainability were actually assessing the true consequences of the NPSs and therefore meeting the requirements of the SEA Directive. It was also concluded by the authors from the evidence of the Appraisals of Sustainability reviewed that they were unlikely to have met the purposes of assessment of influencing the planning process.

Box 1 summarises the significant failings across the draft NPS Appraisals of Sustainability prepared in 2009 which made them potentially challengeable in the courts.[4] One acknowledgement of the environment in the appraisals was in the form of seeking to minimise or avoid impacts through mitigation. However, this approach missed the opportunity to avoid impacts through the appropriate selection of the most environmentally favourable options in the first place (e.g. in terms of energy, a mix of technologies that would have the least adverse impact upon the environment). This is an essential tenet of the SEA Directive which in Recital 5 of the Preamble states that SEA procedures "should contribute to more sustainable and effective solutions".

Box 1. Key failings of the 2009 UK Energy and Ports NPS Appraisals of Sustainability.

- The very nature of the planning processes in these cases was that they, in a variety of ways, sought to circumvent the purposes of the SEA Directive, particularly with respect to the evaluation of reasonable alternatives. In some cases it was because they purported to be high-level strategic planning documents, but were in fact being determined by projects that have already been identified, e.g. sites in the draft Nuclear NPS. In this way the projects are setting the strategic framework for the strategic action and not the strategic action for the projects. In other cases the objectives of the NPS are defined and/or interpreted in such

[4] In 2010 the UK coalition Government subsequently revised some of the draft energy NPSs and their Appraisals of Sustainability partly to address some of the widely recognised shortcomings in the assessments. However, they also put their own stamp on these strategic actions that had been generated originally by the previous Government of a rather different political colour.

Box 1. (*Continued*)

narrow terms (e.g. as simply elaboration of current policy) as to exclude proper consideration and evaluation of "*reasonable alternatives*".

- The Ports NPS Appraisal of Sustainability which was considered consistently inadequate in its approach and appeared to substantially fail to meet the requirements of the SEA Directive.
- The Overarching Energy Appraisal of Sustainability (and technology-specific appraisals) raised some issues similar to those for the Ports NPS appraisal. The issue of alternatives in particular was highly debateable and so it was questionable whether the alternatives considered could be considered "reasonable".
- For the Nuclear NPS Appraisal of Sustainability, the main failing related to the inappropriate appraisal objectives against which the appraisal was undertaken and that there were no proposals for monitoring, proposing instead to outline those in the post-adoption statement. This appeared to be non-compliant with the SEA Directive since the monitoring was therefore excluded from the consultation process on the Appraisal of Sustainability report and draft NPS.

Assessment of the Scottish National Planning Framework

The Scottish second National Planning Framework (NPF2) sets out the strategy for Scotland's spatial development to 2030. The NPF2 focus is on promoting "sustainable economic growth" while also having clear environmental objectives. It sets out a broad vision for planning in Scotland and creates a class of priority national development projects (NDPs), the inclusion of which in the NPF2 signals the desire to see those projects delivered. The process of arriving at the those NDPs however appears far from logical, given that the list started out with only nine projects, and then became a wish-list of some 52 projects before finalising at 14. As part of the preparation of the NPF2, the Scottish Government carried out an SEA (Scottish Government, 2008). Compliance with the SEA Directive, along with the NDP selection process itself, was subject to scrutiny by a judicial review in November 2010 of the proposed Hunterston coal-fired power station and transhipment centre (one of the final 14, but not the original nine NDPs) (Sheate and Twigger-Ross, 2010). The outcome of that judicial review is eagerly awaited.

Given the NPF2's focus on promoting sustainable economic growth it is not surprising therefore that there were a number of conflicts between the SEA objectives, seeking to protect the environment and people, and the programme itself. The key area of conflict was in trying to reconcile climate change and economic growth aspirations due to a large number of proposed major energy and transport projects. Many developments would potentially have significant negative effects on biodiversity, particularly on coastal and marine habitats and species.

The issues with the SEA stem directly from the split nature of the strategic action process itself, which while national was not particularly spatial in its approach to alternatives, and in the creation of a class of national development projects which had the effect of turning strategic thinking on its head. A large amount of time and effort went into determining the national development projects, whereas a more spatially relevant strategy at the national level, followed by a series of programmes or sectoral plans also subject to SEA, would have been a far more logical way to proceed and would have ensured that the most appropriate level of assessment was applied at each stage. As it was, national development projects were created by the NPF2 without adequate assessment of whether they were the most appropriate or environmentally acceptable projects to take forward.

Two types or levels of alternatives were considered, neither of which appeared particularly coherent (or indeed "reasonable"), given the nature of the strategic action the SEA was assessing:

- high-level "alternatives" representing four thematic scenarios: economy, sustainability, communities and connectivity, each of which represents a selection of policy options. "Business as Usual" scenario was also included in this "category" as a fifth scenario; and
- National Development "alternatives", representing a number of energy, transport or environmental infrastructure projects, although as the NPF2 process and SEA unfolded there appears to have been even less logic to the way these projects might have been considered "alternatives".

A fundamental problem illustrated by this approach to alternatives is the gap between "strategic scenarios" and "national development projects", i.e. there is no sectoral or programme level of assessment, nor assessment of spatial alternatives. The NPF2 SEA attempted to do both high-level policy assessment and low-level (but superficial) project assessment. It is therefore not surprising that it did not really accomplish either very satisfactorily. The reality here is that the strategic planning process is being driven by the projects, rather than the other way around, constraining the nature of what might be considered to be "reasonable" alternatives and potentially raising a question as to whether the SEA Directive is being circumvented by such an approach, as well as serious issues of the lack of accountability in the decision-making process when a project is labelled a "national development" project with no effective assessment of alternatives to that project. By the time EIA is undertaken on the project, strategic alternatives to it have already been ruled out by the strategic decision and the public effectively have been excluded from participating in an "early and effective" manner in that process (Sheate and Twigger-Ross, 2010), contrary to the requirements of both the SEA Directive and the Aarhus Convention.

Effective Environmental and Sustainability Assessment?

Table 1 summarises the key problems identified across the cases examined, suggesting that these experiences have not delivered effective SEA, either in terms

Table 1. Key problems in the implementation of high-level SEA cases in the UK.

SEA area	Summary of problem
Screening	Many organisations lack the skills and capacity to screen strategically across the whole organisation for which strategic actions might require SEA and what the relationship to each other should be. Consequently what might be better as separate plan and programme assessments can end up being undertaken as one assessment, even though that creates problems regarding scale, detail, participation and accountability.
Scoping	Inappropriate scoping-out of key environmental topics can be a problem as responsible authorities try to minimise the workload of the SEA process. The rationale for scoping out is often unjustified or simply not provided.
Consideration of alternatives	Very limited consideration of alternatives (e.g. "strategic action" or "no strategic action", and the latter, which is effectively evolution of the environment without the plan, is a separate requirement of the SEA Directive anyway). The consideration of reasonable alternatives is the strongest lever for influencing the strategic action-making process.
Integration of SEA with strategic action-making processes	Often poor integration is observed, e.g. alternatives considered as part of the strategic action process often are not returned to or used in the SEA.
Assessment of significant effects	Often inconsistent approaches to the assessment of significance, and particularly a danger that assessors take an overly positive view of how and whether strategic actions will be implemented, often without justification.
Assessment of cumulative effects	Often poor attempts at cumulative effects assessment can be observed, which take a rather one-dimensional view of cumulative effects. Exacerbated by an unjustified belief in mitigation and a general failure to see alternatives as part of the "avoid, reduce or remedy" hierarchy of the SEA Directive.

Table 1. (*Continued*)

SEA area	Summary of problem
Mitigation and monitoring	Often there is little evidence provided as to how mitigation measures will be delivered. At the strategic level there is no means of guaranteeing or securing (a) that mitigation measures will be delivered and (b) that they will be successful. A large amount of blind faith in mitigation pervades the assessment reports. There is a distinct lack of recognition of the precautionary principle being applied in such cases, which would be more consistent with the aim to avoid damaging impacts rather than mitigating them.
Presentation and public participation	Presentation of SEA environmental reports and sustainability appraisal reports often does little to promote engagement and participation in the assessment process. These reports, and the non-technical summaries in particular, often offer little to anyone without some knowledge or understanding of the issues. Considerably more effort is needed on non-technical summaries to focus on the issues that matter and properly convey a summary of the environmental report in a non-technical manner. This issue has been raised in the courts (e.g. in Seaport, and another hearing in Northern Ireland (Sheate, 2009)), where the opposite problem was observed — a one-page non-technical summary that was neither non-technical nor a summary.

of influencing the strategic action or in terms of changing the mindset of the actors involved, which is likely to take a longer period of time to achieve. That is not a criticism of SEA itself, more of its application and the way it is perceived by responsible authorities. In Scotland, for example, where SEA is being applied to a wider range of strategic actions, some responsible authorities would seem to have attempted to minimise the work involved through scoping out wherever possible, or through combining what might more usefully be separate assessments (Phillips and Sheate, 2010). The Scottish Government, however, is addressing this by developing and increasing its SEA capacity and improving resourcing of the SEA Gateway.[5] It will be interesting to see if other responsible authorities enhance their SEA capability in a similar manner.

[5] The SEA Gateway, a unique initiative in the UK, co-ordinates SEA activity throughout Scotland. The Gateway provides advice and information about the SEA process, co-ordinates consultation, liaises with Scotland's Consultation Authorities: SNH, SEPA and Historic Scotland, and undertakes Scottish Ministers' functions under the Environmental Assessment (Scotland) Act 2005. See http://www.scotland.gov.uk/Topics/Built-Environment/planning/National-Planning-Policy/themes/enviro-assessment/sea/Q/editmode/on/forceupdate/on.

On the other hand, from the authors' own experience of undertaking SEA at the strategic level, one of the key challenges to delivering effective SEA is the limitation of time and resources. Where sufficient time is available, and SEA is considered from the outset, the SEA team and process have the opportunity to engage more fully with the plan-making team and stakeholders, and SEA and Sustainability Appraisal can be used to help structure the strategic action itself and properly consider alternatives.

If the UK Government is to meet the objectives of the SEA Directive (Article 1), its application would need to ensure it is delivering substantial compliance with the specific requirements for an environmental report as laid out in Article 5 and Annex 1. In the authors' view the cases described in this paper showed considerable variation in quality of the assessment, including the quality of baseline data descriptions and evaluations undertaken. This stems partly from variability in the objectives set for the strategic actions themselves which had consequences for the nature of the alternatives considered. Almost across the board the issue of alternatives has been a problem and it is highly questionable whether the alternatives in many cases can be considered to have been "reasonable" as required by the SEA Directive. The constraint on alternatives seems to have been imposed partly because of the political nature of the strategic actions at this national level.

It is difficult to conclude whether the assessment processes have influenced the strategic actions and their planning processes. Only if *real* alternatives (e.g. technology mixes in the case of the energy NPSs or sector-specific programmes for NPF2) had been considered, might the assessment processes have been able to influence the direction of the strategic action. An assessment of such alternatives could valuably create an opportunity to further specify government policy. This would help provide a strategic policy framework for lower level strategic actions while ensuring sufficient accountability and legitimacy of strategic decisions.

What does this say about the application of the SEA Directive in the UK to national level planning processes? The strategic case examples discussed in this paper suggest that either the approach to the assessments has been poorly conceived by Government in terms of applying the SEA Directive, or that it has been well conceived in order to circumvent (or at best barely meet) the purposes of the Directive. Either way the application of strategic environmental and sustainability assessment by the UK Government appears disingenuous since there seems to be little real intention that the assessments should influence the planning process. Why might this be? There may be a strong degree of institutional inertia at very strategic levels, where strategic actions are seen to be the preserve of political considerations. Such actions may also be hampered by an "administrative rationalism" (Dryzek, 2005) among civil servants with little practical or "expert" knowledge of assessment, unconvinced from a bureaucratic perspective that SEA/Sustainability Appraisal is worth the cost and administrative effort involved.

The very purpose of speeding up the planning process goes against effective environmental assessment since the assessment process needs to influence the planning process from the earliest possible opportunity and should seek to ensure public participation during that process. Is this a peculiar feature of Appraisals of Sustainability and/or Sustainability Appraisals (given the experience with Energy and Ports NPSs and Eco-towns Planning Policy Statement at the national planning level and concerns over Sustainability Appraisals of Regional Spatial Strategies) rather than SEA, or a wider pattern of the approach by Government in the UK to the implementation of the SEA Directive? Is there a potential for a return to a minimalist approach (e.g. CPRE, 1991; Sheate, 1994; Wood, 1995), similar to that seen in the 1980s on EIA and the early developments of a draft SEA Directive?

There are a number of important considerations to take into account in seeking to answer these questions. The Sustainability Appraisal/Appraisal of Sustainability approach is one rather peculiar to England and Wales, and has not been pursued in Scotland (where SEA has been further enshrined in primary legislation) or elsewhere in Europe. The principle of assessing all aspects of sustainability together is an attractive one, but depends very much upon the perspective of sustainability being adopted by policy makers. An approach that adopted the full Brundtland definition would include a clear recognition in decision-making of the environmental and social imperatives that underpin the whole concept. That would also be consistent with seeing SEA as a means of ensuring the environment (and many social aspects such as populations and human health) are fully integrated into decision-making. However, this paper argues that the view is a more equivocal one, which sees sustainability as a simple balance between economic, social and environmental factors, and therefore trade-offs between those factors become more acceptable and a high level of protection for the environment is not necessarily guaranteed (Sadler, 2005). In practice, Sustainability Appraisal and Appraisals of Sustainability can appear to reinforce that, claiming sustainability while accepting loss in one or more factors is acceptable, even though there could be alternative ways of enhancing economic growth that would also enhance other factors.

The experience to date before the UK courts (e.g. Weatherup, 2007; Mitting, 2009) supports the perception that the Government is somewhat ambivalent about the purpose of the SEA Directive; otherwise it could, for example, have avoided judicial review over the East of England Regional Spatial Strategy. This also suggests a misunderstanding of the fundamental purposes of SEA with which the case study assessments purport to comply. Spatially-relevant NPSs and assessments, for example, would have allowed a much better assessment of the strategic and cumulative effects of different levels of strategic actions at different broad locations and so enable an approach that might actually help to maximise environmental benefits and avoid adverse environmental and social impacts as much as possible. By the time projects come forward there is far less scope for avoidance, and mitigation becomes the main focus of EIA. There is also a possible pattern emerging from these examples as the inadequate consideration of alternatives and the over-reliance on mitigation were seen as raising serious issues of potential non-compliance with the SEA Directive.

The EU experience suggests that it is still too soon to judge the real effectiveness of SEA even in the limited terms discussed here, but UK research has highlighted some similarities with the early days of the EIA Directive (Sheate, 1994), i.e. variable to poor understanding and application of SEA in practice.

Conclusions

The analysis presented in this paper suggests that the current performance by the UK Government in implementing the SEA Directive for national level strategic actions is far from exemplary. In this case it is not failure to transpose the Directive for these classes of decisions, but poor application. A serious question is raised as to whether this is simply symptomatic of the early learning phase of SEA in Europe or whether it is due to the approach taken to SEA implementation in the UK at national level. The former argument is a weak one, given the UK, and England and Wales in particular, has a long history prior to the SEA Directive being introduced of undertaking forms of environmental appraisal for a range of land use planning documents (e.g. the good practice guide of environmental appraisal of development plans published by Department of the Environment, 1993).

Undertaking SEA that delivers the objectives of the SEA Directive requires time and resources to be invested early in the strategic decision-making process. It also requires a receptive process to avoid unnecessary expenditure and delay later on. It is inevitable that the more consultation and assessment are squeezed, the more recourse will be sought through the courts in the form of judicial review. Delays in past infrastructure planning because of long public inquiries (e.g. Heathrow Airport Terminal 5 inquiry lasted for some four years) were not caused by sheer bloody-mindedness, but because they were controversial proposals that would have significant impacts on the environment and local communities. The Aarhus Convention is clear in promoting access to information, public participation and access to justice that if the first two are not forthcoming the last has to be available as a measure of last resort. Where time is curtailed, options severely constrained or strategic assessment is only introduced late in the process, the SEA inevitably becomes a bolt-on exercise with little real influence on the policy- or decision-making process. The ability of SEA to influence the strategic action process or the mindset of those responsible for the strategic action or stakeholders is severely limited in such circumstances, and there is little real opportunity for exchange of knowledge or understanding. The core objective of the SEA Directive, contained in Article 1 (see Introduction section), clearly is not achievable under these circumstances.

Despite investing considerable time and effort in the strategic assessments described in this paper, if there is little integration with the actual planning process then it is hard to see how these assessments are significantly influencing decision-making. This is particularly true if reasonable alternatives are not being properly addressed, since this is the main way changes to a strategic action might actually be brought about to avoid adverse environmental impact, enhance environmental benefits and deliver more sustainable solutions.

There is a risk that where environmental or sustainability benefits are identified as part of the objectives of the strategic action, then it is automatically seen as somehow a "sustainable plan". What is revealed very clearly by the evidence of the draft NPS Appraisals of Sustainability, for example, is the yawning gap at the policy level above the NPSs where there is no formal SEA undertaken. The problems surrounding the alternatives at the Overarching Energy NPS level stem directly from the lack of any meaningful environmental assessment of options and mix of options of energy technologies at the policy level.

There has always been a need for policy level SEA (Sheate *et al.*, 2003), but as SEA is applied at higher strategic levels it will continue to expose the lack of SEA at the highest government policy level. Without a proper assessment of alternative options at the highest level, options at lower levels will continue to be constrained, as will SEA's role in promoting more sustainable development and improving public participation, scrutiny and accountability of high-level decision-making. At the root of the problems with the application of Sustainability Appraisal and Appraisals of Sustainability in the UK is the apparently weak conception of sustainability adopted by the UK Government and the perception that having to undertake an assessment and comply with the SEA Directive is a hurdle rather than a useful mechanism for helping to deliver better and more sustainable evidence-based policy making. SEA, in these cases, is therefore being allowed to perform neither an effective advocative nor an integrative role.

Acknowledgements

This article is partially based on work commissioned by the RSPB and WWF and a full version of the report upon which it draws is available from http://www.rspb.org.uk/ourwork/policy/planning/statements.asp.

References

Cashmore, M, A Bond and B Sadler (2009). Introduction: the effectiveness of impact assessment instruments. *Impact Assessment and Project Appraisal*, 27(2), 91–93.

Cashmore, M, T Richardson, T Hilding-Ryedvik and L Emmelin (2010). Evaluating the effectiveness of impact assessment instruments: theorising the nature and implications of their political constitution. *Environmental Impact Assessment Review*, 30, 371–379.

CEC (Commission of the European Communities) (2009). Report from the Commission to the Council, the European Parliament, The European Economic and Social Committee and the Committee of the Regions on the application and effectiveness of the Directive on Strategic Environmental Assessment (Directive 2001/42/EC).

CLG (Communities and Local Government) (2008a). Sustainability Appraisal and Habitats Regulations Assessment of the Draft Eco-towns Planning Policy Statement.

CLG (Communities and Local Government) (2008b). Sustainability Appraisal and Habitats Regulations Assessment of the Eco-towns Programme.

CLG (Communities and Local Government) (2010). Towards a more efficient and effective use of Strategic Environmental Assessment and Sustainability Appraisal in spatial planning. Final Report.

COWI and Milieu (2009). Study concerning the report on the application and effectiveness of the SEA Directive (2001/42/EC). Final Report. European Commission DG-ENV.

CPRE (1991). Submission by the Council for the Protection of Rural England (CPRE) to the European Commission's five-year review of EC Directive 85/337/EEC on environmental assessment: the environmental assessment directive-five years on.

Dalal-Clayton, B and B Sadler (2005). *Sustainability Appraisal — A Review of International Experience and Practice*. London: Earthscan Publications.

Department for Transport (2009). Ports: National Policy Statement for England & Wales — Appraisal of Sustainability Report.

Department of Energy and Climate Change (2009a). Planning for new energy infrastructure. Appraisal of Sustainability for the Draft National Policy Statements for: Overarching Energy (EN-1), Fossil Fuels (EN-2), Renewables (EN-3), Gas Supply and Gas and Oil Pipelines (EN-4), and Electricity Networks (EN-5).

Department of Energy and Climate Change (2009b). Planning for New Energy Infrastructure. Appraisal of Sustainability of the Draft Nuclear National Policy Statement.

Department of the Environment (1993). Environmental Appraisal of Development Plans — A Good Practice Guide, Department of the Environment (DoE). London: HMSO.

Dryzek, J (2005). *The Politics of the Earth: Environmental Discourses.* 2nd ed. Oxford: Oxford Universisty Press.

Eales, RP and WR Sheate (2010). Opportunities missed, and challenges to come? *Town and Country Planning*, 79(3), 138–143.

Fergusson, M, C Coffey, D Wilkinson, D Baldock, A Farmer, RA Kramer and AG Mazurek (2001). *The Effectiveness of EU Council Integration Strategies and Options for Carrying Forward the "Cardiff Process"*, IEEP/Ecologic, March 2001.

Fischer, T and P Gazzola (2006). SEA effectiveness criteria — equally valid in all countries? The case of Italy. *Environmental Impact Assessment Review*, 26, 396–409.

Fischer, TB (2005). Having an impact? Context elements for effective SEA application in transport policy, plan and programme making. *Journal of Environmental Assessment Policy and Management*, 7(3), 407–432.

Fischer, TB (2010). Reviewing the quality of strategic environmental assessment reports for English spatial plan core strategies. *Environmental Impact Assessment Review*, 30, 62–69.

Gazzola, P (2008). What appears to make SEA effective in different planning systems. *Journal of Environmental Assessment Policy and Management*, 10(1), 1–24.

Gibson, RB (2006). Beyond the pillars: sustainability assessment as a framework for effective integration of social. Economic and ecological consideration in significant decision-making. *Journal of Environmental Assessment Policy and Management*, 8(3), 259–280.

Gibson, RB, S Hassan, S Holtz, J Tansey and G Whitelaw (2005). *Sustainability Assessment: Criteria, Processes and Applications.* London: Earthscan Publications.

Hertin, J, J Turnpenny, A Jordan, M Nilsson, D Russel and B Nykvist (2009). Rationalising the policy mess? Ex ante policy assessment and the utilisation of knowledge in the policy process. *Environment and Planning A*, 41(5), 1185–2000.

IAIA (International Association for Impact Assessment) (2002). *SEA Performance Criteria*, International Association for Impact Assessment (IAIA), available at http://www.iaia.org/publicdocuments/special-publications/sp1.pdf [accessed 10 January 2011].

IEMA (Institute of Environmental Management and Assessment) (2004). *SEA Review Criteria.*

João, E (2007). A research agenda for data and scale issues in Strategic Environmental Assessment (SEA). *Environmental Impact Assessment Review*, 27(5), 479–491.

Kørnøv, L and WAH Thissen (2000). Rationality in decision and policy-making: implications for strategic environmental assessment. *Impact Assessment and Project Appraisal*, 18(3), 191–200.

Mitting, J (2009). Judgment in, *City and District Council of St Albans and Hertfordshire County Council v Secretary of State for Communities and Local Government* [2009] EWHC 1280 (Admin), 20 May 2009.

Morrison-Saunders, A and T Fischer (2006). What is wrong with EIA and SEA anyway? — A sceptic's perspective on sustainability assessment. *Journal of Environmental Assessment Policy and Management*, 8(1), 1–21.

Neumayer, E (2003). *Weak Versus Strong Sustainability*. 2nd ed. London: Edward Elgar Publishing.

Nilsson, M and H Dalkman (2001). Decision making and strategic environmental assessment. *Journal of Environmental Assessment Policy and Management*, 3(3), 305–327.

Noble, B (2000). Strategic environmental assessment: what is it and what makes it strategic? *Journal of Environmental Assessment Policy and Management*, 2(2), 203–224.

Noble, B (2002). The Canadian experience with SEA and sustainability. *Environmental Impact Assessment Review*, 22, 3–16.

Offical Journal (2001). Directive 2001/42/EC of the European parliament and of the council of 27 June 2001 on the assessment of the effects of certain plans and programmes on the environment, *OJ* L 197, 21 July 2001, pp. 30–37.

Owens, S, T Rayner and O Bina (2004). New agendas for appraisal: reflections on theory, practice and research. *Environment and Planning A*, 36, 1943–1959.

Partidário, MR (2000). Elements of an SEA framework — improving the added-value of SEA. *Environmental Impact Assessment Review*, 20, 647–663.

Partidário, MR, WR Sheate, O Bina, H Byron and B Augusto (2009). Sustainability assessment for agriculture scenarios in Europe's mountain areas: lessons from six study areas. *Environmental Management*, 43, 144–165.

Phillips, P and W Sheate (2010). A new SEA pathway: reflecting on strategic environmental assessment in Scotland. *The Environmentalist*, 104, 19–22.

Pope, J and W Grace (2006). Sustainability assessment in context: issues of process, policy and governance. *Journal of Environmental Assessment Policy and Management*, 8(3), 373–398.

Retief, F (2007). A performance evaluation of strategic environmental assessment (SEA) processes within the South African context. *Environmental Impact Assessment Review*, 27, 84–100.

Runhaar, H (2009). Putting SEA in context: a discourse perspective on how SEA contributes to decision-making. *Environmental Impact Assessment Review*, 29, 200–209.

Sadler, B (2005). *Strategic Environmental Assessment at the Policy Level: Recent Progress, Current Status and Future Prospects*. Report prepared by Regional Environment Centre on behalf of Ministry of the Environment, Czech Republic.

Sadler, B and R Verheem (1996). *Strategic Environmental Assessment: Status, Challenges and Future Directions*. Ministry of Housing. Spatial Planning and the Environment of the Netherlands.

Scottish Government (2008). *National Planning Framework 2 SEA Environmental Report*.

Scottish Government (2010). Strategic Environmental Assessment Pathfinder Project Stage 1 and 2: Combined Summary Report.

Sheate, WR (1994). *Making an Impact: A Guide to EIA Law and Policy*. 1st ed. London: Cameron May.

Sheate, WR (2008). *Letter to the BARD Campaign Reviewing Eco-Towns Sustainability Appraisal Reports*, available at http://www.bardcampaign.com/press/pressreleases/198064/the_sheate_report.html [accessed 6 January 2011].

Sheate, WR (2009). Affidavit on behalf of the Applicant in *Irwin Glenbank Ltd v Department of the Environment, Northern Ireland.*

Sheate, WR, S Dagg, J Richardson, R Aschemann, J Palerm and U Steen (2001). *SEA and Integration off the Environment into Strategic Decision-Making* (3 Volumes). Final Report to the European Commission, DO XI, Contract No. B4–3040/99/136634/MAR/B4 available at http://ec.europa.eu/environment/eia/sea-support.htm, Office for Official Publications of the European Communities, Luxembourg, 438 p, 2003.

Sheate, WR, S Dagg, J Richardson, R Aschemann, J Palerm and U Steen (2003). Integrating the environment into strategic decision-making: conceptualizing policy SEA. *European Environment*, 13(I), 1–18.

Sheate, WR and MR Partidário (2010). Strategic approaches and assessment techniques — potential for knowledge brokerage towards sustainability. *Environmental Impact Assessment Review*, 30, 278–288.

Sheate, WR and C Twigger-Ross (2010). A report to Patrick Campbell and Company Solicitors *'Early and Effective' Participation in SEA, in the Case of McGinty v Scottish Ministers, November 2010.*

Stookes, P (2011). EIA — a precautionary tale. *The Environmentalist*, 109, 16–17.

Sustainable Development Research Network (2008). *Issues for the Practice of Sustainability Appraisal in Spatial Planning — A Review*. Final Workstream Report. Prepared by Land Use Consultants and the Royal Town Planning Institute (RTPI).

Theophilou, V, A Bond and M Cashmore (2009). Application of the SEA directive to EU structural funds: perspectives on effectiveness. *Environmental Impact Assessment Review*, 30(2), 136–144.

Therivel, R, G Christian, C Craig, R Grinham, D Mackins, J Smith, T Sneller, R Turner, D Walker and M Yamane (2009). Sustainability-focused impact assessment: English experiences. *Impact Assessment and Project Appraisal*, 27(2), 155–168.

UNECE (1998). *Aarhus Convention on Access to Information, Public Participation in Decision-Making and Access to Justice in Environmental Matters*, available at http://www.unece.org/env/pp/.

Wallington, T, O Bina and W Thissen (2007). Theorising strategic environmental assessment: fresh perspectives and future challenges. *Environmental Impact Assessment Review*, 27, 569–584.

WCED (World Commission on Environment and Development) (1987). *Our Common Future*. Report of the Brundtland Commission, Oxford University Press.

Weatherup, J (2007). Judgment in (1) An Application by Seaport Investments Limited (2) An Application by Magherafelt District Council, FP McCann (Developments) Limited, Younger Homes Limited, Herron Bros Limited, G Small Contracts and Creagh Concrete Products Limited in *Northern Ireland: No. [2007] NIQB 62*, 7 September 2007.

Weiland, U (2010). Strategic environmental assessment in Germany — practice and open questions. *Environmental Impact Assessment Review*, 30(3), 211–217.

Wood, C (1995). *Environmental Impact Assessment: A Comparative Review*, Longman 1st ed.

CHAPTER 5

Change Agents in the Field of Strategic Environmental Assessment: What Does it Involve and What Potentials Does it Have for Research and Practice?

Lone Kørnøv*, Ivar Lyhne, Sanne Vammen Larsen and Anne M. Hansen

Department of Development and Planning
Aalborg University
Fibigerstraede 13, 9220 Aalborg Ø
Denmark
**lonek@plan.aau.dk*

One of the challenges facing strategic environmental assessment (SEA) is finding ways to work in research and practice allowing critical interrogation and appropriate action to support sustainability. The point of departure for this article is the hypothesis that a certain kind of cooperative knowledge production, called Mode 3, where SEA researchers interact with the societal milieu as change agents, provides a potential for SEA research and practice to further sustainability. Based on literature and three cases, this paper seeks to contribute to two questions: 'what does acting as a change agent within the field of SEA involve?' and 'what potentials does it have for research and practice?' The three cases illustrate how Mode 3 can support reflective SEA practice and practice-based SEA research. Theoretically the current understanding and discussion on change agents is sharpened through the focus on real-life linkages, putting forward the contextual influence and the unpredictability related hereto.

Keywords: Change agent; Mode 3 research; strategic environmental assessment; knowledge production.

---

*Corresponding author.

Introduction

The point of departure and underlying assumption behind this paper is that different types of cooperation between research and practice influence the knowledge production and the possibility to influence decision-making.

During the last decade science and technology have increasingly been harnessed in the quest for a transitioning towards sustainability, among other things grounded in the belief that for knowledge to be useful from a sustainability perspective, it generally needs to be coproduced through close cooperation between scholars and practitioners (Clark and Dickson, 2003). The important scholarly discussion about the role and effectiveness of environmental assessment (EA) as a tool to promote sustainable development has simultaneously increased over the past years, and it has been questioned if EA has the wanted impact on the planning and decision-making process. The discussion involves questioning whether EA tools are too often developed from an expert-driven perspective without sufficient attention to contextual circumstances including the practitioners' needs and capacities (Emmelin 2006) and without sufficient understanding and recognition of the actual non-linear decision-making processes (Richardson (2005); Kørnøv and Thissen (2000); Lawrence (2000); Nilsson and Dalkmann (2001); Bina (2001)). The reasons for the experienced gap between EA research and practice can be found in these arguments, and can be due to a non- or low-collaborative knowledge production, with a clear demarcation between research and practice.

The practice of connecting theoretical knowledge with practical problems, including a high personal engagement, is by Jamison (2001, 2008), called "change-oriented research" and refers to a knowledge making which is problem-based with the aim "...to intervene creatively and constructively in an ongoing social or political process: to contribute to change. Rather than the traditional notion of enlightenment, by which is usually meant that the role of the scientist is to provide insights for the broader society, derived from a 'disinterested' pursuit of the truth, change-oriented research is about empowerment, where the researcher applies knowledge gained from experience to processes of social learning, carried out together with those being 'studied'" (Jamison, 2010, p. 9). This engagement of the researcher as a change agent is in different fields of research referred to by other names like, e.g. empowerment and action research. The change agent research, which is closely linked to current societal needs and is undertaken in close cooperation between research and practice, is also termed "Mode 3" (Huff and Huff (2001); Kurek *et al.* (2007)). Kurek *et al.* (2007) provides an analytical framework for studying the strategic positioning of the researcher, which makes it possible to distinguish between modes of research, hereunder Mode 3.

This paper is inspired by both Jamison's normative framework and argument about the need for change-oriented research, and by the analytical framework developed by Kurek *et al.* (2007). These frameworks are used for discussing experiences with connecting research and practice, and thereby approach the mentioned insufficiencies in the field of SEA. The hypothesis, which this paper is based on, is that situated form of knowledge making in which SEA researcher that acts a change agent within a Mode 3 positioning has a potential to improve the connection between research and practice, and promote sustainable development.

Aim

At Aalborg University's Department of Development and Planning, researchers acting as change agents conduct three research projects on SEA. This paper seeks to collect and communicate experiences from these cases. The paper is aimed at contributing to the following questions:

- What does acting as a change agent within the field of SEA involve? and
- What potential does it have for research and practice?

All modes of research have their own justification — whether an independent or interdependent relation between research and practice, and the authors also have experience with, and value all, modes of research being discussed in the next paragraph. The Mode 3 research with its strong cooperative interdependence is though found by the authors to have a distinctive capacity to influence decision-making towards sustainability. The aim of the paper is to explore Mode 3 and make up illustrative analyses of collection of experiences, showing characteristics of Mode 3 and illuminating possible ways of conducting SEA research in Mode 3 and the potential it may have.

With this aim, first an analytical framework is developed through a discussion of different research modes in the following section. In the fourth section, this framework is used for presenting and analysing the three cases, in terms of what it involves to conduct Mode 3 research, and acting as a change agent within the field of SEA. This covers discussions of strategic positioning in relation to the formal and informal frames for the research projects. In the fifth section this is followed by an analysis of the potentials of Mode 3 research, based on the authors' and collaborating organisations' observations and assessments of the research projects. Thus focus is on the potential of conducting Mode 3 research, seen both from the perspective of the researcher and from the perspective of the organisation. This underpins the objective of the paper: to identify if and how this specific setup of research provides potential in terms of practice in the organisation and in terms of research.

The Discussion of Research Modes

When discussing the different modes of research with focus on the connection of research and practice, the contribution of Gibbons and colleagues in *The New Production of Knowledge: The Dynamics of Science and Research in Contemporary Societies* from 1994 is found very relevant and inspiring. This work is an influential contribution to the ongoing discussion of the need to improve research relevance and knowledge flows from research to practice. Gibbons *et al.* distinguish between two modes of knowledge production.

Mode 1, typically produced in universities and named "ivory tower research" by critics, has the characteristics of largely being discipline-based, intra-scientifically produced and not related to a specific context for application (Gibbons *et al.* 1994). In Gibbons words, "This structure provides the guidelines for researchers about what the important problems are, how they should be tackled, who should tackle them, and what should be regarded as a contribution to the field. In its social dimensions, it also prescribes the rules for accrediting new researchers, procedures for selecting new university faculties, and criteria for their advancement within academic life" (Gibbons, 1999, p. 9).

The strength of the structured research in Mode 1 is widely acknowledged. However, when it comes to research aiming at changing practice, Mode 1 research meets criticism, e.g. the risk of limited relevance of research for society. Mode 1 research on SEA does not necessarily take a point of departure in experienced problems in certain contexts, and therefore it may not be relevant and it may not be applied. In line with this criticism, Gibbons (1999) points to a need for knowledge production, which is "socially robust", ensured through a new social contract between research and society. It becomes not just a matter of *how* knowledge is produced but also *what* knowledge is produced. Here Mode 2 research offers a different approach.

In Mode 2 the relationship between research and practice is characterised by interaction and cooperation, which according to Gibbons and colleagues leads to change-oriented research in which "the boundaries between the intellectual world and its environment have become blurred" (Gibbons *et al.*, 1994, p. 81). The characteristics are knowledge produced trans-disciplinarily, jointly and bound to a specific context. Therefore, Mode 2 research is validated by its relevance for practice. Compared to Mode 1, Mode 2 is argued to be "more timely, more practical, more democratic" (Huff, 2000, p. 291).

Huff (2000) criticises Mode 2 research for having limitations "especially as it moves away from science and technology into management" (Huff, 2000, p. 291). According to (Huff, 2000, p. 292), Mode 2 research is too pragmatic and tends to make "big bets on the basis of limited evidence". Another criticism of Mode 2 is the commercialisation of research, e.g. raised by Jamison (2001) in *The Making of Green Knowledge*. Research is defined by market interests in funding organisations rather than by the interest among researchers (Jamison, 2001). Furthermore, Jamison (2001, p. 124) criticises Mode 2 for limited change "…many of the actual practices of the companies they run and/or represent all too often continue to follow a 'business as usual' strategy".

The discussion of research modes and trends in knowledge production has received considerable scholarly attention (Nowotny *et al.*, 2001). In the midst of these discussions the concept of Mode 3 arose.

Strategic positioning and Mode 3

In line with Jamison's discussion of the need for a "change-oriented research", the limitations of Mode 2 lead Huff and Huff to suggest Mode 3 knowledge production with the purpose "…to assure survival and promote the common good, at various levels of social aggregation" triggered by "…appreciation and critiques of the human conditions, as it has been, is, and might become" (Huff and Huff, 2001, p. 53). The researcher within this Mode 3 is closely linked to societal needs and compared to Mode 2 is capable of influencing his milieu by creating demand for the scientific knowledge instead of supplying on an external demand (Kurek *et al.* 2007).

Some characteristics, used in the literature on Mode 3, are multiple stakeholder involvement and interdisciplinarity, conversation and cooperation, community-driven, engagement in study field, high organisational autonomy and strategic interdependence (Huff and Huff, 2001); Kurek (2008)). The normative element of Mode 3 is explicated by the goal of a "future good" (Huff and Huff, 2001) and "giving voice" through science as social advocacy (Jamison, 2009b).

Whereas in Mode 1 the researcher mainly is accountable to oneself, and in Mode 2 accountable to the milieu and financing organisation, the researcher in Mode 3 is mainly accountable to the people and/or environment affected both in the research process and the research outcome. Mode 3 involves not only personal, active engagement and intervention in on-going processes, but also a normative framework within which the researcher works.

The relationship between the change agent and the milieu (researchers, government, industry and NGO) is established through negotiation, and the researcher in Mode 3 must make on-going choices of how much he is willing to let others influence the research. An analytical model of the strategic positioning of the researcher within the milieu is developed by Kurek and colleagues (Kurek *et al.* (2008)). The model is

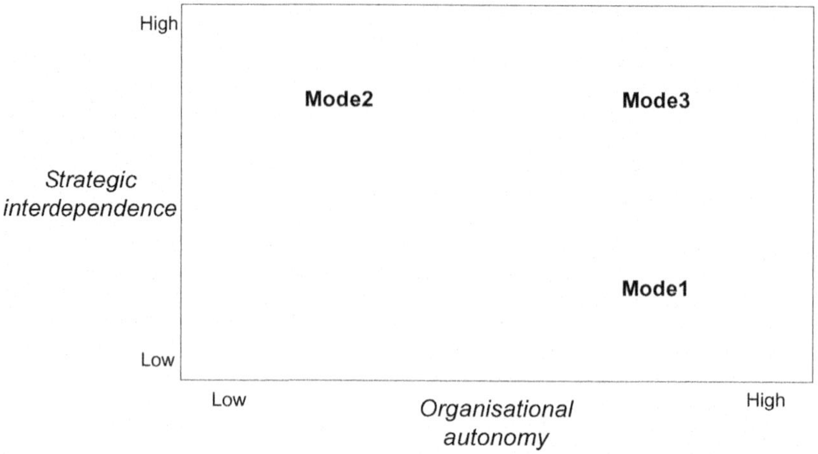

Figure 1. Three modes of strategic positioning. (Based on Kurek *et al.*, 2007, p. 503.)

based upon two dimensions — organisational autonomy and strategic interdependence — and provides a typology with the different modes of researchers positioning, see Fig. 1.

We understand Mode 3 as being characterised by high organisational autonomy and strategic interdependence, and at the same time attributed a normativity guiding the on-going knowledge making and negotiation process taking place between the researcher and the milieu. Mode 3 is understood as flexible in terms of the characteristics of the research activities carried out during the process in the efforts of reaching the goals. The autonomous researchers may in periods choose to do studies independently or choose to engage in research that is defined by the partner. Thus, a Mode 3 collaboration may be a heterogeneous process in terms of the characteristics of its research activities, and the specific activities may have characteristics in common with the other modes. The autonomous researcher in Mode 3 may choose activities that help keeping a critical distance or ease the cooperation with the partners. What make the research activities a Mode 3 collaboration are the overall properties of normative orientation, strategic interdependence and organisational autonomy.

In Mode 3, like in Mode 2, the researcher and milieu share resources (money, time, knowledge) but at the same time the researcher "...autonomously determine directions of research. He retains his responsibility for directing a project" (Kurek *et al.*, 2007, p. 504). So in Mode 3 both the researcher and the milieu are strong enough to sanction each other, and both the strategic interdependence and organisational autonomy is high. This also means that the normative framework,

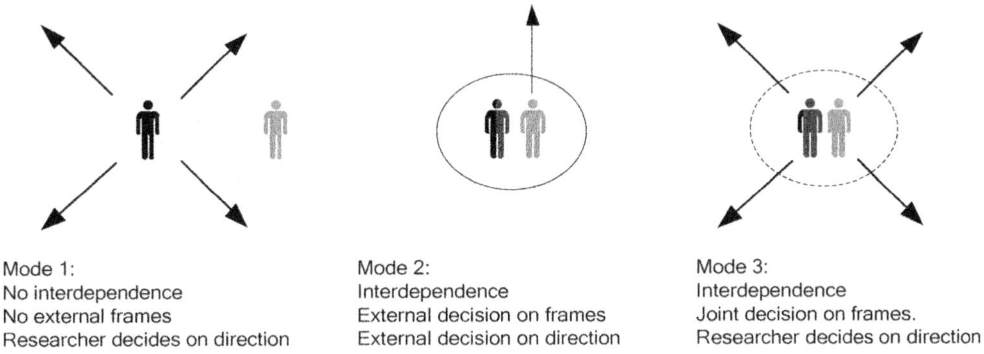

Figure 2. The relation between the researcher and external milieu in the three modes of research. The darker figure represents the researcher, while the lighter figure represents the milieu.

guiding Mode 3 research, is developed by, and acceptable to, both the researcher and the milieu. The difference is visualised in Fig. 2.

All modes exist within the field of SEA research. Examples of SEA research undertaken by a Mode 1 approach are, e.g. theoretical-based work (Kørnøv and Thissen, 2000; Richardson, 2005; Wallington *et al.*, 2007, etc.) and overview articles outlining SEA systems, implementation and experiences (e.g. Therivel and Minas 2002). Several scholars, hereunder Sheate and Partidario (2010) and Seitz *et al.* (2010), are discussing and supporting Mode 3-like cooperation between research and practice, and the need to strength the policy-science interface. An example of doing Mode 3 SEA research is Asplund and Hilding-Rydevik's intervention in on-going SEA and planning processes in Sweden (1996), and writing and publishing taking place in cooperation between researchers and practitioners with examples like Ross *et al.* (2006) and Morrison-Saunders and Arts (2005). An illustration of Mode 2 SEA research is guidance development with a clear term of reference from the contracting institution. These examples illustrate the relevance and importance of all modes of research within the field of SEA.

Thus we are distinguishing between three different modes of research, all with distinct advantages and disadvantages. The focus of this paper is to shed light on experiences with Mode 3 research, answering the questions of what Mode 3 within SEA research involves and what potential it may have. However, before turning to these questions the cases and methods applied are presented in the following section.

Cases and Methods

The analysis in this paper is based upon case studies, from which experiences with Mode 3 research are drawn. In the following the three cases are introduced, and the methods applied in the two analyses are presented. Further information about the three cases is presented continuously in the paper, where it is included in the analysis. The analyses deal with the strategic positioning of the researchers and the potentials for SEA research and practice. The empirical basis for the analyses is document analysis, the researcher's personal observations, and subjective assessments by the researchers as well as the contact person in the organisations.

Cases studied

The study comprises three cases, where PhD researchers are working on their projects in close cooperation with an organisation outside the university. The three research projects have different foci in relation to SEA and different reasoning for the cooperation between SEA research and practice. In all three cases the organisations have co-financed the research projects.

Case 1

Case 1 is carried out in cooperation with the Danish company Energinet.dk, in charge of Danish energy infrastructure. The project is organised with an Aalborg University-based professor as supervisor and the head of the Research and Environment section as main contact person at Energinet.dk.

Aim and methodology: The project concerns the first generation of SEA of plans and programmes in relation to the national energy infrastructure in Denmark (gas and electricity). In this case, Energinet.dk faced implementation of SEA and without sufficient internal professional resources, they initiated cooperation with AAU that ended up with the project aimed at developing and implementing SEA in the energy sector, including SEA methodology targeted at the strategic decision-making processes in the sector. The project has a theoretical basis in decision-making theory and sense-making theory, which are used to understand practice and develop methodology. The project is based on an interactive research approach, in which the researcher is situated at Energinet.dk for a year, participating in meetings and planning processes. To maintain a critical distance, the remaining two years of the project is carried out at AAU, however, still with periodical participation in meetings at Energinet.dk. The research conducted from AAU is widely based on document analysis and interviews.

Case 2

Case 2 is carried out in cooperation with the Greenlandic Self Government and is furthermore co-funded by the independent Alcoa Foundation. The project has an AAU-based professor as main supervisor and the head of the department of physical planning from the Greenlandic Self Government as co-supervisor.

Aim and methodology: The project concerns SEA of mega industry in Greenland in a system with no legislation or guidelines in place. This case is rooted in the environmental and democratic challenge of planning and assessing an aluminium smelter in Greenland (Hansen and Kørnøv, 2010). The aim of the research project was to secure a critical and independent view upon the processes and effect of carrying out SEA. The project is conducted as a case study of the SEA and the planning process of an aluminium reduction plant in Greenland. A theoretical approach is taken, combining power theory with impact assessment theory on the concept of effectiveness. These theories are used to set up an analytical frame for the case study. Document analysis is used to determine the chronology, and thus the backbone of the mapping of decisions in the project. Participant observation and statements are collected primarily by qualitative interviews with key persons from the central actor groups, and by attending meetings as an observer. The interviews supplement the document review concerning the case activities and behaviour, also regarding identification of interests among the actor groups and their access to resources. Based on this, reflections regarding effectiveness and power structures relating to the use of SEA as a decision-making tool when planning new industries in Greenland will be made in terms of development of process and methodology.

Case 3

Finally, the project in Case 3 is carried out in cooperation between AAU and the major Danish engineering consultancy Rambøll. It is organised with an AAU-based professor as main supervisor and a head of department from Rambøll as co-supervisor.

Aim and methodology: The research takes point of departure in the Danish process of preparing river basin management plans (RBMPs), implementing the EU's Water Framework Directive, and preparing SEAs of these plans. Currently, climate change as an environmental factor has been excluded from the planning process, with the reasoning that there is not enough knowledge about climate change to assess its consequences for the water environment and the RBMPs. Concerning this background, the project is aimed at developing the work with climate change in SEA of the RBMPs. A theoretical approach is taken, using sociological risk theory as a framework for research. Document analysis, interviews and a survey is utilised to uncover the attitudes of different actors towards inclusion of climate change in the RBMPs, while a document analysis and interviews are used to assess the experiences with climate change in SEA in Denmark. Based on this, reflections regarding integration of climate change in SEA will be made in terms of development of process and methodology.

Analysing what it involves to be a change agent within the SEA field

The conclusion upon the formal strategic positioning of the researchers in the three cases, and thus whether and how they conduct Mode 3 research, is first and foremost reached by analysing the content of the project contracts. The standard issues like e.g. time schedule is not perceived interesting and relevant for this paper, but the non-standard and unique issues are more interesting and symbolise the negotiated parts of the cooperation. The analysis of the contracts is focused on the explicated objectives and the clauses. Both are used to indicate the strategic interdependence and organisational autonomy, and thereby map the research mode. In addition informal positioning and negotiation takes places in an ongoing dialogue between the SEA researcher, the university and the collaborating organisation. The analysis of the informal process, influencing the research intention, the methods applied and the output of research, is based upon the researcher's observations and experience.

Analysing what potential acting as a change agent has for SEA research and practice?

As stated previously, the hypothesis behind this paper is that Mode 3 research can support SEA and sustainable change via its potential for connecting research and practice. This constitutes the point of departure for the analysis of what potential Mode 3 research has. Two sources form the basis for the analysis: the first part is assessments from the researchers that point at potential for research. These assessments are substantiated by examples from the projects. The second part is based upon open questions related to the potential for influencing practice. The contact persons at the organisations answer the questions: (1) "How has the involvement of the researcher and his/her research influenced the organisation?" (2) "How has the involvement influenced the broader society?" (3) "In which way has the involvement and cooperation influenced the SEA (understanding of SEA, the SEA process, the documents)?" and (4) "How would you characterise the strengths and weaknesses of the set up of the cooperation between your organisation and the researcher?"

In respect to the premature concept of Mode 3 research, the sources are (intentionally) not constrained by mode classifications or characteristics. The sources are instead held open to any impact of the research and this inductive approach may support a refinement of the Mode 3 concept. As the three cases are on-going research projects, the analysis is primarily focused on the process rather than the outputs. The cases do, however, outline a picture of the potential of the research mode.

What Does Acting as a Change Agent Within the Field of SEA Involve?

The Mode 3 research is analysed in terms of the strategic positioning of the researchers in the three cases, and thus it is assessed whether and how they carry out Mode 3 research. Focus is both on formal and informal frames for the research, and these frames will show what is involved in Mode 3 research.

The analysis begins with the strategic interdependence and the organisational autonomy in accordance with the model of strategic positioning proposed by Kurek *et al.* (2007, 2008). The analysis presented in Tables 1 and 2 are inspired by, and to a large extent based upon, the work of Kurek *et al.* (2007, 2008) who build upon Talcott Parsons' theories on social systems. Table 1 gives an overview of the parameters chosen to describe and analyse the strategic interdependence and organisational autonomy. These parameters are inspired by Parsons' model of social systems in which four media can function as exchange means: inducement (e.g. money); deterrence (negative sanctions); commitment; and persuasion (Parsons, 1963).

Common for the research projects is that most of the strategic positioning is happening in an on-going and informal process between the researcher and the cooperating organisation. This will be analysed and discussed in the following, where the strategic interdependence and the organisational autonomy are analysed separately.

Formal and informal strategic interdependence

Table 2 shows the analysis of whether and how the researchers and organisations have strategic interdependence. The analysis shows an economic interdependence in all three cases. This is partly evident from the contracts and partly evident from

Table 1. Parameters chosen as basis for describing and analysing modes of research.

Strategic interdependence	Organisational autonomy
— Understood as the deliberate sharing of heterogeneously distributed resources, assets and capabilities between the partners in order to achieve a joint goal.	— Understood as the researcher's degree of self-governing the research. It is analysed in relation to the researchers autonomy to decide upon:
Economic interdependence	Research goals
Interdependence on exchange of information sources	Acquiring information
Interdependence on engagement	Working place and working balance
	Writing and publishing research results

Table 2. Analysis of the SEA researcher's strategic interdependence in relation to the collaborating organisation. Whether the dependence is explicated formally (in the contract) or informally in the process is indicated in the left column.

	High interdependence		Low interdependence
Economy	Researcher is either fully or partly funded by the organisation and the organisation must get return of their investment in the project		Researcher is economically independent and the organisation is not dependent on return of their investment.
Formal	*Case 1, 2 and 3*		
Exchange of information sources	The organisation is an essential source of information for the researcher and the organisation needs information from the research society		Researcher is not dependent on information from the organisation and vice versa.
Informal	*Case 1*	*Case 2 and 3*	
Engagement	The researcher and the organisation are mutually dependent on the other parts' engagement in the project		Neither the researcher nor the organisation is dependent on engagement from the other part in the project.
Informal	*Case 1 and 2*	*Case 3*	

the informal negotiations. The economic interdependence gives both parties a possibility for sanctioning.

The analysis of the second parameter, dependence on exchange of information sources, as shown in Table 2, reveals some differences. Only Case 1 is really highly dependent upon the collaborator. This has to do with the nature of the SEA research. This project has a focus of getting the right environmental information to the right people at the early stage in decision-making, and to do so the researcher is very dependent on understanding the processes within the collaborating organisation. The contract in Case 1 is a standard contract added restrictions on confidential data that may only be used after approval by Energinet.dk. However, both Cases 2 and 3 do experience some dependence upon information from other actors in the milieu, which the collaborating organisation either informally hinders or supports access to.

Another kind of interdependence is engagement in the project. The researcher is dependent on engagement from the organisation, since it is necessary that the organisation continues internal activities relating to the research and is able and willing to consider and use the research to achieve change in these activities. If the organisation is not engaged, the researcher cannot change anything. The organisation is likewise dependent on the engagement of the researcher to fulfil the expectations of changes. In Case 1, the researcher is dependent on the engagement of the collaborating organisation developing its SEA system, since this is the object of study and change. At the same time, the company relies on engagement from the research in this process of development, e.g. by securing adequacy in terms of regulation. In Case 2 the interdependence is similar, since it also revolves around change in the collaborating organisation. Case 3 is different from this, because the change, which is aimed for, is not restricted to the collaborating organisation, but to a wider range of actors.

Formal and informal organisational autonomy

Table 3 shows the analysis of whether and how the researchers in the cases have organisational autonomy. Regarding to what extent the researchers set research goals autonomously, the analysis shows both high and medium organisational autonomy for all cases. Formally, based upon the contracts, the autonomy is assessed as high/medium as all cases include a loosely formulated goal for the research. In Case 2, the contract emphasises the need for an autonomous researcher, providing critical and independent guidance based on "insider" knowledge/understanding. It is furthermore emphasised that the researcher must work independently and with high validity in relation to the second co-funder Alcoa Foundation. Differing from this, in Case 3 the consultancy expects the PhD-study to "enter directly into Rambøll's work with developing services and having

Table 3. Analysis of the SEA researcher's organisational autonomy in relation to the cooperating organisation. Whether the dependence is explicated formally (in the contract) or informally in the process is indicated in the left column.

<table>
<tr><th></th><th>High autonomy</th><th>Medium autonomy</th><th>Low autonomy</th></tr>
<tr><td>Autonomy to decide on research goals</td><td>Researcher sets research goals within a negotiated overall frame.</td><td>Research goals are based upon the problems of the organisation involving the researcher.</td><td>The organisation set specific research goals.</td></tr>
<tr><td>Informal and formal</td><td colspan="2">Case 1, 2 and 3</td><td></td></tr>
<tr><td>Autonomy in the acquisition of scientific knowledge</td><td>Researcher decides on how and what data is collected.</td><td>Joint decisions are made.</td><td>Decisions on data collection are made by the organisation.</td></tr>
<tr><td>Informal</td><td>Case 1, 2 and 3</td><td></td><td></td></tr>
<tr><td>Autonomy to decide on working place and working balance</td><td>Researcher decides upon where to work and to what extent he will do research-related work with the organisation.</td><td>Joint decisions are made continuously.</td><td>The organisation decides upon the working conditions.</td></tr>
<tr><td>Informal and formal</td><td>Case 2</td><td>Case 1 and 3</td><td></td></tr>
<tr><td>Writing autonomy</td><td>Researcher suggests the content of publications and gives argument why certain theories etc. are chosen.</td><td>Researcher edits or re-writes publications partly or fully.</td><td>Researcher comment on drafts.</td></tr>
<tr><td>Informal</td><td>Case 1, 2 and 3</td><td></td><td></td></tr>
</table>

dialogue with costumers", which is limiting the autonomy for setting research goals. Within the broadly stated research goals, the researcher informally decides on the research in negotiation with the collaborating organisation.

The contracts do not mention methods of data collection, besides the data collected through interaction between researcher and collaborating organisation. In all cases the researchers thus have a high autonomy in the acquisition of scientific knowledge.

For Cases 1 and 3, the organisational autonomy regarding working place and working balance is assessed as medium. For both cases this is due to informal negotiation between the collaborating organisation and the researcher, but also due to the researcher's own interest in being close to what is being studied. Additionally, for Case 3, the contract is more explicit and includes the expectation that the researcher "...spends the main part of the time at our office in Virum." For Case 2, the organisational autonomy is assessed as high, as there are no restrictions or expectations from the collaborating organisation regarding working place and working balance.

Writing autonomy is high in all cases, as the researchers decide on what should be included in publications, and in which journals to publish their results. In all cases, the milieu has interests in certain media, however, which media to use remains the researcher's decision.

The two analyses presented in Tables 2 and 3 show that the cases represent predominantly Mode 3 research, which for the researchers involves high and/or medium strategic autonomy, and primarily high organisational autonomy. The Mode 3 research carried out involves a high engagement in the study field and cooperation with exchange of sources and views. At the same time the researchers retain the responsibility for directing the research and freedom to be critical. For the researcher it thus involves freedom to govern the project within a broadly given frame, which differs from the other modes of research, as shown in Fig. 2 and discussed in the following.

Despite the categorisation of all three projects as predominantly Mode 3, the analysis reveals that in practice there are differences between what this involves. The differences observed are e.g. different levels of how much the researcher identifies with the study field at a personal level, as well as different levels of critical participation in the processes studied. These differences indicate that within Mode 3 many nuances exist, and that Mode 3 research does not lead to one specific research design and practice. Mode 3 research can be undertaken in various ways, depending upon the specific context, including personal preferences, timing, resources etc. After having clarified what conducting Mode 3 research involves in the examples of the three cases, the next step is to analyse the potential for research and practice.

What Potential does Acting as a Change Agent have for Research and Practice?

The second part of the aim of this paper is to investigate the potential of Mode 3 research for research and practice. Investigating two issues does this: (1) if and how being a change agent in relation to SEA influences the research process and content; and (2) if and how the research and cooperation influence the organisation and its work with SEA. These two questions are treated in the following by interpreting the Mode 3 research cases in terms of influences enabled by the combination of high autonomy and high interdependence. The interpretation is based on the experiences and observations of the researchers and contact persons respectively.

Potentials for research: the researchers' experience

The first analysis of the potentials for research of Mode 3 research is based on the researchers' experiences from the three cases. This section is organised around the main issues of access, dialogue on direction and ownership of the research.

Access to people, processes and information

The researchers point to the potential of access in the close association with the organisations. Access to the right person at the right time and place makes it possible for the researcher to make suggestions that test hypotheses or theories. With high strategic interdependence, the researcher is provided with insight and access to follow processes in the organisation. At the same time, the researcher has high autonomy, which means that the researcher potentially can make suggestions that are relevant for practice and, at the same time, tests hypotheses or theories as part of the research process. An example of this potential is from case 1, where the researcher has continuously taken part in organisational processes, which has given possibilities for testing hypotheses, e.g. about the timing of decision aid put forward in theories of organisational decision-making.

At the same time the researcher is allowed to use the information independently, which may improve the research, e.g. by getting feedback on the research from a wider research community. An example of this potential is from Case 2, in which the researcher was allowed to use confidential documents on assessment practice as the basis for research. The confidential data was a key source for research, which included recommendations on how to improve practice. These recommendations would not otherwise be made, as no one else has an interest in using this material for this purpose. The combination of interdependence and autonomy thus made it possible to publish research with a highly relevant content.

The close association with the cooperating organisation through the high strategic interdependence has also been experienced as limiting the research, when the researcher is trying to gain access in areas with opposition towards the associated organisation. For example in case 3, the task of performing SEA of the river basin management plans, which is the topic of the research project, was tendered and won by a competing consultancy. This meant that the researcher being closely associated with a competing consultancy was excluded from studying the process. In other situations, the high organisational autonomy may make it possible for the researcher to go beyond the organisation and interact with competing organisations. Such an act may be validated by a belief that the result of it is (more) beneficial for the research project and the collaboration. This has been possible in case 2, in which the researcher has experienced being excluded from access because of her association with the respective organisations. The researcher used her autonomy and built her own relationships beyond the cooperating organisation, emphasising her relative independence from it.

Dialogue on direction of research

The researchers point to dialogue concerning the direction of the research as an important potential of the Mode 3 set up. The dialogue is seen as an opportunity for enhancing the relevance to practice and society.

The high interdependence in the cases is likely to ensure a dialogue with the organisation as the organisation has an interest in the output of the research. In the three cases, the dialogue has given valuable input from a practical angle to keep the project relevant to practice. The organisational autonomy means that the researcher is still free to develop the research design and secure a scientific rigour independently of the practical wishes of the organisation. In cases 1 and 3 this influence has been experienced through the fact that the research results are continuously being "reality-checked" by practitioners from the organisation. In this way the researcher gets a valuable input on whether suggestions are relevant for practice.

This dialogue also poses a challenge for researchers because the researcher constantly has to balance between the interests of the organisation, scientific demands and the researcher's own interest. In case 3, for example, the organisation has clear wishes for immediately usable methodology, while the scientific community expects more time to be spent on issues such as theoretical angle and research methodology.

Ownership of outputs of autonomous research

The last influence identified by the researchers is connected to the utilisation of the results of the research projects. The Mode 3 set up is experienced to give the organisations ownership of the output of the autonomous research, meaning that the output is more likely to be used in the organisations. This support is especially relevant as the researcher — retaining the organisational autonomy — may have chosen approaches and theories that the organisation would not have preferred at first although the researcher found these more beneficial. The combinations of interdependence and autonomy may in such situations make it possible to improve research and practice by double-loop learning processes (Argyris, 1977) in the organisations. For example, case 1 is aiming at this by using theory that is not previously related to the field, and the organisation has supported the researcher's choice.

The experience from the case studies is that for the organisations, the sense of ownership is related to getting a return on their investment, cf. Table 2. The organisations have invested in the research projects and have had influence on the direction of the research, so that it has relevance, and they will, if at all possible, try to benefit from it in their organisations. The organisations may even work as platforms for disseminating the research results to society and other practitioners. Case 3 is an example of this, because Rambøll will strive to implement any methodology developed, in their subsequent consultancy work, thus communicating it to their clients. The ownership and backing from the collaborating organisation is in case 2 furthermore experienced to give the output of the research a higher status among related institutions.

Potentials for practice: the organisations' experience

The organisations' responses to the questions of potential shed light on the cooperative mode of research seen from practitioners' experiences. This section is organised around the main issues arising in the written response: the importance of linking research and practice closely; the influence observed and assessed; and the risk and weaknesses.

The importance of close linkages between SEA research and SEA practice

The respondents in general stress the importance of a close relationship between research and practice. The respondents from Energinet.dk and Rambøll, e.g. express the value for SEA research as:

> "The strength is that SEA theory is challenged by reality's diversity of asymmetrical courses and sudden political and strategic changes." (Head of Section, Energinet.dk)

> "Sanne gets input for understanding everyday life and problems of the practitioners. Thereby the research study adjusts to a more societal beneficial approach." (Head of Department, Rambøll)

The contextual aspects of practice are hereby put forward as important for enhancing the relevancy of SEA research, even though this does not guarantee an easy implementation in practice. The importance for SEA practice is also raised and related to the organisations' motivation for entering a Mode 3 set up. Energinet.dk chose to initiate the cooperation with Aalborg University because they wanted research input on how to practise SEA, on which plans and especially how to integrate SEA into decision-making: "It has always been — and still is — the attitude in Energinet.dk, that SEA shall not be a shallow paper exercise. SEA shall enter the decision-making processes at a time and with content that makes SEA an active element." The same line of motivation is found in the Self Rule who puts it this way:

> "I like to see the units' cooperation with Anne as an expression of a greater openness to external challenges than some other units' ...Whether it can be said to be evidence that we to a higher degree operate with 'governance' administration principles, I will leave for others to objectively assess — but it is what I as manager of the unit strive for as a principle." (Head of Department, The Greenlandic Self Rule)

While Energinet.dk and the Self Rule emphasise both the short, and long-term perspectives in their views upon the importance of a close relationship between SEA practice and SEA research, Rambøll especially stresses the motivation as short-term business expansion through a competency development. In the long term Rambøll views the importance of cooperation with research for the SEA practice in general:

> "Rambøll gets access to Aalborg University on a more personal level and thereby easier access to future sparring and development of other cooperations." (Head of Department, Rambøll)

The researchers' high engagement in practice is by two respondents underlined as important for the cooperative mode and the content of the research. The following statements from Energinet.dk and the Greenlandic Self Rule exemplify this and point to the importance of grounding research in an understanding of specific contextual circumstances:

> "Ivar has from the first day shown genuine interest in the dilemmas of Energinet.dk, and has very thoroughly acquainted himself with the

> atypical decision processes behind a decision on large scale infrastructure projects." (Head of Section, Energinet.dk)

> "In relation to the societal perspective, it has been an unconditioned benefit — supposedly a precondition — for Anne, that she is an integrated part of the Greenlandic society." (Head of Department, The Greenlandic Self Rule)

The physical affiliation, involving staying in the environment for periods, is part of the high engagement by the researchers and is stressed as an important basis for the influence on their SEA work. The first-hand acquaintance with the actual projects and issues are mentioned as a positive consequence of physical affiliation — in addition to the possibility of involving the SEA knowledge in the processes and to challenge the work undertaken continuously. The researcher becomes integrated and "not just an external consultant or observant" (Head of Section, Energinet.dk).

The influence observed and assessed

A general observation in the answers from the respondents is the conclusion that the close cooperation has influenced the respondents' competences through the developed understanding and actual work on SEA:

> "On the concrete and praxis-related level, it has had great impact for progress and development of the specific SEA, that Anne has 'wafted over the water' in different matters. Anne has through the whole process been a really good sparring partner for me being responsible for the SEA." (Head of Department, The Greenlandic Self Rule)

Rambøll who also refers to the personal competency development, but finds it difficult to assess the direct competency development for others and the company in general, supports this. The reason put forward is that the application-oriented part of the research is not yet finished. This may have to do with the character of the company being a consultancy, and the expressed need for tool making. Energinet.dk raises the influence on the competences on a more institutional level:

> "It has qualified the research project and brought valuable knowledge on SEA from Ivar. Several internal workshops have been held to qualify key employers within SEA. Ivar has participated in the development of internal and external minutes on SEA to be used for establishing a proper SEA policy." (Head of Section, Energinet.dk)

And continues to stress the influence for other actors and society in general: "Energinet.dk and other authorities have a need to get the SEA processes defined and coordinated properly — in that case the project has already been of great importance".

The hidden influence, or indirect influence, for which it is difficult to establish a clear causal relationship between the research and changes in practice, is discussed as important. The respondent from Greenland explains this indirect influence — due to publication, involvement of informants and just general presence by the researcher — through examples like these:

> "In relation to The Bureau of Minerals and Petroleum (BMP) and Anne's insistence on getting access to the (so-called) SEA's written by BMP, I think that this insistence has had an impact on the decision that BMP in January 2010 for the first time has started to publish their SEA's."

> "It is difficult to express but it has to do with a small society, and here Anne's contribution to the debate has made the media image a bit more nuanced — not on the axis advocate versus opponent, but on the axis unreflective versus reflective."

These influences are from the authors' perspective related to Mode 3 research, with the normative sight on, e.g. democratic SEA processes, supplemented at times with research carried out more independently by the researcher to secure the necessary distance to keep a critical stance.

Risks and weaknesses

Working as closely as it has been the case in the three research projects can also be associated with different risks. One is that researchers do not use the synergies between the three modes of research and get too involved in the specific contextual setting with a risk of not keeping enough distance to be critical. The respondent from Energinet.dk raises this risk:

> "A potential weakness in the cooperation model is if Ivar is not capable of getting the necessary distance to the experiences in Energinet.dk. If he becomes part of the processes because they are interesting, it might be difficult to keep the appropriate academic distance to the experiences... Energinet.dk has in general not experienced these weaknesses...more to consider as observation points." (Head of Section, Energinet.dk)

Another risk put forward by the respondents is the unpredictability in the research process and thereby the actual possibilities of creating synergies between practice and research. Rambøll experienced a lower degree of synergies due to lack of jobs of relevance to the research project:

> "We tried to get jobs within the core of the research field, but unfortunately failed. Had we won just one of these jobs, and especially the environmental assessment of the river basin management plans, it would presumably have meant a greater involvement of Sanne in the production." (Head of Department, Rambøll)

The opposite situation was the case for Energinet.dk, since, during the research period, they experienced massive intake of large projects, which has given a large empirical base for the research project. These experiences raise the need to acknowledge the unpredictability in having cooperative processes, and that the benefits for SEA and the organisation as such might appear later than assumed. For Rambøll it was also an unexpected experience that the close cooperation between Rambøll and Aalborg University limited the access to the process of preparing SEA of the new RBMPs: "We were very surprised to experience, that the process was so closed, and that Rambøll's cooperation with the university and Sanne in that respect was hindering the openness of the authorities" (Head of Department, Rambøll). Still the research has a role to play, but the influence is more on the societal level than for the company as such: "...the research project can give the Danish approach to integration of climate in environmental assessments a lift..." (Head of Department, Rambøll).

Another risk mentioned is the lack of engagement from the organisation in general. It is experienced by the respondents that a risk with the cooperative model is that only the key person is fully engaged in bridging SEA research and practice:

> "Rambøll only benefits from the cooperation, if individuals in Rambøll have time/interest/will in getting involved in the cooperation — our conditions for this has actually not been the best." (Head of Department, Rambøll)

In the Self Rule the cooperation has also been solely coupled to the key person, which has not given beneficial and automatic access to other parts of the organisation:

> "Some specific conditions have meant that I have right of disposal over necessary resources and at the same time taken the necessary decision competence for the cooperation to become a reality, but I do not hold a sufficiently high position to personally spread 'the happy message' to other parts of the Self Rule. This work should have been done by others, but unfortunately no one else has taken on this task." (Head of Department, The Greenlandic Self Rule)

Through the examples above it is stressed by the respondents that the members of the organisations need to be open and accessible to make a bigger difference. This is in line with the emphasis on interdependence in the Mode 3 set up.

Conclusion and Discussion

The article has raised the potential of SEA research being involved in engaged knowledge making starting with the environmental problem. The point of departure has been the international questioning whether SEA is effective in influencing planning and decision-making processes in the quest for sustainable development. The authors further question whether the experienced gap between SEA research and SEA practice can be due to a scientific non- or low-cooperative knowledge production. The article, based upon theories on knowledge production and empirical analysis of three cases of SEA research intervention in on-going processes, reveals results presented and discussed in the following.

What SEA research as Mode 3 involves

The cases analysed show that Mode 3 research involves predominantly high interdependence between the researcher and the organisation, mainly in terms of economy and engagement. Also a predominantly high organisational autonomy is present, mainly related to acquisition of scientific knowledge and writing. Also there is a measure of autonomy in deciding on research goals, where in Mode 3, research goals are set through a negotiation. The cases also show that doing SEA research can involve different issues, such as different degrees of involvement. Borrowing terminology from Jamison (2009a), three roles for SEA researchers in the process of inclusiveness are shown:

(1) "*Taking side*": The researcher identifies with the field of study (The Greenlandic case in which the researcher develops a kind of partisanship with the Greenlandic society possibly impacted by the drive for implementing new mega industries).
(2) "*Helping out*": The researcher becomes a "critical friend" (The Energinet.dk case in which the researcher critically participates in the processes in the organisation to find ways for SEA to influence decision-making).
(3) "*Giving advice*": The researcher keeps an academic distance in advising the organisation (The Rambøll case in which the researcher gives professional input to the development of SEA of water plans and incorporation of climate change in SEA).

The three cases indicate that Mode 3 researchers work in a variety of ways. This variety may be triggered by different situations that the researchers adapt to in the process of doing research.

Potentials for Mode 3 to influence SEA research and SEA practice

The empirical analysis, based upon the experience and reflection of both the researcher and the key person in the cooperating organisations, shows that in the three cases Mode 3 influences SEA research and practice in other ways than Modes 1 and 2.

The engagement and involvement in what is being studied has developed a timely and real-life correlated understanding of the processes in which we are trying to integrate and use SEA as a means for sustainable development. The context is being brought to the forefront, which is assessed by all parties in the three cases as positive and important for research to increase relevance for SEA practice and influence this. Some of the main potential experienced by researchers and organisations in the three cases are:

- The research mode renders possible a critical review of planning, assessments and decision-making processes, as well as of research.
- The research mode furthers development of attitudes towards SEA and development of specific assessment skills within the organisations.
- The research mode assists in building bridges among actors within the organisation, and between the organisation and external actors, and eases the communication of SEA results to, e.g. the public.

By cooperating on knowledge making, the researchers have also gained benefits by getting increased access to information and processes. This is assessed as improving both the quality of research, and ongoing dissemination of knowledge and research results in a non-academic forum.

The high autonomy in Mode 3 means that the suggestions of the researcher are likely to go beyond the assumptions and rules that govern practice in the milieu.

The overall conclusion from the study is that potentially a researcher, with high autonomy and interdependence, functions as a change agent for more environmentally sustainable decisions by being part of, and influencing, the field studied — without devaluing or compromising the traditional scholarship.

The challenges for Mode 3 SEA researchers and the organisations involved

Being part of Mode 3 knowledge making is experienced as challenging in different aspects. First, the researcher is putting himself "in game". One needs to know and recognise own knowledge, values and delimitations — and at the same time recognise others. Second, Mode 3 research is, and needs to be, personally driven, based upon a high engagement and clarification of own values. An overall pitfall of Mode 3 research is also the balance of having a close cooperation and at the same time retaining the critical approach of a researcher. It is a challenge to have a high interdependence and, at the same time, maintain high autonomy, i.e. without compromising slightly with your ability or willingness to be critical to the organisation with which you are associated. For the organisation the study especially shows challenges in getting a broader organisational engagement and commitment in the SEA research.

The analysis of the Mode 3 cooperation shows research activities with characteristics similar to Modes 1 and 2. From time to time, the researcher's work resembles a consultancy for the benefit of the cooperation and in other periods the researcher's efforts resemble traditional research in detailed studies of a specific subject. In addition to autonomy and interdependence, what distinguishes the Mode 3 researcher from these other modes is also the reflexivity that precedes and follows the activities resembling other modes: the Mode 3 researcher consciously chooses such research activities when these are beneficial for the cooperation and goal.

The point of departure for the article is that if the SEA research society is to make a difference for practice, we need a wide and deep form of cooperation between researchers and practitioners. This cooperation can be achieved through Mode 3 research entailing co-funding, co-formulation of research questions and co-production of results. We as SEA researchers can choose to be close to the SEA practitioners, decision makers and affected parties and at the same time create temporary space of distance to the relevance demands coming from the co-operators to safeguard rigour. The contextually based Mode 3 research, and the appertaining critical pragmatism, can give us one way to minimise the gap between SEA research and SEA practice. Preconditions for this to happen prove to be personal engagement, a shared wish for research to make a difference for SEA practice and dialogue with a confrontation of own research intention listening to the intentions of the society.

Acknowledgements

The authors wish to acknowledge the work on change agents by Professor Andrew Jamison and would like to thank him for helpfully commenting on a draft of the work.

References

Argyris, C (1977). Double-loop learning in organizations. *Harvard Business Review*, 55(5), 115–125.

Asplund, E and T Hilding-Rydevik (1996). SEA: integration with municipal comprehensive land-use planning in Sweden. *The Practice of Strategic Environmental Assessment*, eds. R. Therivel and M. Partidario. London: EarthScan, 130–140.

Bina, O (2001). Strategic environmental assessment and decision-making processes: new challenges ahead. Conference paper, IAIA'01, Cartegena, Colombia.

Clark, W and N Dickson (2003). Sustainability science: the emerging research program. *Procedings of the National Academy of Sciences of the United States (PNAS)*, 100(14), 8059–8061.

Emmelin, L (ed.) (2006). Effective environmental assessment tools — critical reflection on concepts and practice. Report No. 1 from the MiST-programme, Blekinge Institute of Technology, Sweden.

Gibbons, M (1999). Science's new social contract with society. *Nature* 40, C81–C84.

Gibbons, M, C Limoges, H Nowotny, S Schwartzman, P Scott and M Trow (1994). *The New Production of Knowledge: The Dynamics of Science and Research in Contemporary Societies*. London: Sage.

Hansen, A and L Kørnøv (2010). Challenges for IA of mega industry in a Greenlandic planning, policy and cultural context. *Impact Assessment and Project Appraisal*, 28(2), 135–145.

Huff, A (2000). Changes in organizational knowledge production. *Academy of Management Review*, 25(2), 288–293.

Huff, A and J Huff (2001). Re-focusing the business school agenda. *British Journal of Management*, 12, 49–54.

Jamison, A (2001). *The Making of Green Knowledge*. Cambridge: Cambridge University Press.

Jamison, A (2008). To foster a hybrid imagination: science and the humanities in a commercial age. *NTM — Zeitschrift für Geschichte der Wissenschaften, Technik und Medizin*, 1.

Jamison, A (2010). In search of green knowledge: a cognitive approach to sustainable development. In *Pragmatic Sustainability — Theoretical and Practical Tools*, S Moore (ed.), Abingdon: Routledge.

Jamison, A (2009a). *The Role of the Researcher in Society — Lecture Notes*. Denmark: Aalborg University.

Jamison, A (2009b). *Change-Oriented Research, or Mode 3*. Seminar *The Researchers Role in Society* held 20 October 2009 at Aalborg University, Denmark.

Kurek, K, P Geurts and H Roosendaal (2007). The research entrepreneur: strategic positioning of the researcher in his societal environment. *Science and Public Policy*, 34(7), 501–513.

Kurek, K, P Geurts and H Roosendaal (2008). The impact of the strategic positioning of researchers on their production of knowledge. In *Strategies in the Production and Dissemination of Knowledge*, K Kurek, pp. 68–103. Unpublished doctoral dissertation, University of Twente.

Kurek, K (2008). *Strategies in the Production and Dissemination of Knowledge*. Unpublished doctoral dissertation, University of Twente.

Kørnøv, L and W Thissen (2000). Rationality in decision- and policy-making: implications for strategic environmental assessment. *Impact Assessment and Project Appraisal*, 18(3), 191–200.

Lawrence, D (2000). Planning theories and environmental impact assessment. *Environmental Impact Assessment Review*, 20(6), 607–625.

Morrison-Saunders, A and J Arts (2005). Learning from experience: emerging trends in environmental impact assessment follow-up. *Impact Assessment and Project Appraisal*, 23(3), 170–174.

Nilsson, M and H Dalkmann (2001). Decision-making and strategic environmental assessment. *Journal of Environmental Assessment Policy and Management*, 3(3), 305–327.

Nowotny, H, P Scott and M Gibbons (2001). *Re-Thinking Science: Knowledge and the Public in an Age of Uncertainty*. Cambridge: Polity Press.

Parsons, T (1963). On the concept of influence. *The Public Opinion Quarterly*, 27, 37–62.

Richardson, T (2005). Environmental assessment and planning theory: four short stories about power, multiple rationality and ethics. *Environmental Impact Assessment Review*, 25(4), 341–365.

Ross, WA, A Morrison-Saunders, R Marshall, LE Sánchez, J Weston, E Au, RK Morgan, R Fuggle and B Sadler (2006). Round table: common sense in environmental impact assessment: it is not as common as it should be. *Impact Assessment and Project Appraisal*, 24(1), 3–22.

Seitz, NE, CJ Westbrook and BF Noble (2010). Bringing science into river systems cumulative effect assessment practice. *Environmental Impact Assessment Review*, 31(3), 172–179.

Sheate, WR and M Partidario (2010). Strategic approaches and assessment techniques — potential for knowledge brokerage towards sustainability. *Environmental Impact Assessment Review*, 30, 278–288.

Therivel, R and P Minas (2002). Ensuring effective sustainability appraisal. *Impact Assessment and Project Appraisal*, 20(2), 81–91.

Wallington, T, O Bina and W Thissen (2007). Theorising strategic environmental assessment: fresh perspectives and future challenges. *Environmental Impact Assessment Review*, 27, 569–584.

CHAPTER 6

GIS in Environmental Assessment: A Review of Current Issues and Future Needs

Ainhoa González Del Campo
Girobi Environmental Services, Spain
and
Trinity College Dublin, Ireland
agonzal@tcd.ie

The generation and use of spatial information has significantly increased in recent years. Its importance has been magnified by the INSPIRE Directive, which has subsequently led to specific requirements within several legislative frameworks, such as the Water Framework and Noise Directives, for generating spatial data and spatially-specific outputs, as well as encouraging the creation of spatial data infrastructures at European level. The increased availability of spatial datasets resulting from these initiatives facilitates and promotes the application of Geographic Information Systems (GIS) in planning and environmental assessment, among other sectoral applications. Arguably, the inclusion of geographic information through GIS supports and enhances environmental planning processes, enabling a rapid and objective analysis of environmental issues, and presenting information in a spatial and graphical manner. However, there are a number of considerations for the successful application of spatial data and GIS, and a number of issues remain affecting consistency and validity of results. This paper provides a critical review of current issues affecting spatial dataset management and use (such as format, scale, completeness, timeliness and metadata), and discusses common GIS methods and existing constraints to their application in Strategic Environmental Assessment (SEA) and Environmental Impact Assessment (EIA). It concludes outlining existing opportunities to optimise the contribution of spatial data to environmental planning, and establishing future research and practice needs to enhance GIS use in SEA and EIA practice.

Keywords: Spatial data; GIS; environmental assessment; issues; opportunities.

Introduction

It has been estimated that approximately 80% of information used in the preparation of development plans, programmes and projects is of a geographic nature (Chan and Easa, 2000). Although the magnitude of such a statement may not be universally applicable, as spatial planning elements and approaches vary from country to country, land use planning has an intrinsic spatial nature. Urban and rural planning, and the associated environmental assessment processes that accompany them, entail the consideration of both spatial and temporal aspects which, in turn, establish specific needs for information management and analysis tools. The requirements for spatially-specific approaches to data management and assessment are within the capabilities of Geographic Information Systems (GIS). GIS enable the spatial (and temporal) examination of critical considerations, increasing the objectivity and transparency of planning and environmental assessment.

The application of GIS to Strategic Environmental Assessment (SEA) — a plan-making requirement in accordance with Directive 2001/42/EC (CEC, 2001), or to Environmental Impact Assessment (EIA) — applied to projects as per Directive 97/11/EC (CEC, 1997), can positively contribute to informing decision-making in environmental matters. Although there is no legal requirement to incorporate spatial information in SEA or EIA, their inclusion provides a range of additional benefits to those provided by traditional analytical methods, such as matrices, which commonly lack spatial and temporal dimensions. GIS address such shortcomings by identifying and defining any geographical and/or seasonal variability of, and among, potential impacts (Patil *et al.*, 2002; Vanderhaegen and Muro, 2005; González, 2010b; González *et al.*, 2011b). Moreover, GIS have the potential to provide significant advantages to current SEA/EIA reporting methods by: increasing the speed of information generation; enabling the integration of multiple sources of information to provide new insights; providing the means for a systematic analysis that facilitates the rapid and objective prediction of potential (cumulative) environmental effects; and conveying the results in graphical form that, in turn, enables a visual and comparative examination of the aspects under scrutiny (João, 1998; Vanderhaegen and Muro, 2005; González *et al.*, 2011b). Therefore, it can be argued that GIS support transparent decision-making in environmental planning as results are demonstrably based on spatially-specific and objective evidence (Skehan and González, 2006; Geneletti, 2008; González *et al.*, 2011b). It is anticipated that the potential benefits of GIS in environmental assessment will continue to increase as the full implementation of the INSPIRE Directive (CEC, 2007) materialises. INSPIRE promotes the creation of metadata or "data about data" which enables determining whether a given dataset is fit for purpose and, as a result, enhances standardisation, accessibility and usability of spatial datasets. The submission of spatially-specific information as required by, for example, the Water Framework and Noise Directives (CEC, 2000, 2002), or the consideration of spatial distribution patterns in order to determine population exposure to high concentrations of air pollutants as required by the Air Quality Directive (CEC, 2008a) are, in some way, offshoots of the INSPIRE initiative. In the light of this, the adoption of spatial approaches for data gathering and submission in future EU legislative frameworks is anticipated.

As existing and future environmental legislation is leaning towards geographically referenced methods, the application of Geographic Information Systems (GIS) in SEA and EIA can be easily justified. Nevertheless, there are a number of critical considerations for the successful application of spatial data and GIS (Fig. 1). Appropriate data management, clear outlining of any data limitations, selection and implementation of adequate GIS techniques, and correct interpretation of spatial assessment outputs are necessary in order to obtain coherent and valid results.

This paper critically examines the applicability of GIS in SEA/EIA with regards to:

(a) Potential limitations associated with "raw" or baseline spatial datasets, with particular emphasis on the factors that influence information management which can, in turn, affect the validity of results;
(b) Constraints posed by specific GIS techniques applied; and
(c) Their ability to improve information generation and dissemination in decision-making processes.

The paper draws on examples mainly from Ireland and the Basque Country, Spain, two western European countries with historical and cultural similarities yet

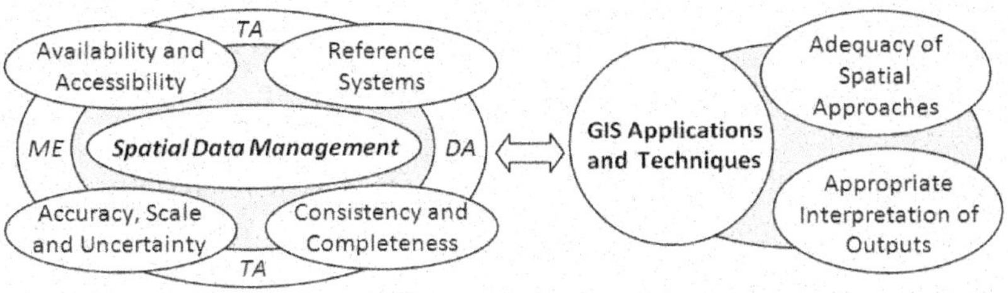

Figure 1. Critical considerations for effective spatial data management and GIS implementation. Note that GIS applications and techniques are dependent on spatial data, and spatial data relays on GIS for their management. The grey circles embody the critical factors affecting each of these components: in the case of spatial data, metadata applies across as they address all such considerations.

differing planning systems, where SEA- and EIA-based GIS applications have resulted in significant advantages but have also revealed current impediments to their effective implementation. Such examples are illustrative and aim to support or challenge observations made in international literature.

Spatial Data Management

Geographic Information Systems (GIS) involves raw field observations/measurements linked to a location — also known as spatial data or geographic information. These raw data can acquire added value and provide new insights when placed in a framework of reference systems in GIS (i.e. spatial, temporal and attribute reference systems providing location and units of measure, time scale/period for which the data is relevant, and qualitative characteristics or quantitative physical properties associated with a given feature or location). Consequently, such raw data can be overlaid to obtain new insights on behavioural relationships within or with other datasets, or incorporated into models to simulate, and in this way anticipate, future changes in field observations and thus support environmental assessment. However, to ensure that all relevant data are incorporated into GIS and to provide an appropriate assessment framework, a number of pre-requisites need to be fulfilled, including: availability and accessibility of datasets; uniformity of reference systems; accuracy, scale, consistency, completeness and timeliness of datasets and their attributes; and provision of detailed metadata (González, 2010a).

Availability

Availability of comprehensive, accurate and relevant data is central to impact assessment effectiveness (João, 1998; Chavan and Ingwersen, 2009; González, 2010b; González *et al.*, 2011b): a pre-condition for ensuring appropriate and efficient examination of relationships between all relevant environmental factors resulting from changes in resource management (e.g. land use). Nevertheless, and despite the European efforts to generate spatial data (e.g. CORINE land uses or WFD water catchment status), and some comprehensive environmental databases being developed at national level (e.g. O'Dea *et al.*, 2004; Herberg, 2008), there are still significant gaps in geographically referenced information throughout Europe, which is affecting the comprehensiveness of GIS-based environmental assessments. Although the availability of national datasets varies from country to country (for instance, governmental bodies in the Basque Country contain a wider set of environmentally relevant and publicly available spatial datasets than in Ireland), a number of datasets that are critical for SEA and EIA types studies have not been produced yet in many countries (e.g. habitat or flood risk mapping are to be prepared in Ireland). Although additional relevant datasets may be available, the general lack of a national inventory or central repository of geographic information renders many datasets unavailable and undiscoverable, particularly those created by the private sector. Several attempts have been made both in Ireland and the Basque Country, and indeed Spain, as well as other EU member states such as Germany, the Netherlands or the UK, to create Spatial Data Infrastructures (SDIs). In the Basque Country, the Gipuzkoa Provincial Council has collated a significant array of land use planning, transport and environmental datasets from public and private sources. However, many of these, particularly those from private sources, remain internal to the organisation due to copyright and licensing issues, their unpublished status constraining common knowledge on available datasets. In Ireland, SDIs are commonly linked to data creators, as no national or regional inventory of environmental and planning datasets exists. Several organisations (such as the National Biodiversity Data Centre — NBDC in association with the National Parks and Wildlife Service, BirdWatch Ireland and An Taisca among others, or the Environmental Protection Agency in association with the Geological Survey of Ireland and Teagasc) have launched theme-specific online inventories, e.g. all Ireland biodiversity-relevant datasets are available in the NBDC Website. Maintaining such inventories requires a commitment for regular updating of available datasets.

Many of the available datasets at European level have been generated at small-scale (i.e. high resolution for national/regional geographical extents) and are, therefore, best applied at SEA level. Datasets containing the appropriate spatial scale/resolution and level of detail are often not readily available at EIA level, requiring additional data collation and generation efforts. Time and financial constraints can often limit extensive data gathering exercises in environmental assessment, particularly in SEA. As a result, practitioners tend to reuse existing datasets — possibly in ways differing from the original intent and thus with a risk for bias, gaps and analytical limitations (Gioia, 2010). Data availability considerations are often acknowledged in environmental reports as data gaps, but are often overlooked when interpreting assessment outputs.

Accessibility

In some cases, spatial information may be available, but not accessible due to copyright, privacy, licensing and confidentiality issues and/or concerns about misuse of data (Kuula and Borg, 2008; González, 2010b). Although quite a common problem at European level, there are significant differences among member states. In most countries, access to spatial datasets is constrained by licensing agreements and use restrictions. This is the case in Ireland and the UK, where the use of digital mapping products is subject to Ordnance Survey purchase agreements and to the payment of annual licensing fees. In contrast, any digital maps (including rectified aerial photographs, satellite imagery and digital terrain models) collated by public organisations in the Basque Country can be freely downloaded through the relevant Provincial Council's Website (see also Herberg, 2008, for the Federal State of Berlin). In other cases, datasets may be available for viewing through web browsers but not for download (e.g. NBDC's and UK Environmental Agency's websites). This "confined accessibility" may pose a significant burden to data use, as the time and resources needed to source the digital version of relevant dataset/s may impose irresolvable limitations.

In all cases, the planning hierarchy at which environmental assessment is applied (e.g. SEA versus EIA) will set the scope for data requirements. Data access and collation can often be incremental and spread throughout the assessment process. However, readily available and accessible data is, in some cases, a prerequisite for complying with specific timing and requirements of some environmental assessment stages, such as scoping (João, 2002; González, 2010b). If and when relevant information is not freely accessible, their incorporation in the Geographic Information Systems (GIS) approach is inhibited, subsequently affecting the comprehensiveness of the assessment and its outcomes (González, 2010b; González *et al.*, 2011b). In such cases, the relevant parameters need to based on other sources of information (e.g. statistical data) and other assessment methods (e.g. matrices), constraining the applicability of GIS.

Reference Systems

A common spatial reference system is critical in a GIS project: datasets in different reference systems or projections do not appropriately overlap in space. Changes in reference systems have occurred worldwide due to the increased use of global positioning systems (GPS). In the Basque Country the new official Spanish geodetic reference system (i.e. ETRS89) has been adopted, which differs by an average of 200 m in X (latitude) and 100 m in Y (length) from the previously applied reference system (i.e. European Datum — ED50). In the case of Ireland, the change from Irish National Grid — ING to Irish Transverse Mercator — ITM (which is consistent with the official geocentric system in Europe) contains a difference in the false origin of 400 km in X and 500 km in Y (Cory *et al.*, 2001). During the implementation of changes in spatial reference systems potential difficulties to data integration may occur. Although current GIS software packages have embedded tools to enable transformations between reference systems,

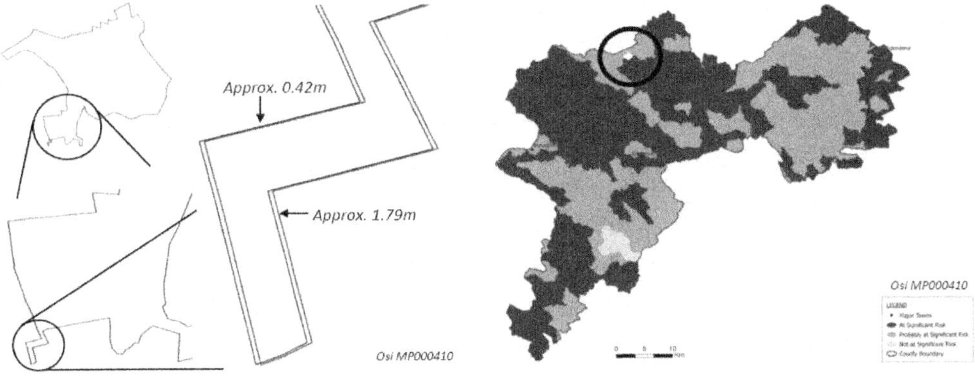

Figure 2. Inaccuracies resulting from applying different layer transformations methods for converting between Irish Transverse Mercator and Irish National Grid (left). WFD surface waters risk assessment dataset illustrating a data gap in north-west County Offaly, Ireland (right).

projection changes require additional data manipulation time and can lead to small inaccuracies in latitude and longitude (González-Matesanz *et al.*, 2003), as illustrated in Fig. 2. Such inaccuracies can have significant implications for assessment outputs, particularly at EIA level.

The temporal scope in SEA and EIA is commonly set by the time-frame of the plan, programme or project. In this context, temporal reference systems also depend, in principle, on the period covered by the proposed intervention/s. However, environmental assessment often relies on available and accessible datasets, particularly in SEA where no additional data gathering or creation is required, and consequently the temporal reference system is frequently compromised by the timeliness of datasets utilised. Although historical data may provide significant insights on trends and patterns of change, more commonly, outdated information can lead to inconsistent results as they fail to reflect the current state of environmental resources. Therefore, it can be argued that spatial datasets need to be current for their effective application in environmental assessment. However, this may be easier to achieve in EIA than in SEA, as EIA processes make provisions for field surveys and additional data collation. Data updates generally depend on the responsible agency. At European level periodic (albeit irregular) updates are commonly established (e.g. CORINE was first created in 1990, and updated in 2000, 2006 and 2009). In the Basque Country, digital base maps and cartography are reviewed on an ongoing basis, while other information types (e.g. socio-economic activities and environmental data) are reviewed biannually. In Ireland, digital mapping is continuously updated by the Ordnance Survey, but environmental data updates are not undertaken as effectively despite existing monitoring programmes. In all cases, limitations posed by the timeliness of datasets (and their effect on temporal reference systems) needs to be taken into account when interpreting Geographic Information Systems (GIS) results.

Similarly, attribute reference systems need to be defined and agreed when creating environmental spatial datasets. Where referring to qualitative or quantitative characteristics, a scaled or thematic approach needs to be adopted to ensure comparability within the features of a thematic layer. Although standardised approaches are becoming customary (e.g. categorisation of land uses as part of CORINE classification), spelling inaccuracies, categorisation errors and attribute data omissions are not uncommon in public datasets, and data rectification and improvement tasks are often required (Onsrud, 1999; González *et al.*, 2011b).

Accuracy, Scale and Uncertainty

Scale determines the spatial accuracy and resolution of data: the larger the scale, the finer the resolution and the more accurate the dataset. Measurements and location on a map must be precise to ensure spatial accuracy. This is particularly relevant when examining spatial correlations among environmental features and potential for cumulative effects. Due to the lack of specifications in the relevant directives, SEA and EIA practitioners have the responsibility and flexibility to identify and select the relevant scale(s) at which the assessment, and description of the baseline environment, should be performed. In SEA, the geographical scale is, in principle, set according to the level of detail required, which depends on whether it refers to regional planning guidelines or local area plans. However, in practice the level of detail is commonly defined by the scale of available datasets. In most cases, both in Ireland and the Basque Country as well as in the majority of European member states, spatial information is generated at national or regional levels (i.e. small scale). This is the case, for example, in CORINE's land use classification, the status of water bodies required by the WFD (CEC, 2000), or the air pollutant concentration under the Air Quality Directive (CEC, 2008a). The generation of local environmental data (e.g. wildlife and green corridors, flood risk zones, etc.) is often rather limited, particularly in some countries such as Ireland where it is only available in a small number of counties. Adopting the scale of available datasets can potentially compromise assessment detail. This may not represent a significant issue in SEA, where assessments are undertaken at small scales (e.g. national to countywide assessments are commonly based on 1:50,000 base mapping in Ireland and tend to be based on broad policies rather than specific zonings), and where it is generally accepted that environmental aspects cannot be described in great spatial detail due to larger geographic contexts and broader assessment parameters

(Therivel, 2004; Skehan and González, 2006; João, 2007). However, the difficulties that exist when spatially identifying and defining certain features (João, 1998) may significantly affect the accuracy of assessments at local level (commonly undertaken at 1:5,000 or larger scale and entailing the specific demarcation of land uses or interventions which sets a requirement for a higher spatial definition). It is therefore quite possible that the information generated at regional level lacks the level of detail required for local area planning SEAs as well as EIAs.

Boundaries of features need to be set at some point (especially when working in spatially fixed regions for land use planning) and some data, particularly in the environmental arena, will always be prone to uncertainty (e.g. geological boundaries can never be accurately mapped). In addition, when integrating modelling operations in Geographic Information Systems (GIS) or when incorporating model results into the spatial assessment, the effects that assumptions made may have on the reliability of outputs have to be addressed. Evaluation of the quality of model results is a prerequisite for their confident use at scientific and policy levels (Borrego *et al.*, 2008). Such an evaluation is transferable to GIS-based environmental assessments and any limitations need to be clearly acknowledged and conveyed for reliability of results. Combining multiple datasets at various scales and resolutions, and with different levels of uncertainty, in a GIS project has the potential to significantly affect the integrity of results. The scale factor and any associated uncertainty issues introduce new considerations on the applicability of GIS to environmental assessment: spatial cognition is necessary when interpreting GIS outputs in order to understand the impact that scale and uncertainty can have on their accuracy and validity.

Consistency and Completeness

Onsrud (1999) considered it inevitable that errors and gaps are contained in any practical database (Fig. 2): no general-purpose datasets will ever be complete for all potential purposes, nor will data accuracy meet the needs of all uses. Even though significant progress has been made across Europe, including in Ireland and Spain, in developing guidelines to standardise data collation and generation methods and ensure completeness, consistency and comparability (e.g. EJ/GB, 2008; Smith *et al.*, 2011), practitioners have to invariably deal with data access, scale and quality issues (João, 2002; Gioia, 2010; González, 2010b). In fact, many of the problems reported by Rybaczuk and Mac Mahon (1995), including completeness and scale, remain issues in Ireland today (González, 2010b). In addition, a significant and yet overlooked issue in both Ireland and Spain, and many other European countries, with regard to data completeness is the general lack of indicator values as attribute information associated with environmental datasets, which impedes establishing the status of environmental resources and their vulnerability, thus reducing their usability (González, 2010b; González *et al.*, 2011a). Attribute values commonly describe qualitative and, sometimes, quantitative characteristics of the elements in a dataset (e.g. *salmonid* river). However, the lack of indicators impedes the determination of the quality status of those elements (e.g. lack of biotic index values in the attributes table hinders the establishment of the river's quality that may have an effect on salmonid populations).

Although it is acknowledged that data must be as current, complete and accurate as is reasonable (Scott and Marsden, 2003; Vanderhaegen and Muro, 2005; González, 2010b), this is not always easily achievable. Data inconsistencies and gaps can be, in most cases, rectified by updating, correcting, and/or completing the relevant information, but these processes require additional time and effort and may hamper the timely incorporation of spatial information into the assessment (González, 2010b; González *et al.*, 2011b). This is particularly relevant when incorporating, for example, AutoCAD-based drawing files which are still commonly applied in both planning (SEA) and engineering (EIA) sectors. The provision of such "legacy data" in the form of AutoCAD files (which lack attribute information and, in most cases, topology) affects their integration and reduces the validity of information for spatial assessment purposes. In all cases, the absence of such rectifications entails incomplete or inaccurate datasets which can lead to an incorrect or deficient assessment of potentially significant environmental issues in SEA/EIA (Vanderhaegen and Muro, 2005; González, 2010b). Practitioners have the responsibility to either undertake such rectifications or acknowledge any limitations in the quality of applied datasets.

Metadata

Metadata allow discovering and inventorying datasets, as well as establishing their fitness for use (i.e. quality) and their fitness for purpose (i.e. usability). Different data sources are utilised and interrogated in both SEA and EIA studies. The existence of metadata for environmental and planning datasets can help to rapidly identify their availability and determine their readiness for integration in the study. The rapid identification of suitable datasets is particularly relevant in the initial stages of environmental assessment (e.g. scoping) where key environmental considerations are set through preliminary assessment. The International Standards Organisation's 19115:2003 Regulations (ISO/TC211, 2003) for standardisation of metadata have allowed consolidation of the metadata requirements initially set by the Dublin Core Metadata initiative launched in 2000. Moreover, and based on these standards, metadata Regulations (CEC, 2008b) have been launched as part of the INSPIRE Directive. Such binding rules establish requirements for the creation and maintenance of metadata for the themes listed in the Annexes of the Directive to ensure that the SDIs of European Member States are compatible and usable in a community and transboundary context.

Following the gradual implementation of the INSPIRE Directive's requirements (CEC, 2007, 2008b), the increasing availability of metadata enables determination of the relevance, validity and quality of datasets for the purpose of a given environmental assessment. SDIs for the publication of metadata are well advanced in the majority of Spanish regional authorities, including the Basque Country. However, in Ireland there are still significant gaps in metadata availability, which can affect the prompt establishment of their applicability (and thus limitations) for the purpose of SEAs and EIAs. Nevertheless, the majority of national public organisations already have (e.g. France, Germany, Italy, the Netherlands, Switzerland or the UK) or are currently putting mechanisms in place to ensure creation of metadata for all newly created datasets.

GIS Applications and Techniques

A review of the 41 SEA environmental reports published in Ireland between 2001 and 2008 revealed that the majority of environmental reports included maps or geographic figures, but that these were solely used for informing the description of the baseline environment (González, 2010b). Although more advanced Geographic Information Systems (GIS) techniques are being currently developed and applied in SEA/EIA, particularly for simulation of future scenarios and impact assessment/prediction (e.g. flood risk modelling, habitat fragmentation) as discussed next, the findings of the review support the observations made by João and Fonseca (1996), and Vanderhaegen and Muro (2005), in that GIS is yet to be applied to its full potential. Nevertheless, a number of GIS techniques are being applied in SEA and EIA (González, 2010b), in particular, thematic mapping and overlays and, occasionally, modelling and 3D visualisations. The outputs of these techniques are commonly used as visual aids for public participation, but GIS-based interfaces are also being developed and explored as meaningful ways for collating public perceptions — although a very few real-life case studies have been published (e.g. Jordan and Shrestha, 2000; Kingston *et al.*, 2000; Weiner and Harris, 2003; González, 2010b).

Spatial Approaches to Environmental Assessment

GIS have been widely applied in resource mapping as the interface for storing and representing field surveys and inventories. Gathering data on-site (by means of

GPS) or collating information from third-party data sources and subsequently creating composite thematic maps that enable the simultaneous assessment of multiple factors have been widely recognised as supporting planning and environmental assessment processes (Kalogirou, 2002; Therivel, 2004; Vanderhaegen and Muro, 2005; Geneletti, 2008; González, 2010b; González *et al.*, 2011b). Such basic Geographic Information Systems (GIS) operations for the creation of spatial and thematic illustrations of environmental and planning considerations support the description of the baseline environment, impact prediction and preparation of environmental reports, inherent to SEA and EIA processes. However, more advanced GIS techniques could enhance their applicability as support tools in environmental assessment.

Overlays and spatial assessments present the next step in GIS techniques, which are typically applied to deal with site selection or site suitability assessment (e.g. Antunes *et al.*, 2001; Kalogirou, 2002; Geneletti, 2008; González *et al.*, 2011b). Spatial analysis for urban and rural planning, and development control is one of the main applications of GIS in Ireland, the Basque Country and many other countries. In the context of SEA, GIS provide the means to integrate and spatially assess multiple environmental and planning considerations in a single interface. Weighted overlays present an alternative GIS technique which enables the integration of public perceptions into SEA/EIA, and has been widely applied for evaluating development and land use suitability in environmental terms (e.g. Ceballos and López, 2003; Chen *et al.*, 2008; Geneletti, 2008). Different approaches to spatial assessment can be found in the international literature, adapted to suit the purpose of each study. All applications are based on the ability of GIS to combine multiple datasets in a spatially-specific manner, as well as on the capacity to integrate relative values of significance or change into each of the datasets. Although limited documented empirical knowledge exists (e.g. Blaser *et al.*, 2004; Bragagnolo, 2011), it is considered that such applications (resulting, for example, in specific environmental indicators or composite environmental vulnerability maps) have the potential to facilitate the assessment of cumulative impacts in SEA.

GIS-based modelling for simulating and, in some cases, visualising in 3D the potential impacts resulting from plan implementation is increasingly used to support environmental assessment. Modelling land use changes, hydrological and ecological processes, such as flood risk or habitat connectivity, and air pollution are most widely applied (e.g. Borrego *et al.*, 2008; Gontier *et al.*, 2010; Jeroen *et al.*, 2011). However, significant gaps still exist between research and practice. Gontier *et al.* (2006), for example, demonstrated the existence of a gap between GIS-based ecological modelling research and current biodiversity impact assessment practice. Arguably, modelling approaches are unlikely to be widely implemented in environmental assessment as they cannot be systematically applied to the wide range of issues considered (Fedra, 2004), commonly incorporate multiple and complex assumptions that may affect the public understanding of outcomes (Therivel, 2004) and there is a general lack of favourable institutional standpoints to anticipate future changes (Thiel, 2009). Moreover, uncertainty in the data that are used, in the methodologies that are applied, and in the value judgements provided by experts (Geneletti *et al.*, 2003), are critical considerations that need to be addressed for reliability of outputs. In this context, the need to provide more "credible" methodologies and "usable" spatial assessment and modelling results has been advocated in literature (Liu *et al.*, 2008, González *et al.*, 2011b).

GIS for Public Consultation and Participation

Public Geographic Information Systems (GIS) use has significantly increased in the last decade, particularly since Internet-based earth navigation tools were published (such as Google Maps). Maps promote "spatial thinking", articulate documents in geographical terms, and improve information delivery. Public participation GIS tools have been developed in parallel, with the aim to use GIS within more inclusive participatory decision-making processes. GIS-based consultation processes assist the involvement of individual citizens directly in planning processes, avoiding the uneven dominance of a minority of representatives commonly encountered in wider consultation approaches (e.g. public hearings and workshops). Moreover, they help empower community groups when responding to local geographic issues by facilitating spatial comprehension, stimulating debate and encouraging submission of personal perceptions.

Although it is argued that participatory GIS cannot be effective on its own, there is strong international support for the use of GIS as a complementary tool in public participation processes due to their potential to enhance conventional participatory approaches and improve the understanding of potential issues (González *et al.*, 2008). Nevertheless, there is still a significant gap between experimental and practical applications of participatory GIS and few real-life case studies are available (see, for example, Germany: http://fs.mapsailor.de/fs-start.htm (Freising Online, 2011); the Netherlands: http://www.ruimtelijkeplannen.nl/web-roo/?planidn=NL.IMRO.0114.2010001-0002 (Ruimtelijkeplannen, 2011); UK: http://www.ccg.leeds.ac.uk/slaithwaite/ (Slaithwaite, 2003)), of which a limited number have been published, the majority of them for land use and transportation planning (e.g. Kingston *et al.*, 2000; Weiner and Harris, 2003; Tang and Waters, 2005; Wood, 2005; González, 2010b; White *et al.*, 2010). Moreover, it is argued that the vast majority of GIS applications for public participation only deliver information to citizens, presenting a one-way rather than the desirable two-way communication process (Steinmann *et al.*, 2005). A number of issues such as institutional power structures, political standpoints, data disclosure, licencing and transfer restrictions, computer literacy and usability barriers, and distrust in IT-based participation remain constraining factors to a more widespread use of real-life Geographic Information Systems (GIS) applications (Elwood, 2006; Hanzl, 2007; González, 2010b).

Addressing Current Issues

Optimising the use of spatial data and the application of GIS in environmental assessment requires addressing current issues and limitations. Opportunities also exist to enhance spatial data management, which in turn would strengthen the applicability of GIS in SEA/EIA and vice versa (Fig. 3).

Future Needs to Improve Spatial Data Management

Spatial data availability and accessibility for certain environmental and planning aspects is still an issue in Ireland, the Basque Country and worldwide. Such lack of spatial datasets may lead to overlooking significant environmental aspects if a GIS-based assessment was solely adopted. This issue could be partially addressed by documenting identified gaps in order to ensure transparency and credibility of assessment outputs. Nevertheless, all relevant information would ultimately need to be provided for informed decision-making, ensuring evidence-based and reasonable judgement. The SEA Directive states that assessments are to be based on existing data sources and that no additional information needs to be collected for the purpose of SEA. In the light of this, data availability limitations can rarely be resolved within SEA. Data gaps can be better addressed at EIA level, where provisions are generally made for additional data collation. In all cases, much of

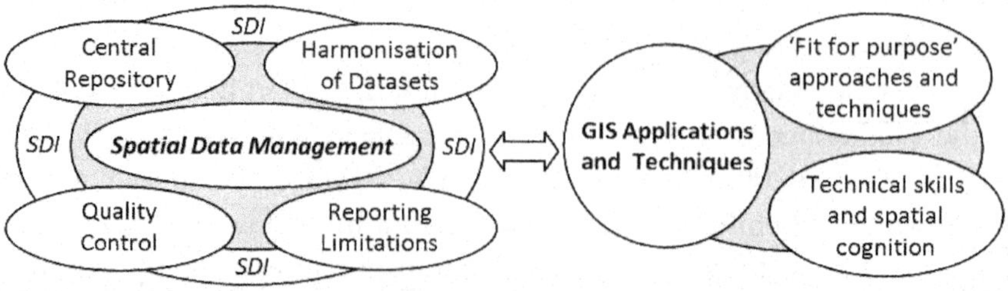

Figure 3. Future needs for improving the effectiveness of spatial data management and GIS implementation. Note that this figure illustrates the opportunities that exist for improving the critical considerations presented in Fig. 1. SDIs enable the creation and distribution of metadata inventories.

the data deriving from individual research and studies remains under-utilised as a result of a lack of data-sharing mechanisms (Chavan and Ingwersen, 2009). Accessibility and licensing issues need to be tackled at institutional level. The preparation of an inventory of all available spatial datasets nationwide or the creation of a central national or regional repository would significantly help improve data discovery and access. Opportunities exist for the establishment of a centralised body at national or European level with the necessary resources for the set up and maintenance of spatial and non-spatial data from different source organisations, or the creation of a distributed network for bringing all available datasets into a single interface (Chavan and Ingwersen, 2009; González, 2010b). Moreover, the inclusion of a binding requirement for data upload into such a repository as part of SEA/EIA submissions would significantly increase the pool of available data as many environmental assessments yield significant sources of information. The dissemination of this accumulable database at a marginal cost would improve accessibility. Improved accessibility would, in turn, significantly enhance data sharing and use, help control duplication, and assist in the assessment of in-combination effects at all planning levels.

Integration of datasets during any Geographic Information Systems (GIS)-based environmental assessment must address format issues of compatibility, spatial reference system and scale. The adoption of European and/or national standards for referencing is being pursued, as recent developments in the adoption of compatible geographical reference systems, the creation of consistent attribute values within legislative frameworks such as the WFD or the implementation of metadata regulations indicate. Several international and national initiatives have already been launched to endorse standardised data collation and distribution in some research areas, such as that promoted by the Global Biodiversity Information Facility or the Irish NBDC. Nevertheless, additional efforts are needed to formally harmonise existing data sources across all research areas and environmental factors. Overall, data sources are inconsistent across the EU, and there is a discontinuity of information across borders and national boundaries (Vanderhaegen and Muro, 2005; Bartley, 2007). The use of multiple datasets in SEA/EIA, created to different standards by different organisations, renders results that cannot be guaranteed for quality or certainty. Guidance on data collation and compilation standards would ensure that datasets from different sources are seamlessly combined and used without undue difficulty in environmental assessment processes and transferable across boundaries (facilitating, in particular, transboundary SEAs). Moreover, the creation of data control mechanisms (e.g. creating an independent body to appraise data quality which could be possibly linked to the proposed centralised data management body), could help ensure that data creators provide standardised and quality datasets.

Shortcomings with regard to accuracy and scale of data could yield uncertain and even misleading assessment results. Scale is particularly relevant in environmental assessment as it can determine and affect assessment detail and accuracy (João, 2002; González, 2010b). More widely available small-scale environmental datasets commonly disclose spatial inaccuracies and render low-resolution results when used in EIA and lower-tier SEAs. In contrast, the appropriateness of large-scale environmental datasets for SEA is affected by lack of spatial detail and "fuzzy" boundaries commonly adopted in higher-tier planning. In order to make optimum use of available datasets at various scales, a multiple-scale approach should be adopted, applying appropriate Geographic Information Systems (GIS) techniques to the scale of assessment, and acknowledging data accuracy and scale limitations and associated uncertainties at each assessment stage. SEA in particular often needs to deal with higher levels of uncertainty (Therivel, 2004; Partidário, 2007). In this context, data uncertainty can be somehow overcome with some acceptance that spatial data and GIS provide indicative results (which may or may not need additional and detailed assessment depending on the relevance of the issue and end purpose of the study), rather than acting as definitive planning tools. Difficulties associated with effectively estimating and communicating uncertainty, particularly when it derives from combining multiple data sources through various GIS techniques, need to be acknowledged. Documenting uncertainty (as well as any limitations with regards to data accuracy, scale and quality) contributes to enhancing transparency and reliability of assessment outcomes.

Datasets must be up to date, have appropriate spatial accuracy and level of detail, and contain comprehensive and complete information — including provision of indicator values where applicable, for reliability of results (João, 2002; CEC, 2007; González, 2010b; González *et al.*, 2011a). Comprehensiveness often depends on the purpose of data collection and accuracy, and on the adopted scale. Therefore, data inconsistencies and uncertainties in both SEA and EIA are not always avoidable. To address this, a spatial data quality control system could be put in place by either public or private organisations. Data quality audits would help ensure that information is current and fit for its purpose. In addition, procedural measures should be put in place to guarantee that, when considered necessary or appropriate, additional site surveys are undertaken and additional spatial datasets are generated for particular environmental considerations at both SEA and EIA level in order to address all significant impacts. Alternatively, and given the current absence of data quality control mechanisms, a data quality checklist could be published to assist practitioners verify, among other aspects, the source of data, its relevance to the study, its spatial accuracy and level of detail (i.e. scale), and the comprehensiveness of its attribute values. This would help identify any potential data inaccuracies/inefficiencies, which could be consequently stated in the SEA environmental report or EIA statement. Additional research is also required to further scrutinise existing spatial datasets in order to determine or estimate their quality, compatibility and reliability in environmental studies, particularly in relation to their scale of collection versus the scale of their application. The documented identification of inefficiencies would also assist in prioritising both imperative data gaps and data improvement tasks.

Spatial data management issues would be better dealt with when INSPIRE metadata requirements are enforced. Comprehensive metadata inventories and their distribution through SDIs would facilitate the prompt establishment of their relevance and quality for the purpose of a given assessment, which would, in turn, help guarantee more reliable and accountable GIS-based assessment outcomes for SEA/EIA.

Future Needs to Enhance the Applicability of GIS in Environmental Assessment

The effective implementation of spatial analysis in both SEA and EIA requires a thorough understanding of the factors that determine the validity of spatial datasets and the applicability of Geographic Information Systems (GIS) techniques in order to ensure accountability of GIS-based assessment outputs. Techniques and tools must be adapted to fit the requirements of each SEA/EIA assessment stage, ensuring that they make optimum use of available data and provide scientifically sound approaches that render reliable results. Therefore, GIS approaches and techniques are likely to vary according to the assessment stage (e.g. basic mapping operations may be applied during scoping while specific modelling techniques may be applied during the assessment of alternatives), scope of the study, assessment parameters (e.g. differing modelling tools would be applied for air quality or green connectivity simulations), and level of detail. Although guidance manuals and reports have been produced for GIS applications in environmental assessment (e.g. Blaser *et al.*, 2004; González, 2009), there is no consolidated pragmatic methodology at EU level. Arguably there is no "one size fits all" solution. However, the adoption of a standard but flexible GIS-based environmental assessment methodology at European or national level would help not only to improve the explicitness and effectiveness of procedures and normalise practice, but also to allow a better comparison between environmental reports and impact statements, and thus to monitor the effectiveness of SEA/EIA implementation

(González, 2010b). In order to develop a fully comprehensive methodological approach, the potential appropriateness and performance of specific GIS techniques for each of the various environmental assessment stages needs to be further examined to ascertain their applicability. Also, additional efforts are needed to bridge the gap between research and practice, and pragmatically apply models and public participation Geographic Information Systems (GIS) approaches that have proven to be useful and valid in research-based applications.

The appropriate application of GIS approaches in assessment is, evidently, subject to "know-how". Technical skills and spatial cognition are pre-requisites for the effective implementation of any spatial assessment technique, the precise interpretation of outputs, and the objective yet truthful communication of findings in environmental reporting and decision-making. In order to achieve this, spatial data management considerations need to be carefully and thoroughly addressed by skilled GIS practitioners. Evaluating the current spatial literacy of practitioners involved in SEA and EIA processes would facilitate devising measures to strengthen GIS approaches in this arena. This would facilitate developing measures to raise the spatial awareness and skills of all personnel involved in planning and environmental management, including stakeholders and the general public, and promote "spatial thinking" throughout impact assessment processes.

Conclusions

Environmental assessment practice derives from, and is informed by, evidence, which in turn supports plan-making and project implementation. One aspect of this evidence is clearly spatial, due to the intrinsic geographic nature of natural resource management and environmental planning. In the light of this, it can be argued that spatial data and GIS have significant potential to support environmental assessment processes. GIS have the potential to augment the quality and quantity of information provided to decision-making.

Nevertheless, in order to optimise the benefits that spatial data and GIS may bring to SEA and EIA, there is a need to improve the management of geographic information as well as the application of GIS approaches to environmental assessment. Although many of the issues outlined in this paper are apparent and could be anticipated in applied-GIS, further insights can be obtained from their contextualisation to environmental assessment research and practice. A number of key areas need to be prioritised, such as adaptation of spatial assessment techniques and promotion of spatial awareness throughout environmental assessment. Usability aspects are also due particular attention: the gap between research and practice in devising appropriate means to effectively engage stakeholders in spatial thinking and GIS-based consultation needs to be tightened. Critical aspects to be further explored include: inventorying availability; political standpoints affecting accessibility; and quantifying and communicating uncertainty associated with scale and quality of data. These fundamental aspects need to be improved, and additional understanding of their implications gained, in order to enhance the operational effectiveness of GIS in SEA/EIA. As GIS outputs largely rely on the quality of inputs, and reliability of GIS-based assessment outcomes largely relies on the validity and transparency of applied methods, it can be concluded that additional efforts are needed in order to optimise the applicability of GIS — if a spatially-specific environmental assessment methodology is to provide fully reliable and accountable results.

References

Antunes, P, R Santos and L Jordao (2001). The application of geographical information systems to determine environmental impact significance. *Environmental Impact Assessment Review*, 21(6), 511–535.

Bartley, B (2007). *Supporting Evidence-Based Spatial Planning and Analysis in Ireland: Towards the Development of All-Island Spatial Databases*. Unpublished paper presented at the National Institute for Regional and Spatial Analysis Workshop, 15 February, Maynooth, Ireland.

Blaser, B, H Liu, D McDermott, F Nuszdorfer, N Thi Phan, U Vanchindorj, L Johnson and J Wyckoff (2004). *GIS Based Cumulative Effects Assessment*. Report No. CDOT-DTD-R-2004-6. Colorado Department of Transportation Research Branch: Colorado.

Borrego, C, A Monteiro, J Ferreira, AI Miranda, AM Costa, AC Carvalho and M Lopes (2008). Modelling uncertainty estimation procedures for air quality assessment. *International Environmental Journal*, 34(5), 613–620.

Bragagnolo, C (2011). *Improving the Consideration of Cumulative Effects in Strategic Environmental Assessment of Spatial Plans: A Case Study in the Peri-Urban Region of Milan*. Unpublished research thesis. Universitá degli Study di Trento: Italy.

Ceballos, A and J López (2003). Evaluating biophysical variables to identify suitable areas for oat in central Mexico: a multi-criteria and GIS approach. *Agriculture, Ecosystems and Environment*, 95(1), 371–377.

CEC (1997). Directive 97/11/EC, of 3rd March, amending Directive 85/337/EEC on the Assessment of the Effects of Certain Public and Private Projects on the Environment. Commission of the European communities. *Official Journal of the European Union*, L0011, 3.1997.

CEC (2000). Directive 2000/60/EC of the European Parliament and Council, of 23rd October, Establishing a Framework for Community Action in the Field of Water Policy.

Commission of the European Communities. *Official Journal of the European Union*, L327/1, 22.12.2000.

CEC (2001). Directive 2001/42/EC, of 27th June, on the Assessment of the Effects of Certain Plans and Programmes on the Environment, Luxembourg. Commission of the European Communities. *Official Journal of the European Union*, L197/30, 21.7.2001.

CEC (2002). Directive 2002/49/EC of the European Parliament and of the Council, of 25th June, Relating to the Assessment and Management of Environmental Noise. Commission of the European Communities. *Official Journal of the European Union*, L189/12, 18.7.2002.

CEC (2007). Directive 2007/2/EC, of 14th March, Establishing an Infrastructure for Spatial Information in the European Community (INSPIRE). Commission of the European Communities. *Official Journal of the European Union*, L108/1, 24.5.2007.

CEC (2008a). Directive 2008/50/EC, of 21st May, on ambient air quality and cleaner air for Europe, Luxembourg. Commission of the European Communities. *Official Journal of the European Union*, L152/1, 11.6.2008.

CEC (2008b). Commission Regulation (EC) No. 1205/2008, of 3rd December 2008, implementing directive 2007/2/EC of the European parliament and of the council as regards Metadata. *Official Journal of the European Union*, 326/12, 4.12.2008.

Chan, Y and S Easa (2000). Looking Ahead. In: Easa, S and Y Chan (eds.) *Urban Planning and Development Applications of GIS*, pp. 253–256. American Society of Civil Engineers: Reston, Virginia.

Chavan, VS and P Ingwersen (2009). Towards a data publishing framework for primary biodiversity data: challenges and potentials for the biodiversity informatics community. *BMC Bioinformatics*, doi: 10.1186/1471-2105-10-S14-S2.

Chen, YW, CH Wang and SJ Lin (2008). A multi-objective geographic information system for route selection of nuclear waste transport. *Omega*, 36(3), 363–372.

Cory, M, R Morgan, C Bray and I Greenway (2001). A New Coordinate System for Ireland. International Federation of Surveyors International Conference Proceedings, 6–11 May, Seoul, Republic of Korea. URL: http://www.fig.net/pub/proceedings/korea/full-papers/pdf/session8/cory-morgan-bray-greenway.pdf.

EJ/GB (2008). Implantación del Sistema de Información Geográfica (GIS) Corporativo. GeoEuskadi; SIG Corporativo del Gobierno Vasco. Eusko Jaurlaritza/Gobierno Vasco.

Elwood, S (2006). Critical issues in participatory GIS: deconstructions, reconstructions and new research directions. *Transactions in GIS*, 10(5), 693–708.

Fedra, K (2004). Sustainable Urban Transportation: A model-based approach. *Cybernetics and Systems*, 35, 455–485.

Freising Online (2011). Interactive statutory master plans in the county of Freising, Germany. URL: http://fs.mapsailor.de/fs-start.htm.

Geneletti, D (2008). Impact assessment of proposed ski areas: a GIS approach integrating biological, physical and landscape indicators. *Environmental Impact Assessment Review*, 28(2–3), 116–130.

Geneletti, D, E Beinat, CF Chung, AG Fabbri and HJ Scholten (2003). Accounting for uncertainty factors in biodiversity impact assessment: lessons from a case study. *Environmental Impact Assessment Review*, 23(4), 471–487.

Gioia, P (2010). Managing biodiversity data within the context of climate change: towards best practice. *Austral Ecology*, 35, 392–405.

Gontier, M, B Balfors and U Mörtberg (2006). Biodiversity in environmental assessment — current practice and tools for prediction. *Environmental Impact Assessment Review*, 26(3), 268–286.

Gontier, M, U Mörtberg and B Balfors, (2010). Comparing GIS-based habitat models for applications in EIA and SEA. *Environmental Impact Assessment Review*, 30, 8–18.

González (2009). *GISEA Manual: Current Practice and Potential on the Application of Geographic Information Systems as a Support Tool in Strategic Environmental Assessment of Irish Land Use Plans*. EPA: Ireland. (URL: http://www.epa.ie/downloads/consultation/name,25835,en.html).

González, A (2010a). Gestión de la Información Espacial en la EAE. In: Casimiro Martínez, MA, Espluga González, AP, Desdentado Gómez, LA, Díaz-Martín, M, García Montero, LG and Sobrini Sagalzeta-de Ilurdoz, I (eds) *Evaluación de Impacto Ambiental: Cooperación, Desarrollo y Sostenibilidad*, pp. 155–163. Print House: Madrid.

González, A (2010b). *Incorporating Spatial Data and GIS to Improve SEA of Land Use Plans: Opportunities and Limitations — Case Studies in the Republic of Ireland.* Lambert Academic Publishing: Germany.

González, A, A Donnelly, M Jones, J Klostermann, A Groot, M Breil and N Chrysoulakis, (2011). Community of practice approach to developing urban sustainability indicators. *Journal of Environmental Assessment Policy and Management*, 13(4), 591–617.

González, A, A Gilmer, R Foley, J Sweeney and J Fry (2008). Technology-aided participative methods in environmental assessment: an international perspective. *Computers, Environment and Urban Systems*, 32, 303–316.

González, A, A Gilmer, R Foley, J Sweeney and J Fry (2011). Applying geographic information systems to support strategic environmental assessment: opportunities and limitations in the context of Irish land-use plans. *Environmental Impact Assessment Review*, 31(3), 368–381.

González-Matesanz, J, A Dalda, R Quirós and J Celada (2003). *ED50-ETRS89 Transition Models for the Spanish Geodetic Network.* Unpublished paper presented at the EUREF Symposium, 4–7 June, Toledo, Spain.

Hanzl, M (2007). Information technology as a tool for public participation in urban planning: a review of experiments and potentials. *Design Studies*, 28(3), 289–307.

Herberg, A (2008). Relevant Baseline Data for Use in SEA — Examples from Germany. In: Fischer, TB, P Gazzola, U Jha-Thakur, I Belčáková and R Aschemann (eds.) *Environmental Assessment Lecturers' Handbook*, pp. 145–152. PENTA Erasmus Mundus: EC. URL: www.twoeam-eu.net.

ISO/TC211 (2003). *ISO 19115:2003 Geographic information — Metadata*. International Standards Organisation Technical Committee for Geographic Information and Geomatics, ISO Published International Standards. Available at: http://www.isotc211.org/.

Jeroen, CJH, WJ Aerts and B Wouter (2011). Climate change impacts on pricing long-term flood insurance: a comprehensive study for the Netherlands. *Global Environmental Change*, 21(3), 1045–1060.

João, EM (1998). Use of Geographic Information Systems in Impact Assessment. In: Porter, A and Fittipaldi J (eds). *Environmental Methods Review: Retooling Impact Assessment for the New Century*, pp. 154–163. The Army Environmental Policy Institute: Atlanta.

João, EM (2002). How scale affects environmental impact assessment. *Environmental Impact Assessment Review*, 22(4), 289–310.

João, EM (2007). The importance of data and scale issues for strategic environmental assessment. *Environmental Impact Assessment Review*, 27(5), 361–364.

João, EM and A Fonseca (1996). *Current use of Geographical Information Systems for Environmental Assessment: a Discussion Document*. Research Papers in Environmental and Spatial Analysis 36, May. Department of Geography, London School of Economics.

Jordan, G and Shrestha, B (2000). *A Participatory GIS for Community Forestry user Groups in Nepal: Putting People Before the Technology*. International institute for environment and development, participatory learning and action notes 39.

Kalogirou, S (2002). Expert systems and GIS: an application of land suitability evaluation. *Computers, Environment and Urban Systems*, 26(2–3), 89–112.

Kingston, R, S Carver, A Evans and I Turton (2000). Web-based public participation geographical information systems: an aid to local environmental decision-making. *Computers, Environment and Urban Systems*, 24(2), 109–125.

Kuula, A and S Borg (2008). Open access to and reuse of research data — the state of the art in Finland. Finnish Social Science Data Archive, 7.

Liu, Y, H Gupta, E Springer and T Wagener (2008). Linking science with environmental decision making: experiences from an integrated modeling approach to supporting sustainable water resources management. *Environmental Modelling and Software*, 23(7), 846–858.

O'Dea, L, V Cummins and N Dwyer (2004). *Developing an Informational Web Portal for Coastal Data in Ireland: Data Issues in the Marine Digital Atlas*. Oceanology International Conference Proceedings, pp. 32–40, 18 March, London, England.

Onsrud, HJ (1999). Liability in the Use of GIS and Geographical Datasets. In *Geographical Information Systems 2: Managing Issues and Applications*, PA Longley, MF Goodchild, DJ Maguire and DW Rhind (eds). New York: John Wiley and Sons, pp. 643–652.

Partidário, MR (2007). Scales and associated data — what is enough for SEA needs? *Environmental Impact Assessment Review*, 27(5), 460–478.

Patil, AA, AP Annachhatre and NK Tripathi (2002). Comparison of conventional and geospatial EIA: a shrimp farming case study. *Environmental Impact Assessment Review*, 22(4), 361–375.

Ruimtelijkeplannen (2011). Online inventory of Dutch spatial plans. URL: http://www.ruimtelijkeplannen.nl/web-roo/?planidn=NL.IMRO.0114.2010001-0002.

Rybaczuk, K and H MacMahon (1995). *Accessing Geographic Information for Ireland*. Dublin, Ireland: Forbairt and the Irish Institution of Surveyors.

Scott, P and P Marsden (2003). *Development of Strategic Environmental Assessment Methodologies for Plans and Programmes in Ireland* (2001-DS-EEP-2/5). Ireland: Environmental Protection Agency.

Skehan, CD and A González (2006). *Spatial Data Management Requirements and Strategic Environmental Assessment*. Unpublished paper presented at the Society of Chartered Surveyors, EuroSDR & Dublin Institute of Technology joint workshop, 31 October, Dublin, Ireland.

Slaithwaite (2003). Virtual Slaithwaite Participatory Planning System, West Yorkshire. University of Leeds, UK. URL: http://www.ccg.leeds.ac.uk/slaithwaite/.

Smith, GF, P O'Donohue, K O'Hora and E Delaney (2011). *Best Practice Guidance for Habitat Survey and Mapping*. The Heritage Council: Ireland.

Steinmann, R, A Krek and T Blaschke (2005). *Can Online Map-Based Applications Improve Citizen Participation*? Lecture Notes in Computer Science, 3416: 25–35, doi: 10.1007/978-3-540-32257-3_3.

Tang, KX and NM Waters (2005). The internet, GIS and public participation in transportation planning. *Progress in Planning*, 64(1), 7–62.

Therivel, R (2004). *Strategic Environmental Assessment in Action*. Earthscan, IncNet Library, ISBN 1-84407-041-7.

Thiel, A (2009). The use of ex-ante modelling tools in european impact assessment: what role does land use play? *Land Use Policy*, 26(4), 1138–1148.

Vanderhaegen, M and E Muro (2005). Contribution of a european spatial data infrastructure to the effectiveness of EIA. *Environmental Impact Assessment Review*, 25(2), 123–142.

Weiner, D and TM Harris (2003). Community-integrated GIS for land reform in South Africa. *Urban and Information Systems Association Journal*, 15(APA II), 61–73.

White, I, R Kingston and A Barker (2010). Participatory geographic information systems and public engagement within flood risk management. *Journal of Flood Risk Management*, 3(4), 337–346.

Wood, SJ (2005). "How Green is My Valley?" Desktop geographic information systems as a community-based participatory mapping tool. *Area*, 37(2), 159–170.

CHAPTER 7

How Clearly is Causality Communicated in EIA?

Anastassios Perdicoúlis* and John Glasson†

**University of Trás-os-Montes e Alto Douro, 5001-801 Vila Real, Portugal and CITTA, FEUP, Porto, Portugal*
**tasso@fe.up.pt*

†Oxford Institute for Sustainable Development (OISD)
School of the Built Environment, Oxford Brookes University
Headington Campus, Gipsy Lane, Oxford OX3 OBP, UK
†jglasson@brookes.ac.uk

Causality, or the relation of cause and effect, is a fundamental notion in impact assessment, indicating how impacts arise from development actions. In the context of Environmental Impact Assessment (EIA), recent research in the UK has indicated that impact statements typically do not contain graphic expressions of causality. This article explores the extent to which causality about environmental impacts is described in the text of the Environmental Impact Statements (EIS). The analysis methodology employs two techniques, semantics analysis and transcription of the text into causal digraphs, while the case study material is sampled from 10 recent UK and US impact statements. The research indicates in many cases that the communication of causality has faults such as uncertainty or absence of important information, and we highlight the implications for the practice of EIA.

Keywords: Environmental Impact Assessment (EIA); Environmental Impact Statement (EIS); Environmental Statement (ES); causality; causal diagrams; semantics analysis.

Introduction

Causality, or the relation of cause and effect, is a fundamental notion in impact assessment, indicating how impacts arise from development actions (Wathern,

*Corresponding author.

1988; Glasson *et al.*, 2005; Therivel, 2004; Perdicoúlis and Glasson, 2006). As such, causality is an issue of concern to everyone involved in impact assessment: academics and practitioners in the roles of authors, readers, and reviewers of technical reports.

In the context of Environmental Impact Assessment (EIA), recent research in the UK has indicated that impact statements typically do not contain graphic expressions of causality (Perdicoúlis and Glasson, 2006). Thus, this article explores how well causality about environmental impacts is described in the text of the Environmental Impact Statements (EIS).

In particular, we sampled material from 10 UK and US impact statements produced between 2004 and 2007, and applied to: (a) semantics analysis (Perdicoúlis and Piper, 2008; Perdicoúlis and Glasson, 2007) of the sample text; and (b) transcription of the text into diagrammatic form (Perdicoúlis and Glasson, 2006, 2007; Perdicoúlis and Piper, 2008), to provide an alternative representation of the same information.

Methodology

Impact statements

On 10 October 2007, using the Google™ search engine, we searched for full text documents using two sets of keywords: "environmental impact statement" for US material; and "environmental statement" for UK material. In the latter case, we had to exclude the homonymous reports of the EMAS context (European Commission, 2001). Using a search engine does not provide random sampling, but we consider it impartial and thus suitable for our sampling purposes.

From the results compiled by the search engine we took those that came up first in the list. The selected case studies are presented in Table 1, and feature a wide range of projects. We cannot tell whether the sampled documents are representative of EISs in their respective countries, but they are likely to be encountered by other people searching with the same keywords (as long as they are maintained in the index of the search engine), thus they gain some projection and importance, and may even serve as "practice models".

Text samples

Having 10 case studies, we opted to work with text samples from each case. The sampling criteria were set to be the following: (a) size: one paragraph; (b) content: the way impacts are presented to arise from the project action (impact analysis, forecasting, etc.); and (c) choice of text: within the page ranges with the defined

Table 1. The 10 case studies of the article, sorted: (a) by chronological order of the reference; and (b) alphabetical order of the locale.

Subject	Locale	Reference
Bypass	Devon, UK	DCC *et al.* (2004)
Mining	Montana, USA	MDEQ and USBLM (2004)
Petroleum	Falklands, UK	RPSE and DP (2005)
Crossrail	London, UK	ERM (2005)
MagLev	Pennsylvania, USA	PAAC and PENNDOT (2005)
Aqueduct	Washington DC, USA	US ACE (2005)
Restoration	Indiana, USA	USDA (2006)
Power station	Montana, USA	USDA-RUS and MDEQ (2006)
Quarry	Montana, USA	MDEQ (2006)
Junction	Essex, UK	MP (2007)

content, select one paragraph at random. The selected text samples are presented in Section 3.

Semantics analysis

An analysis of semantics (i.e. the study of meaning) identifies the special "causality aspects" of the sample text for impact assessment purposes (Perdicoúlis and Glasson, 2007; Perdicoúlis and Piper, 2008): (a) system elements; (b) action; (c) causal relationship; and (d) impacts (Table 2).

The steps of the semantics analysis undertaken in this article are: (a) identification of the four causality aspects in the text samples; (b) interpretation in the

Table 2. Causality aspects and marks of the text. The conventions summarised in this table are used in the semantics analysis of the text, to assist the construction of the corresponding diagrams.

Aspect	Function	Sample expression	In the text	In the digraph
element	system component	fauna, flora, [proponent]	sans serif	node text
action	something done[a]	compaction [of soil]	*italics*	node text
causality	transfer of action	causes (verb)	underlined	arrow (↷)
impact	change or effect	loss, change (noun)	**bold**	text on arrow
comments	explanation	research has shown...	`teletype`	absent

[a]In a broader sense, "something done" to an element is: (i) prescribed by the development proposal; (ii) provoked by a actor alien to the development proposal; and/or (iii) provoked by a change in the element immediately upstream — see illustrations in Perdicoúlis and Glasson (2007) and Perdicoúlis and Piper (2008). In this article we mark only actions (perceived to be) prescribed by the project.

given context, in case of doubt; and (c) marking of the text elements as per conventions (Table 2). Marked text is presented below the original text samples, in Section 3.

Causal diagrams

Among various diagrammatic options for the representation of causality (Perdicoúlis and Glasson, 2006), in this article we opted to use "action-to-outcome maps" (ATOMs), which are well suited to the purposes of impact assessment (Perdicoúlis and Glasson, 2007; Perdicoúlis and Piper, 2008).

ATOMs, chosen to be used for the purposes of this article, are a special category of "graphs" — i.e. knowledge representation structures, consisting of nodes N (or vertices) interconnected with edges E: G = (N, E) (Gentil and Montmain, 2004; Harary, 1969). When the pairs are ordered, as is the case with cause-effect relationships, the graphs are "directed" and they are known as "digraphs" (i.e. opposed to the unordered pairs of "simple" graphs). However, not all digraphs are causal, as Fig. 1 illustrates.

Digraphs are "causal" if the semantics of the edges represent the property of causality: an input node of an edge is one of the causes of the output node (Gentil and Montmain, 2004). In addition to causality, each edge may be assigned extra information, which makes edges a convenient place to mark the likely impacts (Perdicoúlis and Piper, 2008; Perdicoúlis and Glasson, 2006; Rosen, 2003; Sterman, 2000; Canter, 1996).

Case Studies

A380 Kingskerswell Bypass, Devon, UK

Text 1 (Original Text)

[Impact on Farm Holdings] Many factors could contribute to the impact on farm holdings including: loss of land; severance; loss of buildings and fixed equipment; reduced convenience of access; loss of income; and changes in land use and

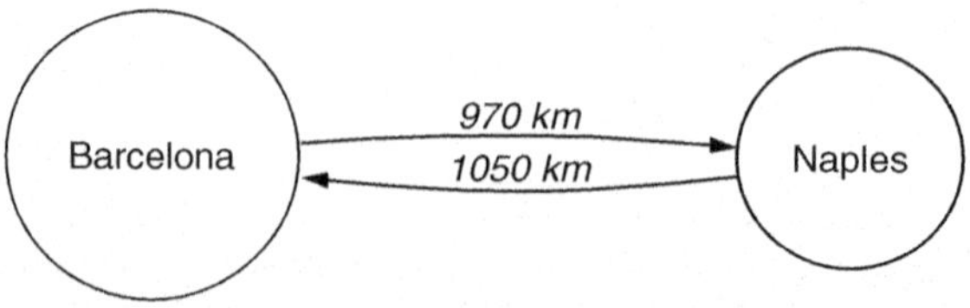

Figure 1. Example of a non-causal digraph representing the distance between two cities, corresponding to two air-traffic corridors.

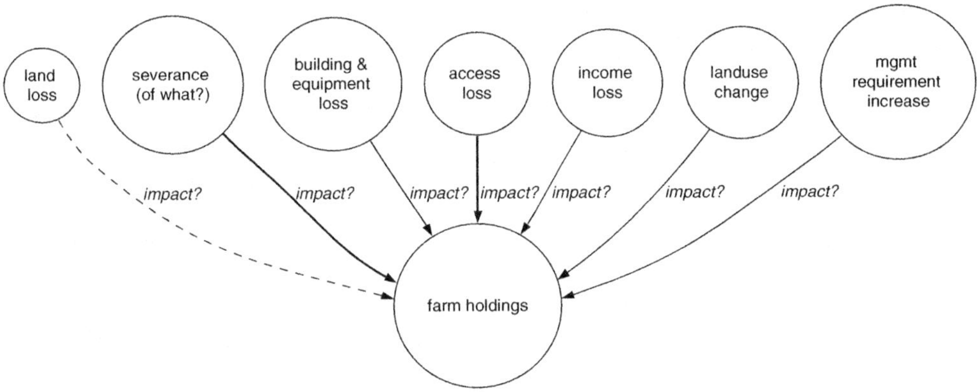

Figure 2. Causal digraph for the DCC *et al.* (2004) case sample text.

increase in management requirement. In terms of land loss, clearly the impact would differ markedly between holdings depending on the use made of the land, though research on the subject has shown that losses of less than 5% rarely, if ever, have a long-term impact on the management of agricultural holdings and that issues related to severance and access are usually of more importance. (DCC *et al.*, 2004, Vol.2, Ch.6, p.9, para. 6.5.6)

Text 2 (Marked Text)
[Impact on Farm Holdings] Many factors could contribute to the **impact** on farm holdings including: *loss* of land; *severance*; *loss* of buildings and fixed equipment; *reduced* convenience of access; *loss* of income; and *changes* in land use and *increase* in management requirement. `In terms of land loss, clearly the impact would differ markedly between holdings depending on the use made of the land, though research on the subject has shown that losses of less than 5 per cent rarely, if ever, have a long-term impact on the management of agricultural holdings and that issues related to severance and access are usually of more importance.`

Observations, doubts, and resolutions

- "Severance" indicates an action or effect, but does not associate an element to it: it is assumed to refer to land
- All the changes that cause impacts to the farm holdings (e.g. loss of land) are assumed to have been prescribed by the project (which could be clarified in the rest of the document)
- The impacts are not explicit in the text: consequently, the impacts are missing from the edges of the corresponding digraph

- The text marked as teletype provides additional comments
- A less important relationship is marked in the digraph with a dashed line, while a relationship of a more-than-average importance is marked with a bold (thicker) line
- All factors that cause impacts to farm holdings are hierarchically in the same rank.

Golden Sunlight Mine Pit Reclamation, Montana, USA

Text 3 (Original Text)
[Pit Impacts to Groundwater] As groundwater enters the pit, it flows through zones of broken and disturbed rock, which contains 0.5–2.0% pyrite. Atmospheric oxygen and dissolved oxygen in water percolating through the broken rock reacts with the pyrite, which leads to sulfide oxidation and generation of ARD. In addition, during precipitation events, water quality is degraded by the flushing of oxidation by-products, such as acid salts that have accumulated on the pit highwall from evaporation [citation] and from heat produced by sulfide oxidation (MDEQ and USBLM, 2004, Ch. 4, pp. 61–62).

Text 4 (Marked Text)
[Pit Impacts to Groundwater] As groundwater **enters** the pit, it **flows through** zones of broken and disturbed rock, which contains 0.5–2.0% pyrite. Atmospheric oxygen and dissolved oxygen in water **percolating through** the broken rock **reacts** with the pyrite, which leads to **sulfide oxidation** and **generation** of ARD. In addition, during precipitation events, water quality is **degraded** by the **flushing of** oxidation by-products, such as acid salts that have **accumulated** on the pit highwall from **evaporation** [citation] and from heat produced by sulfide **oxidation**.

Observations, doubts, and resolutions

- Additional information is required in chemical terms: e.g. knowing that pyrite is FeS_2, we know that it contains iron and sulfur: a chemical reaction, such as $FeS_2 + O_2 + H_2O \rightarrow H_2SO_4 + Fe_2O_2 \bullet H_2O$ (unbalanced), would elucidate the referred process
- Some jargon appears (e.g. ARD), which requires clarification (ARD = acid rock drainage)
- Most of the impacts marked in the diagram were not present in the original text

Offshore Drilling, Falkland Islands, UK

Text 5 (Original Text)
[Discharge of sewage, drainage waters, cooling water and run-off/wash waters] Direct impacts from these emissions will include localised nutrient enrichment,

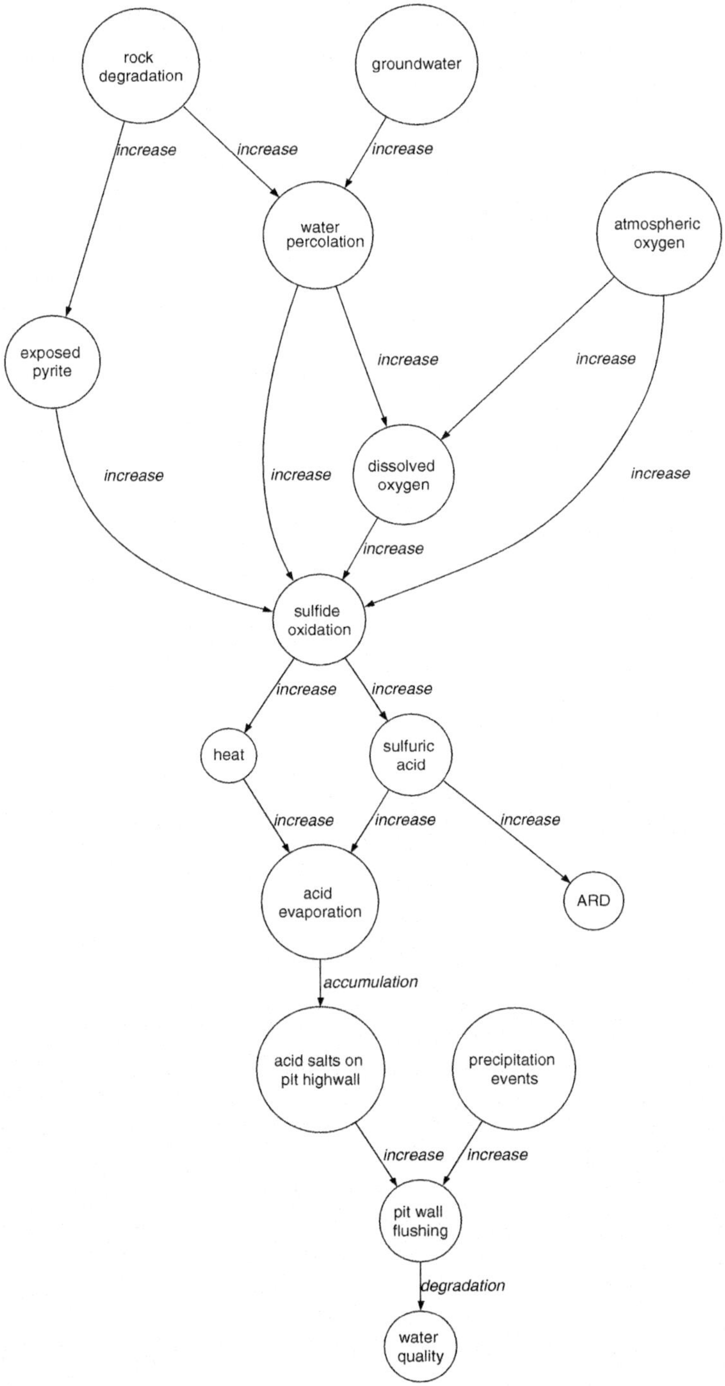

Figure 3. Causal digraph for MDEQ and USBLM (2004) case sample text.

saprogenic effects, temperature increase and low-level pollution from trace oils and chemicals, which will be rapidly dispersed. Indirectly these localised impacts may lead to increased pollution of the ecosystem with a cumulative impact on biodiversity. (RPSE and DP, 2005, p. 92)

Text 6 (Marked Text)

[Discharge of sewage, drainage waters, cooling water and run-off/wash waters] Direct **impacts** from these *emissions* will include localised nutrient **enrichment**,

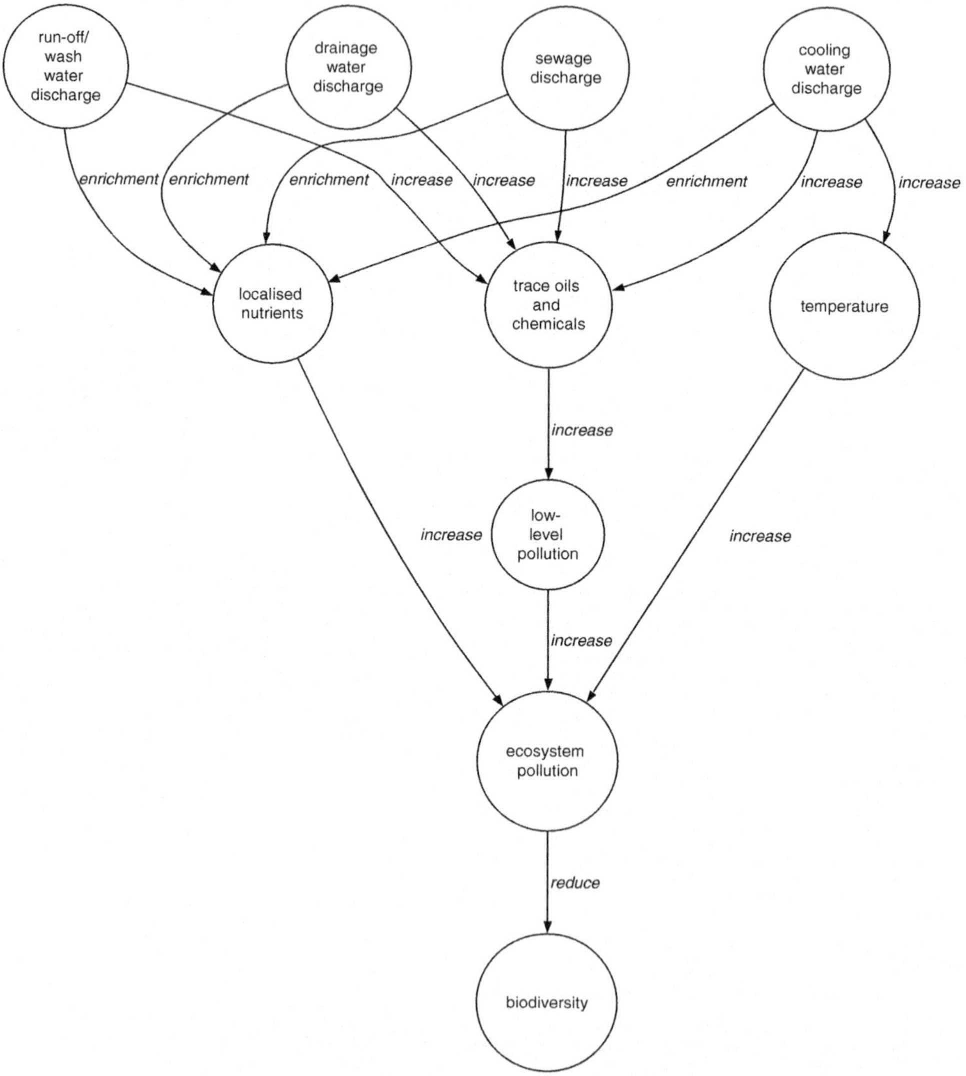

Figure 4. Causal digraph for the RPSE and DP (2005) case sample text.

saprogenic effects, temperature **increase** and low-level pollution from trace oils and chemicals, which will be rapidly **dispersed**. Indirectly these localised **impacts** may lead to **increased** pollution of the ecosystem with a cumulative **impact** on biodiversity.

Observations, doubts, and resolutions

- How do saprogenic effects fit into the system?
- The dispersion of trace oils and chemicals would require yet another level of analysis, but a simpler diagrammatic expression was judged preferable
- The direct impacts of the emissions are identified, but the "receiver" elements are not clearly identified
- Are trace oils and chemicals deriving from the identified emissions?

Crossrail, London, UK

Text 7 (Original Text)
[Direct impacts on listed buildings] Works for the eastern ticket hall could give rise to vibration impacts at the adjacent listed 20 Hanover Square. Appropriate demolition techniques will be used in order to limit vibration impacts and significant impacts will therefore be avoided. However, vibration will be monitored as appropriate during demolition and construction in order that necessary protective measures can be implemented. (ERM, 2005, Vol. 2, Ch. 8, p. 119, Para. 8.6.34)

Text 8 (Marked Text)
[Direct impacts on listed buildings] *Works* for the eastern ticket hall could give rise to vibration **impacts** at the adjacent listed 20 Hanover Square [building]. Appropriate demolition techniques will be used in order to **limit** vibration **impacts** and significant **impacts** will therefore be avoided. However, vibration will be *monitored* as appropriate during *demolition* and *construction* in order that necessary protective measures can be *implemented.*

Observations, doubts, and resolutions

- We had to introduce in the diagram system elements (e.g. vibration, demolition techniques, monitoring registry) that were not explicitly mentioned in the text
- Two negative feedback loops are implied in relation to the function of monitoring: the closing of these loops is presented in dotted lines

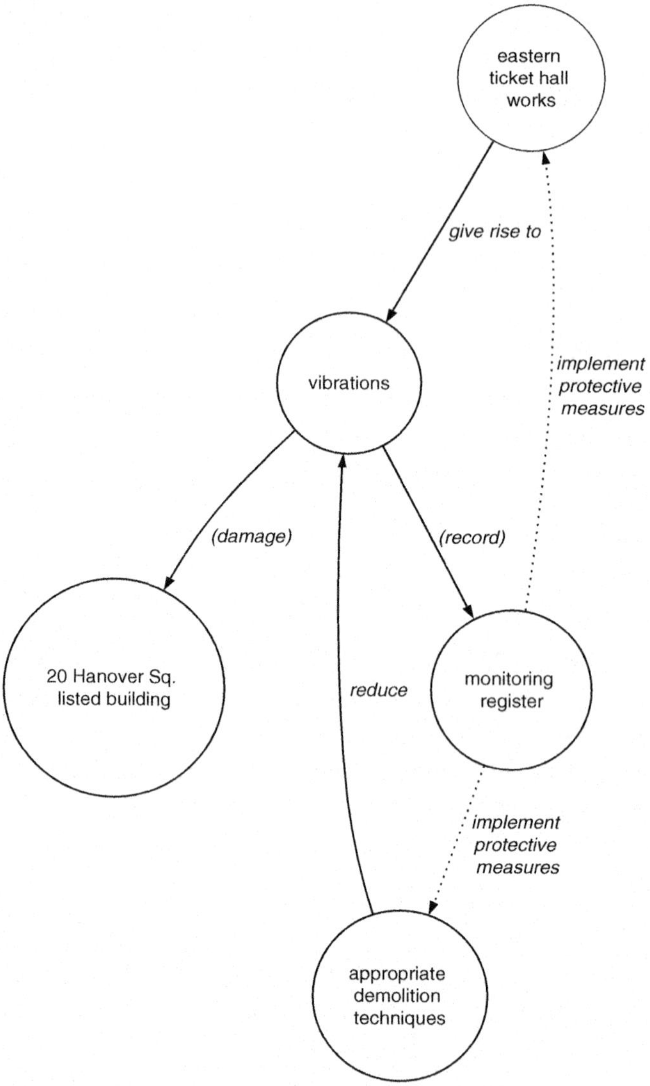

Figure 5. Causal digraph for ERM (2005) case sample text.

High-Speed MagLev Project, Pennsylvania, USA

Text 9 (Original Text)

[Wetlands] Temporary wetland impacts incurred by the project may include disturbance and alteration to the existing vegetation, soils, and hydrology. Wetland vegetation may either be removed or disturbed during project construction. Silt may migrate from exposed areas and enter into wetlands, disrupting vegetation and

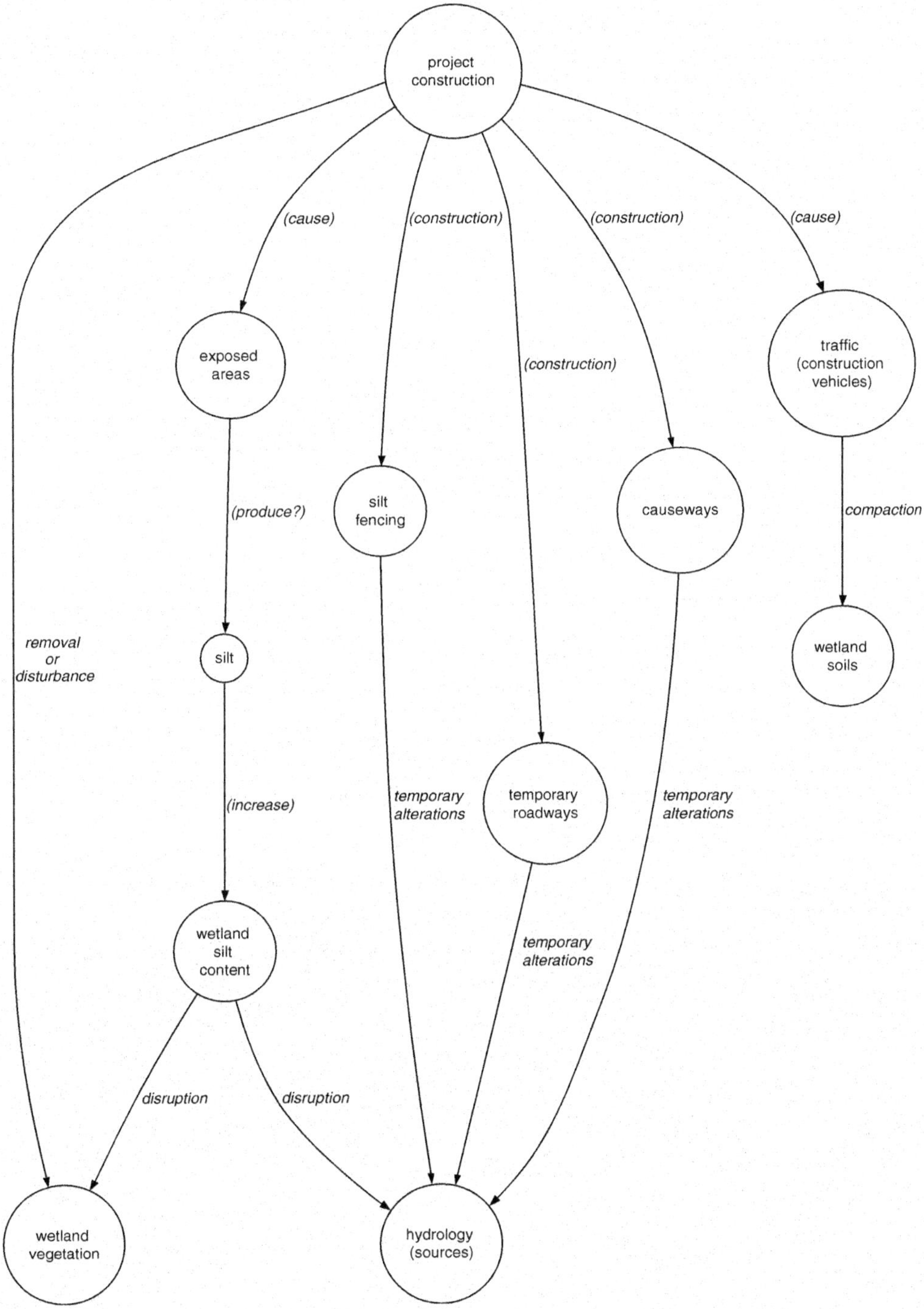

Figure 6. Causal digraph for the PAAC and PENNDOT (2005) case sample text.

hydrology. Soil compaction from construction vehicles may influence wetland soils. Hydrology sources may be temporarily altered by items such as silt fencing, temporary roadways, and causeways. Implementing BMPs and an approved E&S Plan would minimize temporary wetland impacts. (PAAC and PENNDOT, 2005, Ch. 4, p. 56)

Text 10 (Marked Text)
[Wetlands] Temporary wetland **impacts** incurred by the *project* may include **disturbance** and **alteration** to the existing vegetation, soils, and hydrology. Wetland vegetation may either be **removed** or **disturbed** during *project construction.* Silt may **migrate** from exposed areas and enter into wetlands, **disrupting** vegetation and hydrology. Soil **compaction** from construction vehicles may **influence** wetland soils. Hydrology sources may be **temporarily altered** by items such as silt fencing, temporary roadways, and causeways. *Implementing* BMPs and an approved E&S Plan would minimize temporary wetland **impacts**.

Observations, doubts, and resolutions

- Uncertainty in some terms: e.g. hydrology, BMPs, E&S Plan
- Uncertainty in the project action: e.g. are silt fencing, temporary roadways, and causeways standard part of the project, or mitigation measures?
- Uncertainty in impacts: e.g. what does "temporarily altered" or "influence" mean?

Water Treatment Residuals Management, Washington DC, USA

Text 11 (Original Text)
[Rare, Threatened, and Endangered Species (Aquatic Only)] The conclusion of NMFS's (2003) Biological Opinion for the shortnose sturgeon indicated that if a discharge from the aqueduct occurred during the SNS spawning period (1 March–15 May), and SNS eggs and larvae were present, the discharge would adversely affect those life stages within a very small area adjacent to the discharges. It would therefore not jeopardize the continued existence of the Chesapeake Bay SNS population. Because of this NMFS recommendation, the Washington Aqueduct's current NPDES permit does not allow a discharge between 15 February and 30 June to address the broader spring spawning period for resident anadromous species. These time of year restrictions have being implemented, are part of the No Action alternative, and are expected to be protective of the early life stages of SNS. (US ACE, 2005, Vol. 1, Sec. 4, p. 20)

Text 12 (Marked Text)
[Rare, Threatened, and Endangered Species (Aquatic Only)] `The conclusion of NMFS's (2003) Biological Opinion for the shortnose sturgeon`

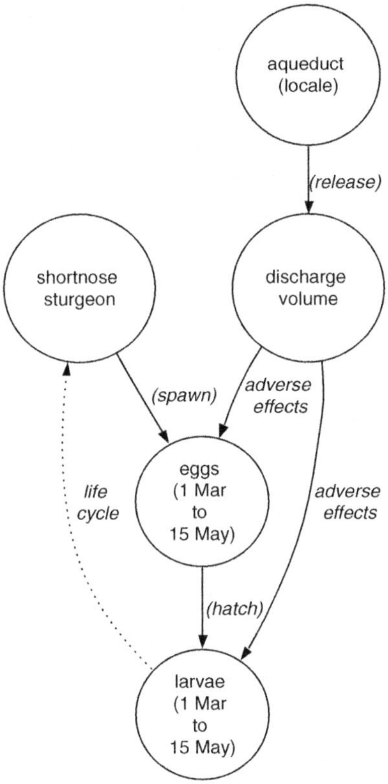

Figure 7. Causal digraph for the US ACE (2005) case sample text.

indicated that if a *discharge* from the aqueduct occurred during the SNS spawning period (1 March–15 May), and SNS eggs and larvae were present, the *discharge* would adversely affect those life stages within a very small area adjacent to the *discharges*. It would therefore not jeopardize the continued existence of the Chesapeake Bay SNS population. Because of this NMFS recommendation, the Washington Aqueduct's current NPDES permit does not allow a discharge between 15 February and 30 June to address the broader spring spawning period for resident anadromous species. These time of year restrictions have being implemented, are part of the No Action alternative, and are expected to be protective of the early life stages of SNS.

Observations, doubts, and resolutions

- Challenge in representing space (e.g. where the discharges were to be made); resolution: mark the space within the element (e.g. "aqueduct (locale)")

- Challenge in representing time (e.g. spawning and hatching seasons); resolution: mark the time within the elements (e.g. "eggs (1 Mar to 15 May)")
- Uncertainty about the life cycle of the shortnose sturgeon (e.g. differentiation of spawning and hatching time, as well as respective locales)

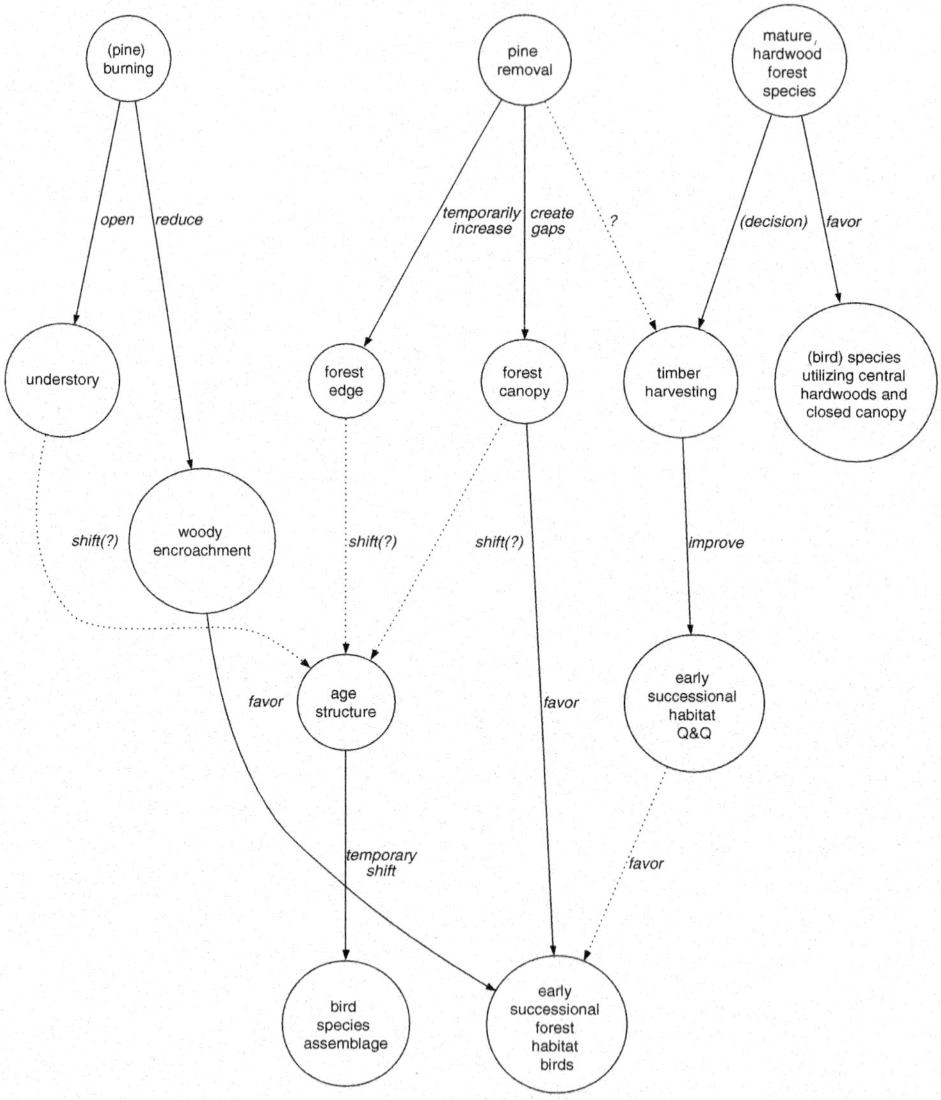

Figure 8. Causal digraph for the USDA (2006) case sample text.

German Ridge Forest Restoration Project, Indiana, USA

Text 13 (Original Text)
[Effects on birds] Many species of birds require specific types of habitat for breeding, nesting, foraging, and rearing of young. Changes in habitat directly affect the health, abundance, and diversity of birds. Pine removal would create gaps in the forest canopy and temporarily increase forest edge, and burning would open the understory. This shift in age structure would also cause some temporary shift in the assemblage of bird species. The increased forest edge would likely favor birds of early successional forest habitats, while the final project goal of mature hardwoods would ultimately favor species that utilize central hardwoods and closed canopy forest. Timber harvesting in conjunction with prescribed burning would improve the availability and quality of early successional habitat, and burning the openings would reduce woody encroachment and benefit bird species that prefer early successional forest habitat. (USDA, 2006, p. 122)

Text 14 (Marked Text)
[Effects on birds] Many species of birds require specific types of habitat for *breeding*, *nesting*, *foraging*, and *rearing of young*. **Changes** in habitat directly affect the health, abundance, and diversity of birds. Pine *removal* would **create gaps** in the forest canopy and **temporarily increase** forest edge, and *burning* would **open** the understory. This **shift** in age structure would also cause some **temporary shift** in the assemblage of bird species. The **increased** forest edge would likely **favor** birds of early successional forest habitats, while the final project goal of mature hardwoods would ultimately **favor** species that utilize central hardwoods and closed canopy forest. Timber *harvesting* in conjunction with *prescribed burning* would **improve** the availability and quality of early successional habitat, and *burning* the openings would **reduce** woody encroachment and **benefit** bird species that prefer early successional forest habitat.

Observations, doubts, and resolutions

- Uncertainties about forest ecology: e.g. is the age structure determined (or influenced) by the understory, forest edge, and forest canopy, as implied?
- Uncertainty about project action: e.g. is pine removal considered as timber harvesting?
- Uncertainty about impacts: e.g. what does "shift" mean?

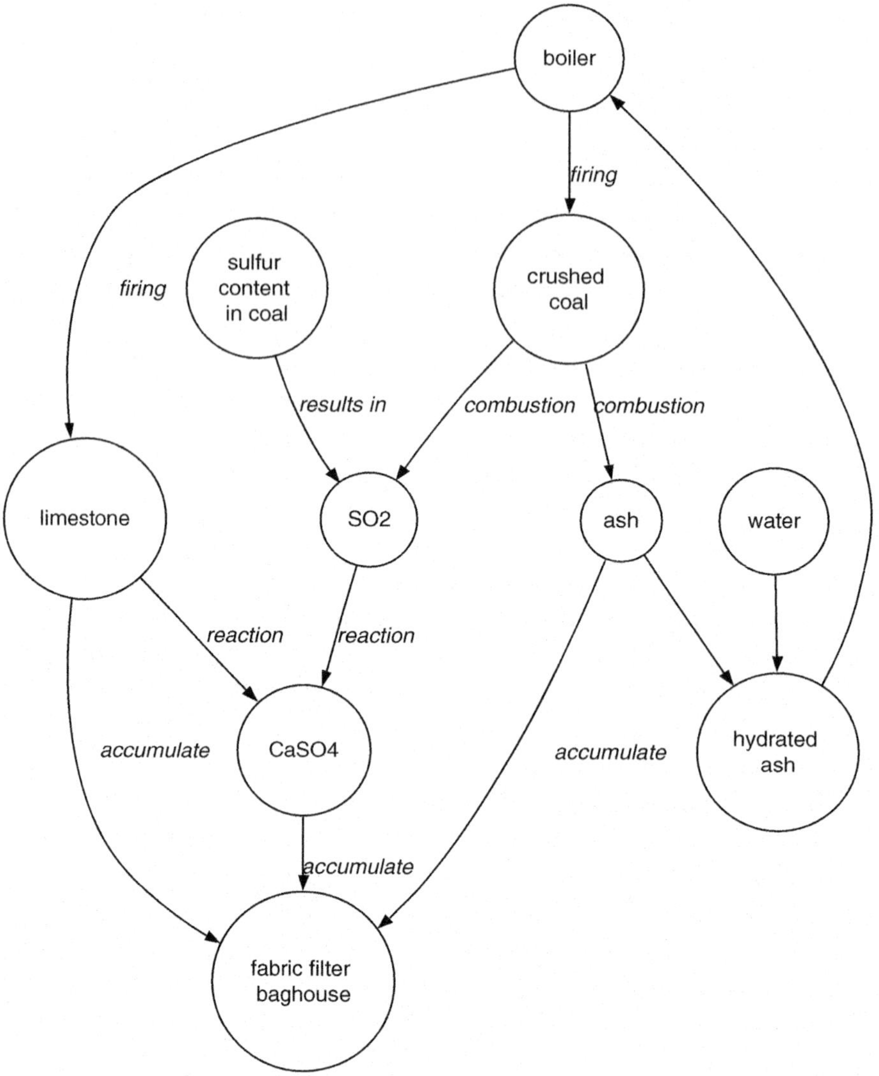

Figure 9. Causal digraph for the USDA-RUS and MDEQ (2006) case sample text.

Highwood Generating Station, Montana, USA

Text 15 (Original Text)

[Air quality] SO_2 emissions in the boiler result from the sulfur present as an impurity in the coal that is fired. The CFB boiler primarily would fire low-sulfur, sub-bituminous coal from the Powder River Basin. This coal varies in sulfur content, but is expected to typically have sulfur contents below one percent by weight. The design of the CFB boiler employs the firing of crushed coal mixed with limestone injected into the combustor. The use of limestone provides control of SO_2 by reacting with SO_2 to form calcium sulfate ($CaSO_4$), which can be removed from the exhaust in the fabric filter baghouse. In addition to this boiler design, the boiler would be equipped with a hydrated ash reinjection system that would take a portion of the limestone and ash collected in the fabric filter baghouse, hydrate it, and re-introduce it into the exhaust in a reaction vessel upstream of the fabric filter baghouse. Hydrated ash reinjection is a type of dry flue gas desulfurization (FGD) system that allows for additional conversion of SO_2 to $CaSO_4$. Overall the use of limestone injection with hydrated ash reinjection would control 97 percent of the SO_2 emissions that would result from an uncontrolled boiler firing low-sulfur coal. (USDA-RUS and MDEQ, 2006, Ch. 4, p. 33)

Text 16 (Marked Text)
[Air quality] SO_2 **emissions** in the boiler result from the sulfur `present as an impurity` in the coal that is *fired*. The CFB boiler primarily would *fire* low-sulfur, sub-bituminous coal `from the Powder River Basin`. This coal `varies in sulfur content, but is expected to typically have sulfur contents below one percent by weight. The design of the` CFB boiler employs the *firing* of crushed coal mixed with limestone *injected* into the combustor. The use of limestone provides control of SO_2 by **reacting** with SO_2 to form calcium sulfate ($CaSO_4$), which can be *removed* from the exhaust in the fabric filter baghouse. `In addition to this boiler design, the` boiler would be equipped with a hydrated ash reinjection system that would take a portion of the limestone and ash collected in the fabric filter baghouse, *hydrate* it, and *re-introduce* it into the exhaust in a reaction vessel upstream of the fabric filter baghouse. Hydrated ash *reinjection* is a type of dry flue gas desulfurization (FGD) system that allows for additional *conversion* of SO_2 to $CaSO_4$. Overall the use of limestone *injection* with hydrated ash *reinjection* would **control** 97 percent of the SO_2 emissions that would **result** from an uncontrolled boiler *firing* low-sulfur coal.

Observations, doubts, and resolutions

- Uncertainty in the process: what are the elements (e.g. boiler, combustor, reaction vessel) that participate in the process described here? Where do injection and re-injection take place? Are references to illustrations (e.g. figures) necessary?
- Further questions: are there any other solid or gaseous byproducts (e.g. from burning limestone, potentially with impurities) that are not mentioned here? Where does the solid $CaSO_4$ end up?

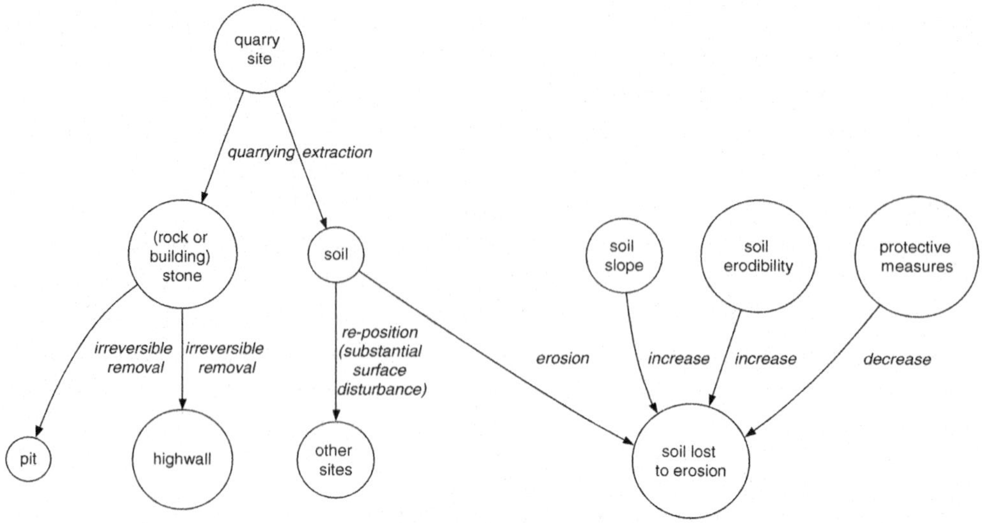

Figure 10. Causal digraph for the MDEQ (2006) case sample text.

Plum Creek Timberlands Quarry Operations, Montana, USA

Text 17 (Original Text)
[Geology and soil quality, stability and moisture] Are soils present which are fragile, erosive, susceptible to compaction, or unstable? Are there unusual or unstable geologic features? Are there special reclamation considerations? Removal of rock or building stone would irreversibly remove the material from the site. A pit and/or highwall may result from quarrying. Soils would be salvaged and replaced at sites proposed for substantial surface disturbance. Additional protective measures would be required on steep slopes and erodible soils to minimize erosion. (MDEQ, 2006, Attachment 1: Supplemental Programmatic Environmental Assessment, p. 9)

Text 18 (Marked Text)
[Geology and soil quality, stability and moisture] Are soils present which are fragile, erosive, susceptible to compaction, or unstable? Are there unusual or unstable geologic features? Are there special reclamation considerations? *Removal* of rock or building stone would **irreversibly remove** the material from the site. A pit and/or highwall may result from *quarrying*. Soils would be *salvaged* and *replaced* at sites proposed for **substantial surface disturbance**. Additional protective measures would be required on steep slopes and erodible soils to **minimize** erosion.

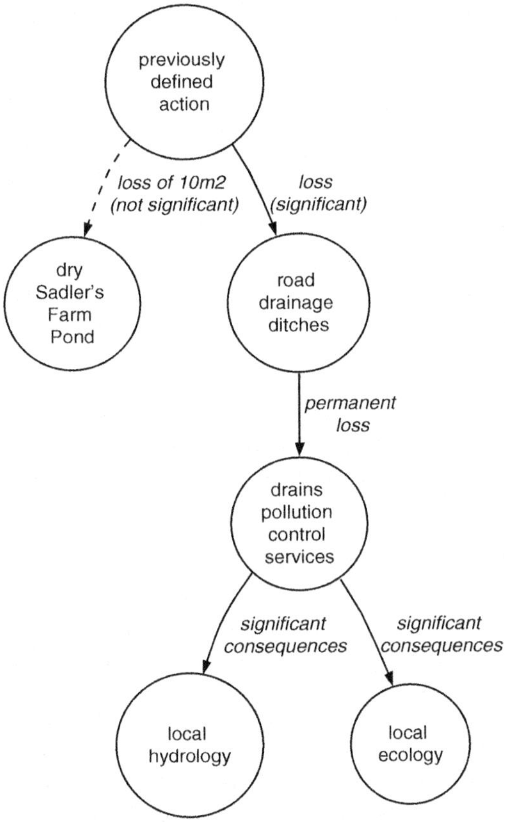

Figure 11. Causal digraph for the MP (2007) case sample text.

Observations, doubts, and resolutions

- Uncertainty about the protective measures: where (in which parts of the system) do they act?

A13/A130 Sadlers Farm Junction Improvement, Essex, UK

Text 19 (Original Text)

[Standing water, direct impacts] Direct impacts are associated with the loss of road drainage ditches and approximately 10 m^2 of the dry Sadler's Farm Pond. The loss of part of the dry Sadler's Farm Pond is deemed not significant, although the loss of the drainage ditches could have significant consequences on the surrounding local hydrology and ecology due to the permanent loss of the drains' pollution control services. In addition, this habitat type on site is known to support protected species such as common reptiles (species are dealt with in later sections below). Consequently, in general and in the absence of mitigation, the loss of drainage ditches is assessed as being significant at no more than the local level. (MP, 2007, p. 123, Para. 8.7.3.4)

Text 20 (Marked Text)
[Standing water, direct impacts] Direct **impacts** are associated with the **loss** of road drainage ditches and approximately 10 m^2 of the dry Sadler's Farm Pond. The **loss** of part of the dry Sadler's Farm Pond is deemed **not significant**, although the **loss** of the drainage ditches could have **significant consequences** on the surrounding local hydrology and ecology due to the permanent loss of the drains' pollution control services. In addition, this habitat type on site is known to support protected species such as common reptiles (species are dealt with in later sections below). Consequently, in general and in the absence of mitigation, the loss of drainage ditches is assessed as being significant at no more than the local level.

Observations, doubts, and resolutions

- Uncertainty about expressions: e.g. hydrology, ecology
- Uncertainty about project action: what causes the two main losses referred to in this paragraph. Is a reference to a project action necessary?

Discussion

Faults and resolutions

The faults regarding causality, observed in the 10 case studies of this article, were either uncertainty in the information provided in the text, or total absence of information (Table 3). Either of these faults is an impediment to understanding, as proved in the creation of the diagrams.

Table 3. Causality faults observed in the 10 case studies. As seen in Section 3, all cases had faults.

Semantics class	Fault	Case studies
Elements	Uncertainty or absence of information	6
Impacts	Uncertainly or absence of information	5
Action	Uncertainty or absence of information	5
Causality	Uncertainly or absence of information	4

The transcription of the information of the text into a diagrammatic form requires knowledge without gaps of what participates (elements), what the actors do (action), what happens after (impacts/ effects), and in which sequence the action takes place (cause-effect relationships). In order to proceed with the diagrams, we had to give some resolutions.

Our selection of a single paragraph from each EIS was based on the premise that EIS authors observe the basic rule of "correct writing" that paragraphs are units with self-sufficient meaning. However, as paragraphs may have dependencies in terms of meaning, our first port of call to resolve our doubts was to consult the adjacent paragraphs and verify whether they were crucial to understanding our focus paragraph. Besides this internal check, sometimes we had to consult external sources such as textbooks to enrich or verify our knowledge about specific processes, such as chemical reactions. Section 3 shows many instances of adding extra information to the diagrams, ranging from expressions of our doubts (e.g. "impact?") to completing the missing information (e.g. new system elements), either based on further reading or as our own assumptions.

Technical issues

Semantics analysis proved to be a limited diagnostic tool. It can capture gross errors, such as the total absence of action, or the generic description of impacts (e.g. as "impacts"). The utility of the semantics analysis, at least here in this article, was mostly to resolve simple doubts in the text such as "which elements exist here?", and thus help in the creation of diagrams — which could be valuable to people new to diagramming. Semantics analysis may proceed and conclude to categorise all the text in causality classes, even when this information contains gaps and omissions, without giving clear warnings. Thus, semantics analysis is not advanced enough as a causality analysis technique, as it fails to highlight the more subtle aspects of information quality in the text. Diagrams, on the other hand, are more sensitive to the quality of information and give clear warnings, typically bringing the diagramming process to a halt.

In the work carried out for this article, we noted that when adequate causality information was present in the text, then diagramming was fast, easy, and added a new visual perspective to the information: the flow of action and the changes induced. When the text presented uncertainty or information gaps, then diagramming needed much more work and a higher degree of intervention (with all the risks of error).

The diagrams created for the purposes of this research were created with a hierarchical orientation, typically from action to outcomes. In the beginning of the diagram (typically at the top, or on the left side) is the rank containing the proposed action: this can be considered as the "beginning" of the system to be depicted. In the end of the diagram (at the bottom or on the right side) there are the final receiving elements, usually associated with the objectives of the development proposal. The value of this hierarchy should be particularly noted and appreciated in the project preparation and revision phases.

Some of the diagrams of Section 3 occupy almost a whole page, even originating from a text of a single paragraph. Thus, for the full extent of impact analysis, the complete diagrams may attain overwhelming dimensions. Hence, with the prospect of diagramming, the issue of focusing on the significant impacts (or information, in general) during impact analysis becomes important. Even if the diagrams are not included in the final version of the EIS/ES (e.g. due to large dimensions), their construction provides a good exercise for focusing and refining the impact analysis text.

Implications for EIA

The two analytic methods revealed causality faults in all 10 case studies of this article (see Section 3 for details and Table 3 for a summary). If we were to judge relatively, then these findings are not alarming — containing causality faults seems "normal". But if we were to judge against the objectives of transparency or integration (IAIA and IEA, 1999), or even to push EIA practice to its best performance, then the situation appears sub-optimal.

The examined EISs/ESs did not provide enough information to explain or indicate why causality is presented in the observed way, but we consider three possible scenarios[1] and discuss their respective resolutions: (a) in case the authors themselves do not fully understand the causality, then they are not ready to write the statement; resolutions include more study at the cost of extra time investment, or pass the case to another team that is ready for the case; (b) in case the authors lack communication skills or resources, they should clearly seek those before committing themselves to producing the statement; resolutions include skill development of the team members by professional courses or self-study, and obtaining the missing resources such as time; and (c) in case the authors do not find it necessary to communicate causality, then they may consider the following reflections.

It is often referred among practitioners as the purpose of EIA to "inform the decision makers". Apparently, the decision-makers only want to know what the

[1] Thanks to the suggestions of the reviewer of a previous version of this article.

effects of the proposed action are likely to be (or, simply, what will happen), and do not ask for how these effects are likely to arise. Logically, if decision-makers do not ask for the extra information, they should not receive it. But if EIA were to merely consider outcomes without considering causal mechanisms, then EIA would descend to the level of older types of practice used to inform decision-makers in other contexts — e.g. consulting oracles. And if EIA undoubtedly informs the decision-makers merely with "what", then the EIA professionals will also adapt to thinking superficially — i.e. only in terms of "what" and not "how" or "why". Although it is easier, faster, and cheaper to think of and write merely "what", going no further would turn the EISs/ESs non-transparent and non-replicable, so it would render the EIA process non-scientific. In the light of this, the purpose of EIA may be slightly modified, to "produce the information with which to inform the decision makers", which has the right to question the quality of the information (product) and of the way it was produced (process).

EIA practitioners may claim that causality must be communicated only when necessary, because that comes at an opportunity cost of doing or not doing something else. Besides being unwise to optimise by eliminating the essential parts — and causality is essential when studying the "effects" of the causes — thinking of and communicating causality is difficult and needs practice before it is "really necessary". These difficulties may be scientific (what sources inform us about this causal relationship?), conceptual (e.g. can this causal relationship be proven, or does it have to be assumed?), or technical (e.g. which kind of diagram is better for my needs?). It is unlikely to communicate causality effectively when necessary, if the EIS/ES authors have not practised enough.

Future work

Of the two analytic methods used in this article to determine the clarity of expression of causality in EISs/ESs, semantics analysis proved to be not useful in the production of such documents. Its value as a diagnostic tool can be used only when necessary for the resolution of doubts. On the other hand, causal diagramming is a viable production technique, which can complement (if not substitute some of) the text with significant advantages such as the flow of action and hierarchy. Diagramming appears to be a field for further experimentation in the conception and communication of causality in EIA, with a possible extension to all impact assessment.

This article discovered the causality faults in several EISs/ESs, and this is what it reports. As a next step for further work, we set to discover the causes of these causality faults. This could be a collaborative work, inviting the authors of EISs/ESs to report on their causal reasoning and its expression, the reviewers to report on methods they use to discover the causal reasoning in the EISs/ESs, and the decision-makers to report on how receiving information on causality could change their work.

Conclusion

Impact analysis text samples from 10 US and UK case studies, processed with semantics analysis and transcription to diagrams, indicated that the expression of causality is often not very clear, and thus difficult to understand. The general faults observed relate to the quality of the transmitted information (i.e. uncertainty or total absence), while their causes remain to be investigated. Authors, reviewers, and even decision-makers may need to re-consider their commitment to causal thinking, in a community-wide call for exchanging ideas from the various points of view.

References

Canter, LW (1996). *Environmental Impact Assessment.* New York: McGraw-Hill International Editions.

European Commission. Regulation (EC) No 761/2001 of the European Parliament and of the Council, of 19 March 2001, allowing voluntary participation by organisations in a Community eco-management and audit scheme (EMAS). *Official Journal of the European Communities*, LI 14, 24.4.2001.

Gentil, S and J Montmain (2004). Hierarchical representation of complex systems for supporting human decision making. *Advanced Engineering Informatics*, 18, 143–159.

Glasson J, R Therivel and A Chadwick (2005). *Introduction to Environmental Impact Assessment.* 3rd ed. London: Routledge.

Harary, F (1969). *Graph Theory.* Boston, MA: Addison-Wesley.

IAIA and IEA (1999). *Principles of Environmental Impact Assessment Best Practice.* International Association for Impact Assessment in cooperation with the Institute of Environmental Assessment, UK.

Perdicoúlis, A and J Glasson (2006). Causal networks in EIA. *Environmental Impact Assessment Review*, 26, 553–569.

Perdicoúlis, A and J Glasson (2007). Action-to-outcome maps in impact assessment. *The Systems Thinker*, 18(10), 7–10.

Perdicoúlis, A and J Piper (2008). Network and system diagrams revisited: satisfying CEA requirements for causality analysis. *Environmental Impact Assessment Review*, 28, 455–468.

Rosen, KH (2003). *Discrete Mathematics and Its Applications.* 5th ed. Boston, MA: McGraw-Hill.

Sterman, JD (2000). *Business Dynamics.* Boston, MA: Irwin-McGraw-Hill.

Therivel, R (2004). *Strategic Environmental Assessment in Action*. London: Earthscan.
Wathern, P (1988). *Environmental Impact Assessment — Theory and Practice*. London: Routledge.

Further Reading: Impact Statements

DCC, TC and PB (2004). A380 Kingskerswell Bypass Environmental Statement. Devon County Council, Torbay Council, and Parsons Brinckerhoff.

ERM (2005). Crossrail Environmental Statement. Environmental Resources Management.

MDEQ (2006). Draft Environmental Impact Statement, Operating Permit Application for Decorative Rock, Building Stone, Rip-Rap and Crushed Aggregate in N. W. Montana. Montana Department of Environmental Quality.

MDEQ and USBLM (2004). Draft Supplemental Environmental Impact Statement: Golden Sunlight Mine Pit Reclamation. Montana Department of Environmental Quality and U.S. Bureau of Land Management.

MP (2007). A13/A130 Sadler's Farm Junction Improvement, Environmental Statement. Mouchel Parkman.

PAAC and PENNDOT (2005). Pennsylvania High Speed MAGLEV Project, Draft Environmental Impact Statement. Port Authority of Allegheny County and Pennsylvania Department of Transportation.

RPSE and DP (2005). Environmental Impact Assessment for Offshore Drilling the Falkland Islands to Desire Petroleum PLC. RPS Energy and Desire Petroleum PLC.

US ACE (2005). Final Environmental Impact Statement for a Proposed Water Treatment Residuals Management Process for the Washington Aqueduct, Washington, D.C. US Army Corps of Engineers, Baltimore District.

USDA (2006). German Ridge Restoration Project, Final Environmental Impact Statement. US Department of Agriculture, Forest Service, Region 9, Hoosier National Forest.

USDA-RUS and MDEQ (2006). Draft Environmental Impact Statement, Highwood Generating Station. US Department of Agriculture, Rural Utilities Service, and Montana Department of Environmental Quality.

CHAPTER 8

The Effectiveness of EIA as an Instrument for Environmental Governance: Reflecting on 25 Years of EIA Practice in the Netherlands and the UK

Jos Arts[*], Hens A. C. Runhaar[†], Thomas B. Fischer[‡,§], Urmila Jha-Thakur[‡], Frank Van Laerhoven[†], Peter P. J. Driessen[†] and Vincent Onyango[‡]

[*]*Spatial Planning and Environment*
University of Groningen, Groningen, The Netherlands

[†]*Copernicus Institute of Sustainable Development*
Utrecht University, Utrecht, The Netherlands

[‡]*School of Environmental Sciences*
University of Liverpool, Liverpool, UK
[§]*fischer@liverpool.ac.uk*

The European Union (EU) Environmental Impact Assessment (EIA) Directive which was introduced some 25 years ago has had a major impact on decision-making practices in EU member states. In the professional literature, this impact has mostly been discussed under the heading of "effectiveness", with an emphasis being given in particular to procedural elements. The extent to which EIA has contributed to objectives, such as raising environmental awareness and leading to an incorporation of environmental values in decision-making has only been rarely investigated. This paper reflects on these latter two aspects of EIA effectiveness in two EU member states: the Netherlands and the United Kingdom. Empirical evidence was compiled mainly on the basis of a comprehensive literature review and online surveys with EIA professionals in both countries. Our results indicate that overall the instrument is about equally effective in both countries with regards to the incorporation of environmental concerns in decision-making. As both countries have different governance mechanisms, further research is needed on why perceived effectiveness does not differ more.

[§] Corresponding author.

Keywords: Environmental Impact Assessment; governance; effectiveness; context; the Netherlands; the United Kingdom.

Introduction

Environmental Impact Assessment (EIA) was introduced more than 40 years ago based on the US National Environmental Policy Act (NEPA). Subsequently, it has had a remarkable career and is now being used worldwide, formally and informally. In the European Union (EU), some 25 years ago the Directive "on the assessment of the effects of certain public and private projects on the environment" (337/85/EEC) was adopted by the Council of Ministers after more than a decade of discussion (see Wood, 2003; Pokorný-Versteeg, 2003). Member states (MSs) then had three years to transpose it into national legislation (Fischer, 2010a). In this context, there was much freedom to adapt EIA to the specific needs of the different MS. As a consequence, there are considerable differences between EIA systems throughout the EU regarding various elements, including, for example, the development of alternatives, quality control and follow-up.

Several countries, including e.g. France, Ireland and Belgium already had some legal requirements for EIA in place by 1985. Other countries were discussing EIA regulation at the time, including the Netherlands, where the focus on that debate was on the Canadian EA system. As a consequence, the resulting Dutch EIA system was modelled much more to this rather advanced system than on the more basic requirements of the EU Directive. In a third group of countries, including the United Kingdom (UK), Germany and Denmark, EIA was established primarily in response to the EU Directive, albeit many elements of the Directive had been present already (see e.g. Hanusch and Fischer, 2011). Since the mid-1980s, much experience has been gained with EIA in practice, and regulations have changed with respect to coverage and procedural elements, in particular through directives 97/11/EC and 2003/35/EC, as well as through Directive 42/EC/2001 (i.e. the Strategic Environmental Assessment (SEA) – SEA – Directive).

Governance and effectiveness

Following Wood (2003), the European EIA-Directive was the first EU intrusion into the planning domain. It is an instrument which contributes to environmental awareness and to environmental protection by requiring the generation of ex ante information of environmental impacts of public and private initiatives. The Directive has had major repercussions on MS decision-making practices. EIA can be considered a governance instrument, as it introduces rules and assigns particular roles and responsibilities to actors. EIA aims to steer actors' behaviours towards greater environmental awareness, leading to the incorporation of environmental values in proposed activities and plans. When EIA was introduced, it was seen as an innovation in environmental governance (see Caldwell, 1982; Taylor, 1984; Sadler, 2004; Richardson and Cashmore, 2011; Kidd and Fischer, 2007). This is connected with various governance mechanisms embedded in EIA, including the formal requirement to provide for environmental information for development alternatives prior to final decision-making, the responsibility of the proponent to prepare an EIA report, formal public participation and the requirement to do follow-up. Various systems also have provisions for conducting strategic level environmental assessment. The Netherlands is one of the few countries that have established an independent expert commission charged with quality review.

The effectiveness of EIA in terms of the extent to which it is actually meeting its objectives is frequently discussed in the professional literature. However, the main focus of empirical research has been on the "procedural" effectiveness of EIA, i.e. the extent to which formal EIA procedures have been followed (Cashmore *et al.*, 2004; Van Doren *et al.*, 2013). Few research projects have focused on the "substantive" effectiveness of EIA. Two commonly recognised indicators of effectiveness are whether environmental awareness is raised through EIA and whether it leads to an incorporation of environmental values into decision-making (Runhaar and Driessen, 2007; Runhaar *et al.*, 2013). The general perception is that in most Western countries, EIA has contributed to greater environmental awareness, as well as to an enhanced environmental protection, albeit generally in a modest way. Initiatives are rarely fundamentally revised (Bassi *et al.*, 2012; Cherp, 2001; Fiddler and Noble, 2012; Gazzola *et al.*, 2011; Fischer, 2002, 2009; Jay *et al.*, 2007; Sadler, 2004; Wood, 2003). It has also been suggested that EIA effectiveness is context-specific. At the project level, effectiveness of EIA is claimed to depend to a large extent on the actors involved, their interests and power positions, and the extent to which the most powerful decision-makers are open to environmental values and to revising their original plans (Fischer and Gazzola, 2006; Hilding-Rydevik and Bjarnadóttir, 2007; Kørnøv and Thissen, 2000; Runhaar and Driessen, 2007). At the system level, EIA effectiveness depends on characteristics of EIA legislation, for example the presence of "follow up" requirements (Sadler, 2004; Wood, 2003). Although various international comparative reviews of EIA systems have been undertaken, we find surprisingly little evidence for how effective EIA systems in different countries are. Furthermore, relevant references often date back to the 1990s (Barker and Wood, 1999; Wood, 1995; Sadler, 1996; EC, 1993).

Aim of paper and research questions

In this paper, a comparative review of the effectiveness of EIA in the UK and the Netherlands is provided. The main reason for the choice of these two countries is that their EIA systems are about the same age and have the same legal basis (both are based on the European Union Directive 85/337/EEC). However, EIA is implemented in different ways, due in parts to different planning traditions, including consensus-oriented Rhinelandic Dutch versus polarized Anglo-Saxon UK administrative cultures (De Jong and Geerlings, 2005). With this paper, we hope to add to the understanding of the importance of context for EIA effectiveness.

Subsequently, the following questions are addressed:

- What EIA governance mechanisms have been identified in the professional literature and what contextual factors are thought to influence the effectiveness of these mechanisms in terms of raising environmental awareness and leading to the incorporation of environmental values in decision-making?
- How have governance mechanisms in the UK and Dutch EIA systems changed since the inception of EIA?
- How effective have the UK and Dutch EIA systems been in raising environmental awareness of the actors involved and in incorporating environmental values in decision-making?
- How can similarities and differences be explained in terms of governance mechanisms and contextual factors?
- What adjustments to EIA legislation may be needed in order to ensure EIA effectiveness in the future?

Outline

In the next section we will elaborate on the theoretical framework used to review governance elements and contextual factors. Subsequently, main components of the UK and Dutch EIA systems will be discussed. We will then briefly reflect on the developments over the past 25 years, as well as on the different contexts within which EIA is being applied. In addition, we will look at differences and commonalties of the two systems. We will then discuss the effectiveness of EIA in both countries. Challenges and potential adjustments of the Dutch and UK EIA systems will be identified and in the final section, conclusions will be drawn.

Theoretical Framework: EIA Governance Mechanisms and Contextual Elements

EIA effectiveness can be described in terms of the extent to which it achieves its goals. These include the incorporation of environmental considerations in decision-making and the enhancement of the environmental awareness among

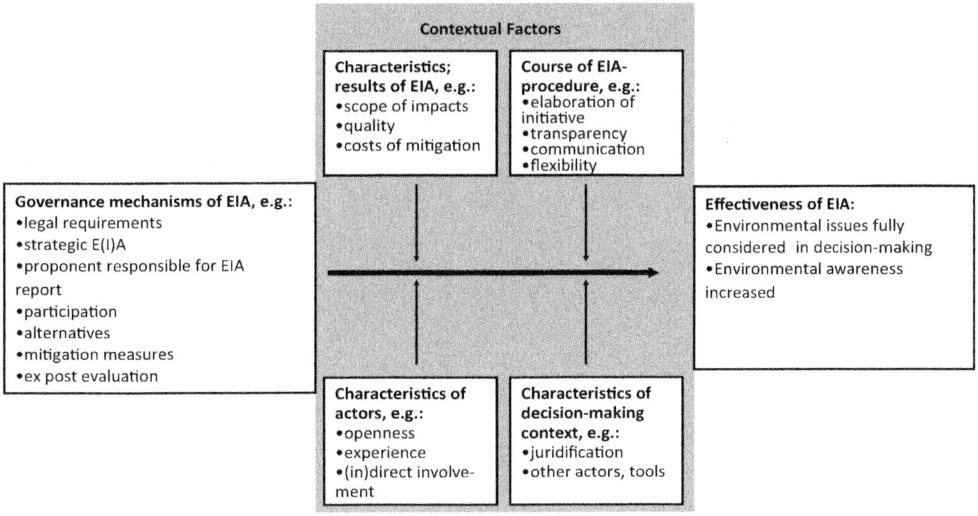

Figure 1. Relationships between governance mechanisms, contextual factors and EIA effectiveness.

proponents and competent authorities (Jesse, 2008; Runhaar and Driessen, 2007; Wood, 2003; Arts, 1998). Effectiveness is connected with existing governance mechanisms and contextual elements. Fig. 1 depicts our conceptual model. In Tables 1 and 2, governance mechanisms and contextual elements identified in the professional literature are specified further.

Governance mechanisms

By assigning particular roles and responsibilities to actors (e.g. to proponents and competent authorities) and by introducing rules (e.g. by formulating requirements to develop alternatives and to justify decisions in the light of an EIA report), EIA aims to steer the behaviour of public and private actors towards a greater environmental awareness and towards the incorporation of environmental values into proposed activities and plans. Table 1 provides for an overview of the various governance mechanisms, the actors being steered, the actors steering, as well as for some related empirical evidence found in the literature for the Netherlands and the UK. It is important to note that the empirical evidence base in the various references provided in Table 1 tends to be small: large-scale evaluations based on large numbers of cases or surveys are scarce (exceptions are Ten Heuvelhof and Nauta, 1996, 1997 and Fischer, 2002).

Table 1. Governance mechanisms in EIA.

Governance mechanism	Who is steered?	In which direction?	By whom?	Empirical evidence? (in the Netherlands: references)	Empirical evidence? (in the United Kingdom: references)
Legal requirements	Proponent, competent authority	Explicit consideration of environmental issues, awareness, "prevention effect", "correction effect", internalisation	Legislator (stakeholders, in the Netherlands the NCEA)	Yes (EC, 2009a,b,c; Scholten and Van Eck, 1994; Ten Heuvelhof and Nauta, 1996)	Yes (EC, 2009a,b,c; Wood, 2003, Pokorný-Versteeg, 2003; Fischer, 2010a,b)
Strategic environmental assessment	Proponent, competent authority	Consideration of impacts at strategic level, which is framework for subsequent decision-making	Competent authority, in the Netherlands the NCEA	Partly (Runhaar and Driessen, 2007; Fischer, 2002)	Partly (Wood, 2003; Pokorny-Versteeg, 2003; Fischer, 2002, 2007; Therivel and Fischer, 2012)
Proponent responsible for EIA	Proponent	Prevention effect, adaptation/(re)design of alternatives, environmental awareness, internalisation	Competent authority, in the Netherlands the NCEA	Yes (in relation to prevention effect) (Ten Heuvelhof and Nauta, 1997; VROM, 1994)	Yes (Wood, 2003)

Table 1. (*Continued*)

Governance mechanism	Who is steered?	In which direction?	By whom?	Empirical evidence? (in the Netherlands: references)	Empirical evidence? (in the United Kingdom: references)
Participation of stakeholders and advisors	Public, in the Netherlands the NCEA	Take into account stakeholder issues, which might enhance support and quality of decisions	Stakeholders (public, in the Netherlands the NCEA, formal advisors)	Yes (however contradictory findings) (Dalal-Clayton and Sadler, 2004; Fischer, 2003; Hildén *et al.*, 2004; Sheate *et al.*, 2003; Scholten and Van Eck, 1994; Ten Heuvelhof and Nauta, 1997; Wassen *et al.*, 2011. NB: Ligtermoet, 2007 finds a negative relationship)	Partly (no formal public consultation at start of procedure) (Wood, 2003; Pokorný-Versteeg, 2003; Glasson *et al.*, 1994; Petts, 1999)
Scoping, advice guidelines for EIA study (until 1 July 2010 in NL)	Proponent, competent authority, consultant	Safeguarding quality of EIA (consideration of all relevant issues and alternatives, go beyond legal requirements)	In the Netherlands the NCEA, government department	Yes (Ten Heuvelhof and Nauta, 1996, 1997; Mostert, 1995)	Partly (Wood, 2003; Fischer and Phylip-Jones, 2008)

Table 1. (*Continued*)

Governance mechanism	Who is steered?	In which direction?	By whom?	Empirical evidence? (in the Netherlands: references)	Empirical evidence? (in the United Kingdom: references)
Requirements to develop alternatives (in NL this included the AMFE until 1 July 2010)	Proponent, competent authority	Reconsideration of initiative, explicit consideration of environmental effects (consideration of possibilities of maximum environmental protection)	Competent authority, in the Netherlands the NCEA	Yes (Based on some 20 in-depth interviews in the Netherlands conducted by the authors in the context of this study)	Partly (Wood, 2003; Pokorný-Versteeg, 2003)
Requirements to develop mitigation measures	Proponent, competent authority	Consideration of environmental protection, correct initiatives	Competent authority, in the Netherlands the NCEA	Yes (Based on some 20 in-depth interviews in the Netherlands conducted by the authors in the context of this study)	Yes (Wood, 2003; Pokorný-Versteeg, 2003)
Quality review	Proponent, competent authority, consultant	Complete and valid information; serious alternatives; legitimacy decision	In the Netherlands the NCEA	Yes (Ten Heuvelhof and Nauta, 1996, 1997; Mostert, 1995; Scholten and Van Eck, 1994)	Yes (Wood, 2003; Pokorný-Versteeg, 2003; Fischer, 2010b)

Table 1. (*Continued*)

Governance mechanism	Who is steered?	In which direction?	By whom?	Empirical evidence? (in the Netherlands: references)	Empirical evidence? (in the United Kingdom: references)
Follow-up (monitoring, evaluation of environmental effects)	Competent authority, proponent	Safeguarding environmental quality after implementation of the activity that has been decided	Competent authority, stakeholders	Yes (though only in some cases done) (Morrison-Saunders and Arts, 2004; Arts, 1998; De Leeuw and Rense, 1991)	Partly (Wood, 2003; Pokorný-Versteeg, 2003; Morrison-Saunders and Arts, 2004; Partidario 2004; Partidario and Fischer, 2004)

NB: NCEA = Netherlands Commission for Environmental Assessment; AMFE = Alternative most friendly to the environment, based on Runhaar *et al.* (2011).

Table 2. Contextual elements relevant for EIA effectiveness as discussed in the international professional literature.

Contextual elements	Mechanisms of influencing	Empirical evidence	References in international literature
(1) Characteristics of EIA results:			
— Scope of impacts	— Bigger (more diverse) scope might enhance acceptance of EIA and decision (though with too big a scope environmental issues might be overlooked)	Yes	Runhaar and Driessen, 2007; Van Kessel *et al.*, 2003; Scholten and Van Eck, 1994; Ten Heuvelhof and Nauta, 1996; Elwell, 2002; Hilden *et al.*, 2004; Fischer, 2002; Glasson *et al.*, 1997; Fischer and Phylip-Jones, 2008; Chadwick, 2002; Wood *et al.*, 2006
— Quality of EIA report	— Higher quality leads to more use of results	Contradictory evidence	
— Costs of mitigation measures	— Higher costs, leads to less use of mitigation measures	Hardly	
(2) Course of EIA procedure:			
— Degree of elaboration of initiative	— The more elaborate, the less scope there is for use of EIA information (relationship with (late) timing of EIA)	Yes	Runhaar and Driessen, 2007; Ten Heuvelhof and Nauta, 1996, 1997; Arts, 1991; Scholten and Van Eck, 1994; Therivel and Minas, 2002; Sheate *et al.*, 2003; Dalal-Clayton and Sadler, 2004; Fischer, 2004; Van Buuren and Nooteboom, 2009; Piper, 2001; Hartley and Wood, 2005
— Transparency of EIA process	— Transparency adds to legitimacy of EIA (how results were produced is clear)	Yes	
— Communication about EIA results	— Good communication meaning results are reaching relevant actors (also enhances acceptance of EIA)	Yes	
— Flexibility of EIA	— Good linkage which decision-making process enhances acceptance and use of EIA (results)	Yes (especially acceptance)	

Table 2. (*Continued*)

Contextual elements	Mechanisms of influencing	Empirical evidence	References in international literature
(3) Characteristics of EIA actors:			
— Openness of proponent and competent authority	— Effectiveness of EIA is better if proponents and authorities are more open to environmental issues	Yes	Bonte, 1996; Partidario, 2000; Elwell, 2002; Dalal-Clayton and Sadler, 2004; Fischer, 2004; Runhaar and Driessen, 2007; Marshall 2001; Glasson *et al.*, 1997; Stoeglehner *et al.*, 2009; Van Doren *et al.*, 2013
— Experience with EIA	— The more experience with EIA, the higher the chances of effectiveness	Yes, indicative	
— (In)directly involved	— The more an actor is directly involved, the more insight they have	Indicative	
(4) Characteristics of the decision-making context:			
— Other actors, tools	— Role of EIA in protecting the environment is taken over by other instruments or actors (such as environmental interest groups)	Indirect (mainly IPPC directive mentioned)	Van Kessel *et al.*, 2003; Arts, 1998; Leroy, 1996 (and in-depth interviews in the Netherlands as part of this study); Weston, 2002; Fischer, 2009.
— Juridification	— Compliance with legal requirements and prevention of juridical procedures are more important than optimising environmental performance	Indirect	

Contextual factors

The effectiveness of EIA is not only determined by the governance mechanisms shown in Table 1, but also by various contextual factors, as discussed by Fischer (2005), Runhaar and Driessen (2007) and Van Doren *et al.* (2013). This relates to the way EIA is applied in practice and includes e.g. the quality of the EIA report (the environmental statement) and the connections of EIA and decision-making processes. Furthermore, it relates to the context in which EIA procedures are implemented, including, for example the extent to which proponents are open to consider environmental issues. We have sub-divided the contextual factors discussed in the professional literature into four categories (Table 2).

Characteristics of the Dutch and UK EIA Systems

In this section, the characteristics of the Dutch and UK EIA systems are outlined. This is followed by a description of differences and similarities between the two systems.

The Dutch EIA system

In the Netherlands, EIA was discussed since the mid-1970s. At the same time, quite a bit of experimenting occurred. Canadian EIA had some great influence on the Dutch system (Arts, 1998; Wood, 2003). Formal regulations were only introduced in 1986 by inclusion of a chapter on EIA in the Environmental Protection (General Provisions) Act (now the Environmental Management Act). The regulations came into force when the EIA Decree was issued in 1987, implementing the European EIA Directive (85/337/EEC).

The formal goal of the Dutch EIA regulations is "to ensure that environmental values are fully considered in decision-making". This objective is in line with the EU EIA Directive. In the Explanatory Memorandum to the EIA regulations (Arts, 1998), two other subsidiary goals of EIA were established by the legislator, including the internalisation of environmental awareness (improving environmental attitudes) and the streamlining of decision-making (coordination and transparency). An increased environmental awareness is also mentioned in the EU EIA Directive (see e.g. Jesse, 2008 and Pokorný-Versteeg, 2003).

The regulations specify initiatives subject to EIA. Next to various specific projects (e.g. application for a licence to build a factory), many strategic initiatives are also identified in the EIA Decree, giving rise to strategic environmental assessment (SEA). The Dutch EIA procedure encompasses the development of alternatives for the initiative, an assessment of environmental impacts of the initiative and its alternatives, and the development of measures to mitigate or compensate for negative impacts. Until July 2010, the development of an "alternative most friendly to the environment" (AMFE) was also mandatory. Proponents of initiatives, which are either governmental agencies or private companies, are responsible for conducting an EIA. The preparation of the environmental statement is usually outsourced to a consultancy. The competent authority decides on how to use the EIA outcomes and has to justify its decision, giving consent to the activity. Proponents may adjust their initiatives voluntarily. Stakeholders have the opportunity to participate in the EIA process during two stages, including the scoping stage (where alternatives and assessment criteria are decided upon) and during the presentation of the assessment outcomes. The independent Netherlands' Commission for Environmental Assessment (the NCEA) is responsible for quality review. The NCEA has its own staff and mobilises specific expertise by involving scientists and other experts for reviewing specific EIA studies. EIA regulations contain a requirement to evaluate the decision subject to EIA while or after it is implemented. This is also called EIA follow-up.

Changes to governance mechanisms in the Netherlands

In the Netherlands, the formal EIA system has been subject to various amendments as a consequence of EU evaluations and regulations changes. In 1994, a system of case-by-case screening was introduced for smaller projects, as the EU noted some imperfections in this area of application in the Netherlands. In 1996, a court decision made it clear that the European EIA Directive prevails over Dutch national regulations. In 2006, the requirements for strategic-level EIA for plans and programmes were amended in order to comply with the SEA Directive (2001/42/EC).[a] In 2010, there was a more fundamental revision of the Dutch EIA system. This was called "modernisation of EIA" and focused on limiting the costs associated with the procedure. These changes mean that there are now fewer requirements that go beyond EU standards. Importantly, AMFE was abolished. Furthermore, more tailor-made procedures were introduced, with a distinction between procedures for plans and complex projects on the one hand and more simple projects on the other. The public is now involved more and competent authorities enjoy more freedom (and responsibility). Yet, despite the various changes made to the Dutch EIA system over the last 25 years, overall governance mechanisms have not been fundamentally altered, apart from the introduction of formal SEA.

[a] Under the Dutch EIA system, many strategic decisions such as (spatial) plans were already subject to EIA (see Sadler and Verheem, 1996; Arts, 1998; Fischer, 2002).

Earlier evaluations of EIA

The Dutch EIA system and practice have been subject to several evaluation studies, including ECW (1990, 1996), Ten Heuvelhof and Nauta (1996), Van Kessel *et al.* (2003) and most recently Runhaar *et al.* (2011). Generally speaking, in the international literature, the Dutch EIA system has been considered as extensive and advanced because of the number of elements that go beyond EU standards (Wathern, 1990; Devuyst, 1994; Glasson *et al.*, 1994; Wood, 1995, 2003; Sadler, 1996; Fischer, 1999; Pokorný-Versteeg, 2003; Morrison-Saunders and Arts, 2004).

Over the past 25 years, some 2500 EIA and SEA studies have been carried out (NCEA, 2011, Runhaar *et al.*, 2011). After 25 years, the formal policy discourse on EIA seems to have changed. Many see "old-style" EIA as an instrument which is delaying decision-making, raising administrative costs, lacking all too often quality and adding little value to decision-making. This criticism was instrumental in the revision of the EIA regulations of 2010 which focused on meeting "just" the basic quality standards set by the EU EIA Directive. As a consequence, the advanced and comprehensive nature of the Dutch EIA system has recently started to be called into question. Increasingly, there has been simplification of regulations and reduction of safeguards. Other countries, on the other hand, have improved their systems.

The UK EIA system

In the UK, interest in EIA started in the 1970s and was connected with oil drilling activities in Scotland. It is worth noting that the concept of EIA in the UK was first introduced by practitioners (Glasson and Bellanger, 2003). In the 1970s and early 1980s, much work was done on the methodological aspects of EIA (Clark *et al.*, 1976). However, central government was hesitant to commit to it (Wood, 2003). As the European EIA Directive started to emerge, discussion in the UK on how to implement it also commenced. Just before the deadline for implementation, EIA was introduced through inclusion in the Town and Country Planning Regulations for England and Wales and in specific Environmental Assessment Regulations for Scotland and Northern Ireland. Hence, unlike the Netherlands where the resultant regulations came under the broad remit of nature conservation, in the UK, they were placed under the planning system (Glasson and Bellanger, 2003). As the regulations do not cover all types of activities mentioned in the European EIA regulations, specific arrangements were made for those activities not covered by the Town and Country Planning Regulations, including e.g. highways, land drainage, electricity, pipeline and forestry projects (see Glasson *et al.*, 1997; Marr, 1997; Wood, 2003; Pokorný-Versteeg, 2003).

As a consequence, there is not one regulation to implement the European EIA Directive in the UK, as there is in the Netherlands. In practice, various governmental circulars and guidelines for EIA are important next to the formal regulations. However, the Town and Country Planning Regulations are the most important piece of legislation under which 70% of the EIAs are undertaken in the UK (Glasson and Bellanger, 2003; Glasson, 1999). Over the past 25 years, more than 10,000 EIAs of developmental projects have been carried out and since 2000 around 600 EIAs each year have been undertaken in the UK (IEMA, 2011).[b]

As far as the EIA process in the UK is concerned, the environmental statement (ES) is usually prepared by or on behalf of the proponent and is submitted alongside the planning application to the relevant Local Planning Authority (LPA). LPAs can be consulted by the proponents for both, screening and scoping, while LPAs can consult statutory bodies, other experts and the public on the ES (RTPI, 2001). In contrast to the Dutch EIA system, the UK system does not contain obligations to assess those alternatives considered for a proposed project. In 2011, a voluntary scheme to observe the quality of EIA was introduced by the Institute of Environmental Assessment and Management (IEMA), equating to a voluntary type of accreditation of consultancies involved in conducting EIAs. This is based on conducting quality checks across EIA activities of an organisation (IEMA, 2011) and is called "EIA quality mark" (http://www.iema.net/eia-quality-mark).

Changes in governance mechanisms in the UK

Since 1988, various changes have been made to the EIA regulations, in particular in order to deal with the amendments of the EU Directive in 1997, which in 1999 led to an "almost to the letter implementation" (Wood, 2003, p. 56) in the Town and Country Planning (Environmental Impact Assessment England and Wales) Regulations 1999 and also to new regulations in Northern Ireland and Scotland. In 2004, an important amendment of the regulations was made in order to implement the EU SEA Directive by inclusion of an umbrella regulation, related to an integrated process for land-use plans (Fischer, 2007; Sadler *et al.*, 2011). Separate regulations apply to England, Scotland, Northern Ireland and Wales. More recently, in 2011, new EIA regulations were introduced in Scotland and later in England and Wales to replace SI293/1999 (as amended). The new set of regulations is seen to consolidate and update the EIA regime.

[b] In addition, over the past eight years, well over a thousand SEAs are likely to have been conducted in England alone (Therivel and Fischer, 2012).

Earlier evaluations of EIA

The UK EIA system and related practice has been subject to various evaluations, for example by Glasson (1999) and Weston (2002). Furthermore, international comparative studies like those by Marr (1997), Barker and Wood (1999), Sadler (1996), Fischer (2002), Wood (2003), Jha-Thakur *et al.* (2009) have shed further light on the UK EIA system and practice (see references in Tables 1 and 2). More recently, in anticipation of a quarter century of EIA in the UK, studies by IEMA (2011) and the University of Liverpool (Jha-Thakur *et al.*, 2012) have been prepared. While the former looks at the current state of EIA practice and its future in the UK, the latter also explores the role of EIA in encompassing health considerations.

EIA practice has developed considerably since it was formally introduced in the UK in 1988. EIA within the UK has frequently been criticised with respect to its somewhat unregulated scoping practice, the lack of consideration of cumulative impacts, as well as the lack of assessment of alternatives and somewhat complex public participation processes (Glasson, 1999; Wood *et al.*, 2006; Weston, 2002; see also Table 1). The existence of more than 20 EIA regulations and numerous amendments imply that the EIA system in the UK is more complex than its Dutch counterpart. The recently introduced 2011 regulations are expected to simplify the process by consolidating the various amendments. Hence, it is unlikely that EIA regulations within the UK will face any further profound changes before 2017 (IEMA, 2011), when the new EIA Directive will have to be implemented.

Differences and commonalties between the Dutch and UK system

Commonalities between the Dutch and UK systems stem from the fact that both have to comply with the same EU directives for EIA and SEA. Differences between the UK and the Netherlands relate to factors such as different legal traditions (common law versus Roman/Napoleonic law), administrative cultures (Anglo-Saxon versus Rhinelandic: polarised, conflict-oriented versus more consensus-oriented) and traditions of fragmented sectoral versus coordinated (integrated) planning. For more in-depth discussions, see e.g. De Jong (1999), De Jong and Geerlings (2005), Pokorný-Versteeg (2003) and Fischer (2001, 2002). In his international comparative study, Wood (2003) concluded that the Netherlands and the UK both scored well on issues such as legal basis, coverage, screening, and mitigation. The Netherlands also did well with regards to the development of alternatives, scoping, EIA report quality, review and consultation. There is room for improvement regarding the influence on decision-making and impact monitoring in both the Netherlands and the UK.

Expectations with regards to the effectiveness of EIA in the Netherlands and the UK

Our assumption is that differences in EIA effectiveness in the Netherlands and the UK may be attributed to differences in governance mechanisms (see Table 1). Furthermore, contextual differences, which may directly or (through differences in institutional contexts) indirectly be expressed in terms of the elements outlined in Table 2, should also play a role. This leads us to the following expectations regarding EIA effectiveness in the Netherlands and the UK:

- In the UK, there are fewer provisions for participation and consultation of stakeholders and advisors than in the Netherlands, which should result in a reduced effectiveness of UK EIAs.
- In the UK, the scoping stage is regulated less than in the Netherlands,[c] which should result in a reduced effectiveness of UK EIAs.
- In the UK, there is no formal requirement to assess alternatives, in contrast to the Netherlands: the effectiveness of EIA should therefore be reduced in the UK compared with the Netherlands with respect to this aspect.
- Quality reviews of EIAs are better institutionalised in the Netherlands than in the UK: EIA reports should therefore be of a higher quality in the Netherlands than in the UK.
- In the Netherlands' EIA regulations, there is a requirement for follow-up in contrast to the UK: EIA effectiveness in the Netherlands with regards to this aspect should therefore be greater than in the UK.
- In the UK, EIA is more directly integrated into main decision-making procedures than in the Netherlands[d]: it can therefore be expected that the (perceived) importance of the linkage of EIA to the dynamics of decision-making is greater in the Netherlands than in the UK.

And, with regards to the institutional context:

- We expect differences in the degree of controversy amongst respondents between the UK and the Netherlands. This is connected with the very different

[c] At least this was the case before the 2010 revisions of the Dutch EIA regulations. Before 2010, the Dutch EIA system included several formal governance mechanisms regarding scoping, such as a requirement to provide for a notification of intent by the proponent, a formal public review and an advice by the Dutch EIA Commission on this, and guidelines for the EIA report by the competent authority.

[d] Apart from decision-making regarding national transport infrastructure in the Netherlands, in which EIA is integrated into the decision-making procedures by the combined EIA/route decision procedure (see e.g., Arts and Van Lamoen, 2005).

administrative cultures, i.e. in the UK more polarised, conflict-oriented versus more consensus-oriented in the Netherlands.

Effectiveness of EIA in the Netherlands and the UK

In order to review EIA effectiveness in the Netherlands and the UK, various methods have been used. The two main methods include a literature review (which also comprises documentation on opinions of experts) and online surveys of EIA actors.[e] Furthermore, in the Netherlands some additional interviews with EIA experts were conducted and in the UK, a recent report on the subject by IEMA (2011) was consulted. In the surveys, EIA actors were asked about their perceptions of EIA effectiveness and the factors that in their view contributed to EIA effectiveness. Both indicators for effectiveness and explanatory factors were derived from the professional literature (see Section 2). The online surveys held in the Netherlands and the UK had a similar set-up and were covering similar questions. In the Netherlands, the survey was conducted between October and November 2010 and yielded 443 responses. Although it is difficult to comment on the make-up of the EIA professions covered, it appears that some groups were underrepresented, including private proponents, political representatives of competent authorities and lawyers/judges. As the share of EIAs carried out in different sectors may change considerably over the years (see Table 3), it is difficult, though, to be more assertive at this point. In the UK, a similar survey was carried out in spring 2011 and yielded 181 responses. We assume that this lower response rate may be due to "survey fatigue". Recently another survey on EIA had been conducted in the UK (IEMA, 2011). However, the results of this survey have also been processed. In the UK survey, consultants, researchers and scientists in particular appear to be overrepresented, with lawyers being underrepresented.

[e] In order to obtain additional information for interpreting Dutch results, some 20 semi-structured interviews were conducted (of which 17 after the survey). In this context, persons with an understanding of various policy sectors which represent a wide variety of initiatives and represent more than two-thirds of all EIAs conducted in the Netherlands in 2010 — including road infrastructure planning, planning of housing and industrial areas, industrial and energy generation licensing, and water planning and management — were interviewed. All interviewees were asked to identify strengths, weaknesses, opportunities and threats for EIA, how EIA effectiveness could be further enhanced and what other improvements are necessary. Five interviews were conducted with representatives of private proponents and six with (advisors of) political representatives, which partly compensates for the low response among this group of EIA professionals in the survey. An additional source for interpreting UK results was IEMA's 2011 comprehensive study on "the state of EIA practice in the UK".

Table 3. Survey participants.

	Survey in the Netherlands (% response)	Survey in the United Kingdom (% response)
EIA professional (main experience gained in this role)		
Proponent (public agency)	12.9%	16.0%
Proponent (private company)	2.0%	7.2%
Competent authority (politician)	0%	See Competent authority (admin)
Competent authority (administration)	16.0%	5.0%
Expert (administrative)	19.6%	9.9%
Consultant	15.1%	34.3%
Scientist	3.2%	13.8%
EIA Commission	21.0%	NA
Interest group	1.8%	4.5%
Lawyer/judge	1.1%	0.6%
Otherwise	5.6%	8.8%
	443	181
Sector (where main experience is gained)		
Infrastructure (road, waterway, railways)	38.4%	29.1%
Water management	11.5%	13.1%
Housing, business sites, urban extensions	13.3%	10.3%
Industry	5.2%	18.2%
Waste management	1.8%	1.7%
Mining	2.7%	5.7%
Energy	2.7%	4.6%
Farming/forestry	4.5%	4.0%
Recreation	1.1%	1.7%
Otherwise	16.5%	11.4%
Total	443	181

The survey was semi-structured, also allowing respondents to raise relevant issues and post remarks themselves. Our assessment of EIA effectiveness is thus based on perceptions of EIA professionals, rather than on the evaluation of EIA projects through case studies. It is important to be aware that such perceptions may be skewed due to: (a) specific interests (e.g. consultants earning money preparing an EIA study might be more positive about EIA than proponents paying for it); and (b) experiences with EIA (both in terms of the number of years and in terms of the number of EIAs produced). In order to deal with these problems, we asked respondents to indicate their primary role in EIA, their experiences and the main policy sector in which they had been working. Correlation analysis of perceived EIA effectiveness and these background characteristics served as a means to explore "biases".

Findings for the two countries

In the survey, the respondents from the Netherlands indicated that EIA contributed to the environmental awareness of both authorities and proponents (see Figs. 2(a)/2(b)). There is, however, no overall consensus. Respondents representing competent authorities perceive this effect to be significantly lower than other respondents. The UK

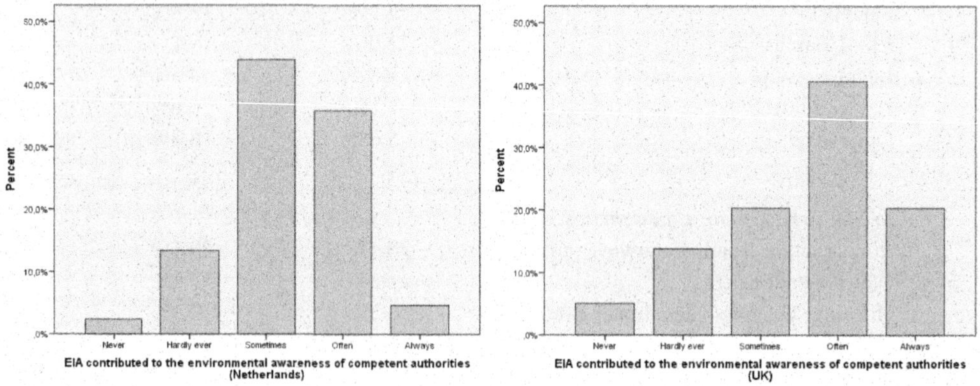

(a)

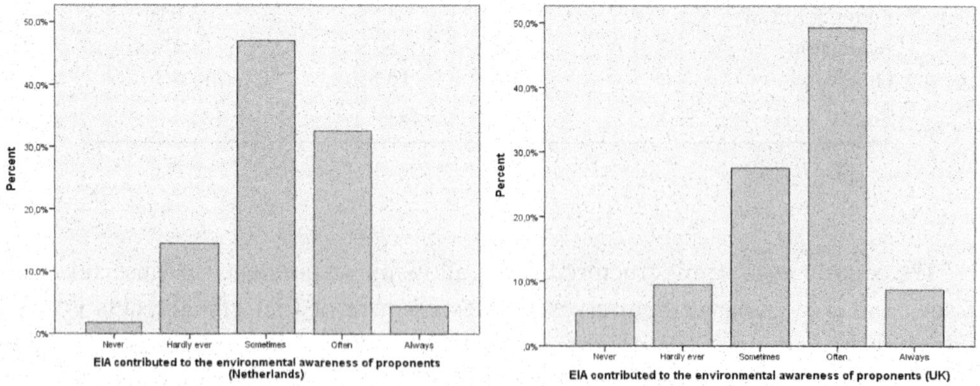

(b)

Figure 2. (a) Contribution of EIA to environmental awareness — competent authorities (left: Netherlands, right: UK). (b) Contribution of EIA to environmental awareness — proponents (left: Netherlands, right: UK).

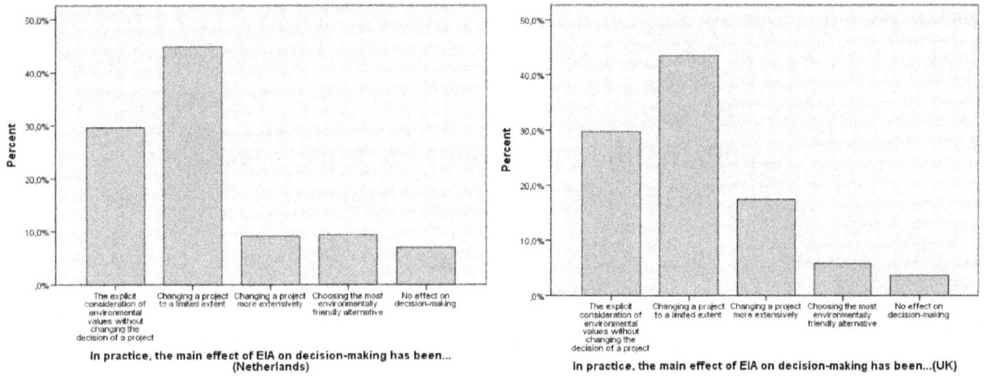

Figure 3. Perceived effect of project EIA on decision-making in the Netherlands (left) and the UK (right).

data indicate that there are similar overall trends in both categories of respondents, except that the "ambivalent" groups in the UK were slightly lower and the "Always" and "Often" groups relatively speaking higher than in the Netherlands (Figs. 2(a)/2(b)). UK respondents were found to be significantly more optimistic than their Dutch colleagues.[f]

Project EIA is perceived to have a modest "greening" impact on projects in the Netherlands. Our respondents thought that EIA mainly resulted in small adjustments of initiatives, not in the choice for other (more environmentally friendly) alternatives (see Fig. 3). A mostly moderate effect of EIA was also observed in the UK survey,[g] however, with 20% of the respondents saying that the impact was more extensive. In just under 10% of the cases, it was said that the most environmentally friendly alternative was chosen, i.e. the impact was greater.

The effectiveness of EIA: comparing Dutch and UK surveys

The above figures and the statistical tests indicate that perceptions of EIA effectiveness differ significantly between the two countries regarding the potential to raise environmental awareness but not in terms of the potential to incorporate environmental values in decision-making. In particular the latter is not what we

[f]The observed difference between the Netherlands and the UK for competent authorities and proponents is statistically significant (for competent authorities: Kolmogorov-Smirnov $Z = 2.031$; Asymp. Sig (two-tailed) = 0.001; for proponents: Kolmogorov-Smirnov $Z = 2.101$; Asymp. Sig (two-tailed) = 0.000). This means that in the UK EIA is perceived as leading to a greater environmental awareness among competent authorities and proponents.

[g]The observed difference between the Netherlands and the UK is not statistically significant (Kolmogorov-Smirnov $Z = 0.700$; Asymp. Sig. (two-tailed) = 0.712).

had expected, based on what had been discussed in the previous section. In the next section, we will provide some further explanations for our observations.

Explaining the Perceived EIA Effectiveness in the Netherlands and the UK

In this section, we first discuss the perceived importance of governance mechanisms and context factors. Subsequently, we make an attempt to explain the similar perceptions regarding the incorporation of environmental values in decision-making through EIA in both the UK and the Netherlands.

The importance of governance mechanisms for EIA effectiveness

Rather than going through all governance mechanisms outlined in Table 1, we focus here on the main findings of our surveys only. In the Netherlands, the greening impact of initiatives is correlated in particular with the so-called "prevention effect" (see Fig. 4). Proponents as well as authorities put this down to the existence of formal EIA and other legal environmental requirements they have to comply with (see Fig. 5(a)). This prevention effect appears to be bigger than the impact EIA reports have on decisions (i.e. what could be called the "correction effect" during/after EIA; see Fig. 3). These findings correspond with earlier studies in the Netherlands (e.g. Ten Heuvelhof and Nauta, 1996). It can therefore be concluded that EIA in the Netherlands is applied in particular as an instrument for appraisal, less as an instrument for active design. Furthermore, the effectiveness of EIA appears to be stable over time, with regulation changes and changes to the

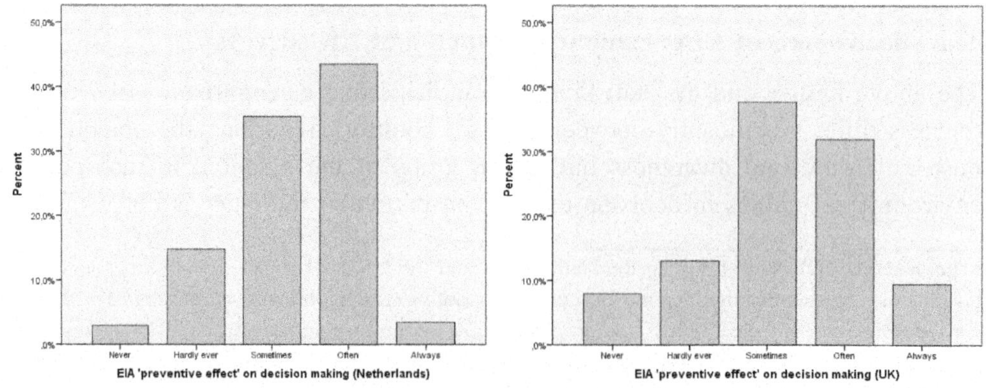

Figure 4. Perceived degree of occurrence of prevention effect in the Netherlands (left) and the UK (right).

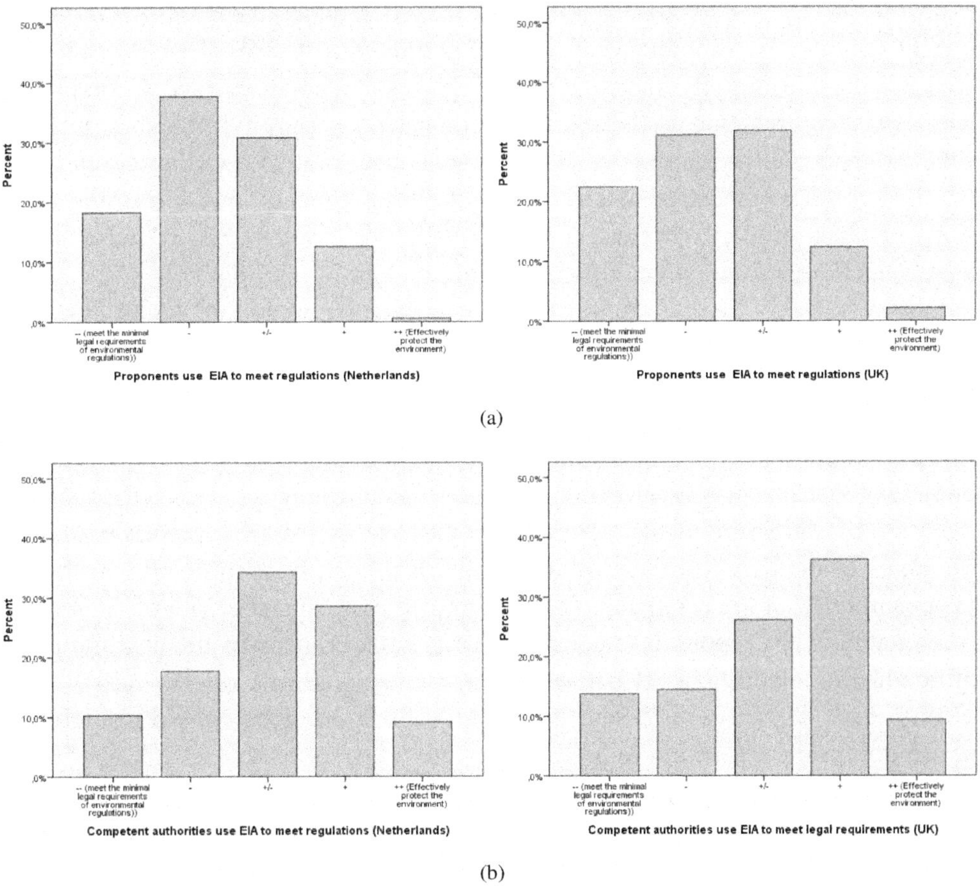

Figure 5. (a) Perceived way in which EIA has been used by proponents in the Netherlands (left) and the UK (right). (b) Perceived way EIA has been used by competent authority in the Netherlands (left) and in the UK (right).

overall context only having minor effects. A very similar picture regarding this prevention effect[h] and the importance of formal EIA requirements for both the proponent[i] and competent authority[j] is also emerging in the UK (Figs. 4, 5(a) and 5(b)).

[h] The observed difference between the Netherlands and the UK is not statistically significant (Kolmogorov-Smirnov $Z = 0.590$; Asymp. Sig (two-tailed) $= 0.877$).

[i] The observed difference between the Netherlands and the UK is not statistically significant (Kolmogorov-Smirnov $Z = 0.367$; Asymp. Sig (two-tailed) $= 0.999$).

[j] The observed difference between the Netherlands and the UK is not statistically significant (Kolmogorov-Smirnov $Z = 0.697$; Asymp. Sig (two-tailed) $= 0.716$).

In the survey, we asked respondents what governance mechanisms and contextual factors they thought were most important for effective EIA. In both the Netherlands and the UK, the majority of the respondents considered complying with legal requirements as the most important reason for environmental responsible behaviour of proponents[k] and authorities[l] (see Table 4 and Fig. 6). In other words, EIAs are conducted and contribute to environmental awareness and revisions of plans primarily because EIA is mandatory, not (or less) because actors want to achieve environmentally responsible outcomes with the help of EIA.

The quality of the EIA report is seen as another important factor for EIA effectiveness. Generally speaking, Dutch EIA professionals are positive about the quality of EIA reports. In this context, scoping and review stages are seen as being of particular importance. According to the respondents, the Dutch EIA Commission in particular plays an important role in safeguarding EIA quality. Other forms of quality control, such as certification, are seen as unwanted. In the Netherlands, a majority of the respondents agreed that the quality of EIA reports is good. Members of the EIA Commission especially, however, were more critical. In the UK, nearly half of the respondents neither agreed nor disagreed with the statement "the quality of the reports was good".[m] In this context, perhaps the new initiative of conducting quality checks across EIA activities by IEMA (the so-called "EIA quality mark" initiative) is a step in the right direction, which could change this perception positively in the future.

Both Dutch and UK survey results underline the relevance of other governance mechanisms of EIA (see Table 4). Public participation is generally seen as important, in particular for gaining public support of the EIA report (Runhaar *et al.*, 2013). Overall, the influence of the public on the quality of EIA is viewed as limited in both the UK and the Netherlands. The consideration of mitigation measures is seen as important and the assessment of alternatives as fairly important. Here, in the Netherlands, the abolishment of the requirement to consider the Alternative Most Friendly to the Environment (AMFE) is expected to have a detrimental effect on the effectiveness of EIA, which has been confirmed in the interviews. Ex post evaluation to EIA (follow-up) is considered to be fairly important for achieving effective EIA, although EIA follow-up practice is limited in

[k] The observed difference between the Netherlands and the UK is not statistically significant (Kolmogorov-Smirnov $Z = 0.539$; Asymp. Sig (two-tailed) $= 0.934$).

[l] The observed difference between the Netherlands and the UK is not statistically significant (Kolmogorov-Smirnov $Z = 0.651$; Asymp. Sig (two-tailed) $= 0.790$).

[m] The observed difference between the Netherlands and the UK is statistically significant (Kolmogorov-Smirnov $Z = 2.993$; Asymp. Sig (two-tailed) $= 0.000$), which means Dutch EIA reports are perceived to be of a better quality.

Table 4. Perceived importance of governance mechanisms and contextual factors on EIA effectiveness.

The importance of factors for the contribution of EIA to more environmentally sustainable decision-making

	The Netherlands: Percentage "very important" or "important"	United Kingdom: Percentage "very important" or "important"	*Independent Samples Median Test*	Null-hypothesis = median scores are the same across countries
Legal requirements	70.7%	75.0%	0.496	Retain null-hypothesis (i.e. insignificant difference)
Quality of the research underlying EA (i.e. validity and comprehensiveness)	68.9%	87.6%	0.000	Reject null-hypothesis (i.e. significant difference)
Transparency of EIA process	61.6%	80.3%	0.000	Reject null-hypothesis (i.e. significant difference)
Extent to which the initiator as well as the competent authority were willing to take into account environmental values	57.6%	83.2%	0.000	Reject null-hypothesis (i.e. significant difference)
The costs of mitigating measures	55.8%	68.6%	0.014	Reject null-hypothesis (i.e. significant difference)
The way the results of the EIA were communicated towards the proponent, competent authority and stakeholders	52.9%	80.3%	0.000	Reject null-hypothesis (i.e. significant difference)

Table 4. (*Continued*)

The importance of factors for the contribution of EIA to more environmentally sustainable decision-making

	The Netherlands: Percentage "very important" or "important"	United Kingdom: Percentage "very important" or "important"	*Independent Samples Median Test*	Null-hypothesis = median scores are the same across countries
Participation of stakeholders in the EIA research	50.7%	73.8%	0.000	Reject null-hypothesis (i.e. significant difference)
Extent to which the EIA process was connected with the dynamics of the decision-making process	48.8%	86.2%	0.000	Reject null-hypothesis (i.e. significant difference)
Extent to which the initiative was elaborated before the EIA was conducted	44.2%	37.3%	0.269	Retain null-hypothesis (i.e. insignificant difference)

Asymptotic significances are displayed. The significance level is 0.05.
Note: Governance mechanisms in normal font; *contextual factors in italics*.

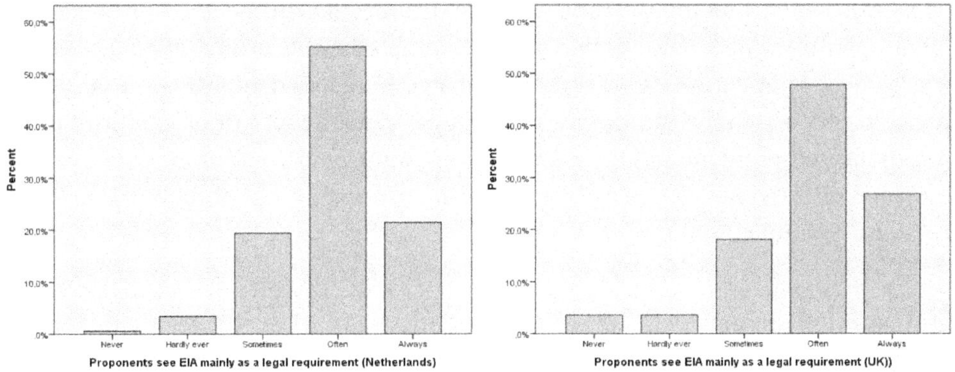

(a)

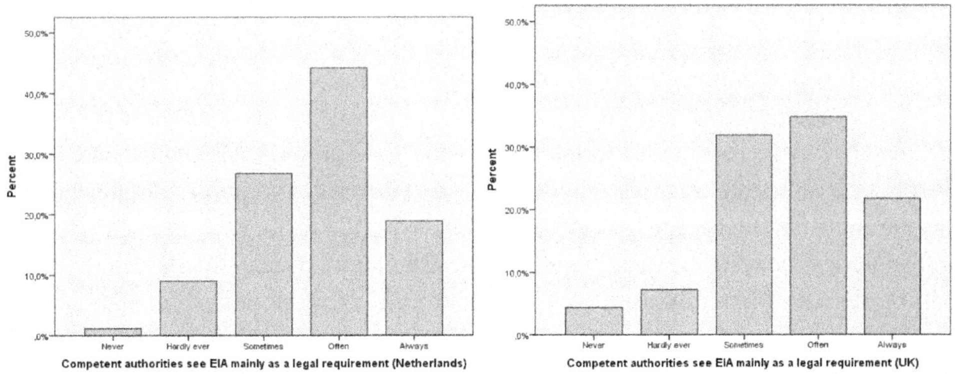

(b)

Figure 6. The extent to which EIA is mainly seen as a legal requirement by proponents in the Netherlands (left) and in the UK (right). (b) The extent to which EIA is mainly seen as a legal requirement by the competent authority in the Netherlands (left) and in the UK (right).

both the Netherlands and the UK (see Arts, 1998; Morrison-Saunders and Arts, 2004; Partidário and Fischer, 2004). The other governance mechanisms seem to be less relevant for effective EIA. For instance, a majority of respondents believed that SEA does not make EIA for projects redundant. In the UK, this opinion is even stronger than in the Netherlands, which is most likely connected with the more strategic nature of many situations within which SEA is applied here (Fischer, 2006). Finally, the responsibility of the proponent to prepare the EIA report is seen as important for overall EIA effectiveness by only a minority, which is striking when considering the importance attached to the "prevention effect" (see discussion above). This might be related to the importance of legal requirements to conduct EIA.

Overall, when looking at Table 4, it is striking that in the UK, all scores are higher apart from one, namely the extent to which there was already an agreement on a project before EIA. This indicates that UK respondents differ less in their opinions and are more outspoken than Dutch respondents, thus confirming the more polarised political and decision-making culture of the former. However, the perceived effectiveness of EIA in terms of incorporating environmental values in decisions is similar.

Contextual elements

Generally speaking, Dutch EIA practitioners consider the quality of EIA reports to be good (see Fig. 7 and Table 4) and consider this to be an important factor for overall EIA effectiveness. However, the scope of EIAs is perceived by many as often being too broad. Proponents in particular have this view, while the EIA Commission and more experienced respondents give this less weight. Furthermore, it is suggested that high costs of mitigation measures will limit the use of EIA results, in the UK even more than in the Netherlands. Both UK and Dutch respondents state that transparency of EIA procedures is good and thought that this was an important factor for EIA effectiveness, with the former being even stronger. Transparency can thus be considered an important added value of EIA in decision-making. In addition, communication with authorities, proponents and stakeholders is seen as a relevant factor. However, this appears to be less so than what might be expected from what has been suggested in the professional literature. In the Netherlands, factors that are of lesser importance include the extent to which a proposal has been elaborated on. This may limit EIA effectiveness (the issue of foreclosure) and the extent to which EIA is connected with decision-making. Here,

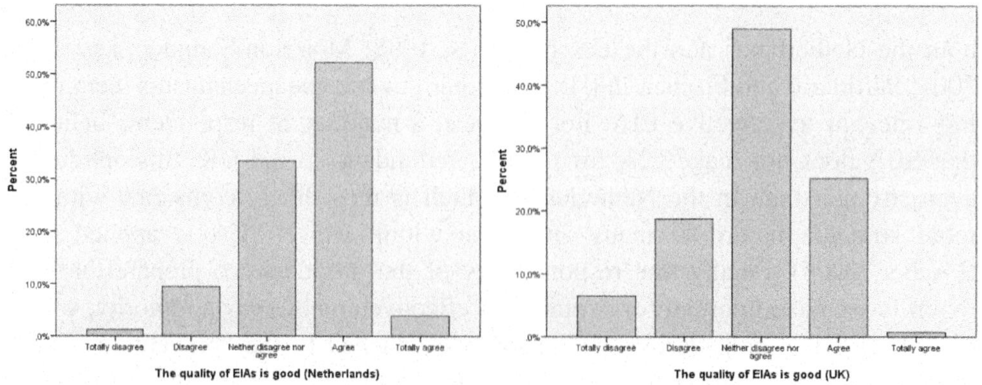

Figure 7. Perceived quality of EIA reports in the Netherlands (left) and UK (right).

opinions of UK respondents are markedly different, given that these connections are of a very high importance. Regarding the characteristics of the actors involved, openness to environmental issues of the proponent and the authority are seen as being important in both the UK and the Netherlands. Also, many respondents indicate that more experience with EIA is an important factor for better EIAs. Finally, with respect to the decision-making context, neither the survey nor the interviews suggest that other actors or instruments are currently replacing EIA or parts of it, with the exception of perhaps the IPPC (Integrated Pollution Prevention and Control) Directive, which limits the added value of EIA for energy and industry projects.

Disagreement and consensus

There is some broad consensus on many issues among EIA professionals from both the UK and the Netherlands. The role of respondents (authority, proponent, consultant, EIA Commission) does not explain any differences in the survey in a statistically significant manner. There is also little difference between sectors (e.g. infrastructure, water management, industry etc.). This is striking, as some authors have suggested that these differences were important (Runhaar and Driessen, 2007). The differences in opinions that do exist seem to be dependent on two factors in particular: experience with EIA (both in the Netherlands and UK); and how closely an actor is involved in the initiative (the Netherlands).[n]

As far as disagreement amongst the UK EIA professionals is concerned, a lack of consensus was observed with regards to the clarity of EIA regulations. Consultants in particular were of the opinion that EIA regulations are not very easy to understand in the UK. This is perhaps owing to the fact that in practice it is the consultants who actually look into the legal requirements for their clients and hence they face this problem more acutely. Some disagreement was also observed with regards to the role of mitigation costs and its importance in leading to more environmentally sustainable decisions.

Revisiting our expectations on the effectiveness of EIA in the Netherlands and the UK

Due to differences in governance mechanisms and context factors we thought that there would be significant differences in the perceived EIA effectiveness of both countries. However, in the previous section, we established that there were some

[n] For a more in-depth discussion for the Netherlands, see Runhaar *et al.* (2013) and for the UK see Jha-Thakur *et al.* (2012).

small differences in the perceived effectiveness regarding the likelihood of raising environmental awareness only and that there were no differences for the incorporation of environmental concerns in decision-making. Below, we discuss these findings further, referring to provisions for participation and consultation, scoping, assessment of alternatives, environmental statement quality review, follow-up and integration of decision-making and EIA.

Provisions for participation and consultation

Regarding participation and consultation of stakeholders and advisors, the UK has fewer provisions than the Netherlands. However, this does neither reflect a lower environmental awareness (Fig. 2) nor does it have a discernible impact on decision-making (Fig. 3). In the UK, the importance of participation is seen to be more important than in the Netherlands (Table 4). This might be related to the fact that the regulations provide for less safeguards. In the Netherlands, participation is seen as being of less importance for influencing EIA, but important for its overall acceptance.

Provisions for scoping

The EIA scoping stage in the UK is less regulated than in the Netherlands.[o] However, this does neither reflect a lower environmental awareness (Fig. 2) nor does it have an impact on decision-making (Fig. 3). In the Netherlands, EIA is perceived as a rather effective instrument, although with respect to scoping there seems to be still room for improvement, which is also the case in the UK (see also next section).

Assessment of alternatives

In the UK, there is no formal requirement to assess alternatives, opposite to the Netherlands. However, this does neither appear to reflect a lower environmental awareness (Fig. 2) nor does it seem to have an impact on decision-making (Fig. 3). In the Netherlands, the assessment of alternatives is seen to function rather well and is seen as an important factor for overall EIA effectiveness.

---

[o] At least this was the case before the 2010 revisions of the Dutch EIA regulations. Before 2010, the Dutch EIA system included several formal governance mechanisms regarding scoping, such as a requirement to provide for a notification of intent by the proponent, a formal public review and an advice by the Dutch EIA Commission about this, and guidelines for the EIA report by the competent authority.

Environmental statement quality review

In the Netherlands, quality reviews of EIA are better institutionalised than in the UK. Indeed, the quality of EIAs is perceived to be better in the Netherlands than in the UK (see Fig. 7). This seems to be reflected in the importance attached to the quality of EIA in the UK, which is much higher than in the Netherlands (Table 4). However, this difference in quality (and quality safeguards) does neither reflect a statistically significant lower environmental awareness (Fig. 2) nor does it appear to statistically significantly impact on decision-making (Fig. 3).

Follow-up

In the Netherlands, EIA regulations contain requirements for follow-up. However, this is not reflected in a statistically significant lower environmental awareness (Fig. 2) nor in a statistically significantly greater impact on decision-making (Fig. 3) in the UK compared with the Netherlands. The importance of follow-up is seen as fairly important for achieving effective EIA, but follow-up practice is limited in both the Netherlands and the UK. This is seen as an area where there is room for improvement.

Integration of decision-making and EIA

In the UK, EIA is more directly integrated with decision-making procedures than in the Netherlands. In the Netherlands, EIA is not well linked, apart from infrastructure planning. However, this is not seen here as a problem. In the UK, functioning integration is seen as a very important factor (see Table 4), although the resulting impact on EIA effectiveness does not appear to be much different from the Netherlands.

The different administrative cultures (more polarised, conflict-oriented versus more consensus-oriented) do not result in many differences in the replies of respondents from the UK and the Netherlands in terms of largely different perceptions. In fact, in both countries, there is some broad consensus on many issues. In the UK, there are some differences regarding the perceived clarity of the regulations, the role of mitigation costs and the relevance for decision-making.

These reflections lead us to conclude that the impact of governance mechanisms and context elements depends more on how these are shaped in practice, rather than on the mere presence or absence of them. In this context, the importance of the quality of environmental statements is illustrative. In the UK, this is considered to be more important for EIA effectiveness because the actual quality is perceived to be lower than in the Netherlands. This suggests an alternative explanation, namely that the importance of specific mechanisms or factors is not perceived realistically. Rather, respondents take it for granted that a particular mechanism or factor functions well. Finally, there may also be some form of compensation between governance mechanisms and context factors. However, our surveys do not allowus to draw any firm conclusions on this.

Conclusions and Discussion: Beyond Appraisal?

In this paper, we assessed and compared the effectiveness of EIA in the Netherlands and the UK. Whilst the two systems have many similarities, there are also differences in terms of the way in which the EU EIA Directive is implemented and in the institutional context. Based on these differences, before our examination of the two systems, we formulated some expectations regarding the effectiveness of EIA in both countries. However, these were only partly confirmed and many aspects of EIA effectiveness were perceived to be similar by EIA actors in both countries. Whereas in the UK, EIA is perceived to contribute statistically significantly more to raising environmental awareness, overall EIA is perceived to have the same (moderate) impact on decisions regarding the incorporation of environmental concerns.

In both countries, some of the governance mechanisms and contextual elements that explain EIA effectiveness are the same, whilst others are different. In the Netherlands, legal requirements and the quality of EIA in particular are important for explaining EIA effectiveness. Furthermore, transparency and environmental attitudes of proponents and competent authorities are also seen as being important.

In the UK, elements that are seen to be of particular importance include the quality of the environmental statement, the linkage with the dynamics of decision-making, environmental attitudes, as well as communication and transparency of the methodology used. In the UK, EIA guidance plays a very important role in improving the quality of EIA reports. In both, the UK and the Netherlands, the mandatory status of EIA, as well as transparency and positive attitudes contribute towards explaining EIA effectiveness. Differences can be explained mainly in terms of governance mechanisms, such as quality control, scoping, and the development of alternatives. These have been more rigorously regulated in the Netherlands than in the UK. Differences in the institutional context related to differences in the administrative culture seem to be limited, which is most likely the result of the moderating influence of the EU.

Our findings suggest that the influence of governance mechanisms and contextual elements is more complex than expected. This requires further analysis, in particular on the extent to which mechanisms and elements reinforce or compensate for each other.

EIA effectiveness was found to be more or less stable across both countries. Earlier evaluations for the Netherlands suggested that EIA effectiveness has also been stable over time (see Runhaar *et al.*, 2012). Despite the various changes to the EIA systems, little seems to have changed with respect to governance mechanisms (apart from perhaps the introduction of SEA). In the Netherlands, EIA regulations underwent a significant revision in 2010, aimed at improving time- and cost-efficiency. Amongst other things, competent authorities were given more responsibility and the use of public participation became more flexible. Subsequently, elements of the EIA system that were going beyond the minimum quality levels of the EU have become vulnerable to downgrading.

In the UK, reforms were last implemented in 2011, under the Town and Country Planning Regulations. The scope of these revisions was more modest, consolidating EIA with amendments in the provisions of the Town and Country Planning (Environmental Impact Assessment) Act. In our view, both revisions do not sufficiently touch upon what our research highlights as barriers to achieving effectiveness. Necessary adjustments appear to be related in particular with the "perennial problems of EIA", as has been discussed in the professional literature for many years (see Ortolano and Shepherd, 1995; Sadler, 1996; Wood, 2003). This includes e.g. reducing the complexity of regulations, better coverage of strategic decisions, clearer screening, more focused scoping (preventing lengthy EIA reports) and more attention to actual follow-up.

The most important challenge to EIA is connected with its primary functions, namely to being an instrument for environmental appraisal and also an instrument for improving project design and decision-making. True effectiveness requires a focus on both environmental appraisal (for which the current instrument appears to work quite well in both countries) and the "greening" of designing and decision-making. EIA currently does not appear to fully provide for an effective platform for an open and creative discussion about plans, programmes and projects. The many formal and complex requirements do not appear to be conducive to this. Neither is the way in which public participation and the review by external advisors are currently organised. In practice, EIA seems to have a design function when the contextual factors are in favour of it, including an open attitude of proponents and authorities, and when there is still room for alternative designs. However, this doesn't appear to arise from the EIA system itself. To improve, this would most likely require other incentives.

After 25 years, EIA has clearly become a "part of life" in the Netherlands and the UK. EIA is viewed as a routine element in decision-making for appraising environmental impacts. In the UK and the Netherlands, EIA is a mandatory test for decision-making and it functions well in this respect, although perhaps it has remained a bit mechanistic. In this respect, EIA certainly does not make decision-making on initiatives easier. However, this might also be considered as a sign that EIA is actually working as a policy instrument. Overall, it appears that in the Netherlands and the UK, EIA is performing quite well, although it is clearly being little loved.

Acknowledgements

For the study of the Dutch EIA system and practice we thank Hans de Vries and Germt de Vries for their input. For the UK study into EIA practice we thank Stephen Ackers for his support.

References

Arts, J (1998). *EIA Follow-Up: On the Role of Ex Post Evaluation in Environmental Impact Assessment*. Proefschrift, Groningen: Geo Press.

Arts, J (1991). *Een kwestie van niveau: Milieu-effectrapportage in De ruimtelijke Ordening*. Rijksuniversiteit Groningen/Ministerie VROM, Groningen.

Art, J and F van Lamoen (2005). Before EIA: Defining the scope of infrastructure projects in the Netherlands. *Journal of Environmental Assessment Policy and Management*, 7(1), 51–80.

Barker, A and C Wood (1999). An evaluation of performance of EIA in 8 EU countries. *Environmental Impact Assessment Review*, 19(4), 387–404.

Bassi, A, R Howard, D Geneletti and S Ferrari (2012). UK and Italian EIA systems: a comparative study on management practice and performance in the construction industry. *Environmental Impact Assessment Review*, 34, 1–11.

Bonte, R (1996). De actoren en hun belangen. In *Milieu-effectrapportage, methoden, effecten en resultaat*, R Bonte, P Leroy and R Mooren (ed.), pp. 95–109, Aeneas: Best.

Caldwell, LK (1982). *Science and the National Environmental Policy Act: Redirecting Policy Through Procedural Reform*. University of Alabama Press, University of Alabama.

Cashmore, M, RG William, R Morgan, D Cobb and A Bond (2004). The interminable issue of effectiveness: substantive process, outcomes and research challenges in the advancement of environmental impact assessment theory. *Impact Assessment and Project Appraisal*, 22(4), 295–310.

Chadwick, A (2002). Socio-economic impacts: are they still the poor relations in UK environmental statements? *Journal of Environmental Planning and Management*, 45(1), 3–24.

Cherp, A (2001). EA legislation and practice in Central and Eastern Europe and the former USSR: a comparative analysis. *Environmental Impact Assessment Review*, 21(4), 335–362.

Clark, B, K Chapman, K Bisset and P Wathern (1976). *Assessment of Major Industrial Application — A Manual*. Department of the Environment, London.

Cmer, Commissie voor de milieu-effectrapportage (Netherlands' Commission for Environmental Assessment) [NCEA] (2011). *Jaarverslag 2010*. Utrecht.

Dalal-Clayton, B and B Sadler (2004). *Strategic Environmental Assessment (SEA): A Sourcebook and Reference Guide on International Experience*. London: IIED. Available at www.iied.org/spa/sea.html.

De Leeuw, B and R Rense (1991). *Evaluatiebepalingen in besluiten met m.e.r.* Amersfoort: Twijnstra Gudde.

Devuyst, D (1994). *Instruments for the Evaluation of Environmental Impact Assessment*. Brussel: Vrije Universiteit Brussel.

De Jong, WM (1999). *International Comparison of Decision-Making on Infrastructure*. Research series of the Directorate for Strategy and Coordination, Ministry of Transport. Public Works and Water Management, The Hague.

De Jong, WM and H Geerlings (2005). *A Roadmap for Infrastructure Appraisal*. Delft: TRAIL.

EC (1993). *Verslag van de Commissie over de tenuitvoerlegging van Richtlijn 85/337/ EEC, incl. bijlage over Nederland*. Commissie van de Europese Gemeenschappen, Brussel.

EC (2009a), *Verslag van de Commissie aan de Raad, het Europees Parlement, het Europees en Sociaal Comité en het Comité van de Regio's over de toepassing en doeltreffendheid van de MEB-richtlijn (Richtlijn 85/337/EEC, als gewijzigd bij Richtlijn 97/ 11/EG en Richtlijn 2003/35/EG)*. Commissie van de Europese Gemeenschappen, Brussel.

EC (2009b), *Toepassing en doeltreffendheid van van de MEB-richtlijn (Richtlijn 85/337/ EEG, als gewijzigd bij Richtlijn 97/11/EG en Richtlijn 2003/35/EG)*. Commissie van de Europese Gemeenschappen, Brussel.

EC (2009c), *Toepassing en doeltreffendheid van de richtlijn betreffende strategische milieubeoordeling (Richtlijn 2001/42/EG)*. Commissie van de Europese Gemeenschappen, Brussel.

ECW (1990). *Naar een volwaardige plaats*. Evaluatiecommissie Wet algemene bepalingen milieuhygiëne, Den Haag.

ECW (1996). *Naar een duurzame milieu-effectrapportage, tweede advies over de regeling milieu-effectrapportage*. Evaluatiecommissie Wet milieubeheer, Den Haag.

Elwell, C (2002). *Sustainability Impact Assessment of the Earth Summit @ 10. A Canadian Perspective*. Ontario: Canadian Institute for Environmental Law and Policy.

Fiddler, C and B Noble (2012). Advancing strategic environmental assessment in the offshore oil and gas sector: lessons from Norway, Canada, and the United Kingdom. *Environmental Impact Assessment Review*, 34, 12–21.

Fischer, TB (1999). Benefits from SEA application — a comparative review of North West England, Noord-Holland and EVR Brandenburg-Berlin. *Environmental Impact Assessment Review*, 19(2), 143–173.

Fischer, TB (2002). *Strategic Environmental Assessment — Transport and Land-Use Planning*. London: Earthscan.

Fischer, TB (2003). Strategic environmental assessment in post-modern times. *Environment Impact Assessment Review*, 23(2), 155–170.

Fischer, TB (2005). Having an impact? — context elements for effective SEA application in transport policy, plan and programme making. *Journal of Environmental Assessment Policy and Management* 7(3), 407–432.

Fischer, TB (2006). Erste Erfahrungen mit der Umsetzung der SUP Richtlinie in der englischen Raumplanung (Experiences with implementing the SEA Directive in spatial planning in England). *UVP Report*, 20(1+ 2), 48–52.

Fischer, TB (2007). *Theory and Practice of Strategic Environmental Assessment — Towards a More Systematic Approach*. London: Earthscan.

Fischer, TB (2009). On the role(s) of (strategic) environmental assessment in "greening" decision making, Copernicus Lecture, *University of Utrecht*, 2 March 2009. www.twoeam-eu.net.

Fischer, TB (2010a), The EU and its regulatory role in environmental policy and assessment, *GeoINova*. Special issue "The Evolution of Integration in Europe, 20 Years after the Fall of the Berlin Wall", pp. 155–167.

Fischer, TB (2010b), Reviewing the quality of strategic environmental assessment reports for English spatial plan core strategies. *EIA Review*, 30(1), 62–69.

Fischer, TB and P Gazzola (2006). SEA good practice elements and performance criteria — equally valid in all countries? — the case of Italy. *Environment Impact Assessment Review*, 26(4), 396–409.

Fischer, TB and J Phylip-Jones (2008). Scoping in environmental assessment. In TB Fischer, P Gazzola, U Jha-Thakur, I Belčákova and R Aschemann (eds.), *Environmental Assessment Lecturers' Handbook*, Road Bratislava, pp. 136–142. www.twoeam-eu.net.

Gazzola, P, U Jha-Thakur, S Kidd, D Peel and T Fischer (2011). Enhancing environmental appraisal effectiveness: towards an understanding of internal context conditions in organisational learning. *Planning Theory and Practice*, 12(2), 183–204.

Glasson, J (1999). The first 10 years of UK EIA system: strengths, weaknesses, opportunities and threats. *Planning Theory and Practice*, 14(3), 363–375.

Glasson, J and C Bellanger (2003). Divergent practice in a converging system? The case of EIA in France and the UK. *Environmental Impact Assessment Review*, 23(5), 605–624.

Glasson, J, R Therivel and A Chadwick (1994). *Introduction to Environmental Impact Assessment.* Natural and Built Environment Series 1. London: UCL Press.

Glasson, J, R Therivel, J Weston, E Wilson and R Frost (1997). EIA-learning from experience: changes in the quality of environmental impact statements for UK planning projects. *Journal of Environmental Planning and Management*, 40(4), 451–464.

Hanusch, M and TB Fischer (2011). SEA and landscape planning. In Sadler B, Aschemann, R. Dusik, J. Fischer, TB. Partidário, M and Verheem R (eds). *Handbook of Strategic Environmental Assessment.* London: Earthscan, pp. 157–173.

Hartley, N and C Wood (2005). Public participation in environmental impact assessment-implementing the Aarhus Convention. *Environmental Impact Assessment Review*, 25(4) 319–340.

Hildén, M, E Furman and M Kaljonen (2004). Views on planning and expectations of SEA: the case of transport planning. *Environmental Impact Assessment Review*, 24(5), 519–536.

Hilding-Rydevik, T and H Bjarnadóttir (2007). Context awareness and sensitivity in SEA implementation. *Environmental Impact Assessment Review*, 27(7), 666–684.

IEMA (2011). Special Report–The state of Environmental Impact Assessment practice in the UK, available at http://www.iema.net/eiareport (last checked on 1 July 2012).

Jay, S, C Jones, P Slinn and C Wood (2007). Environmental impact assessment: retrospect and prospect. *Environmental Impact Assessment Review*, 27(4), 287–300.

Jesse, K (2008). *Een hernieuwd perspectief op milieu-effectrapportage. Over de inhoudelijke invloed van m.e.r. op de besluitvorming en enkele mogelijkheden ter flexibilisering.* Tilburg: Proefschrift, Universiteit Tilburg.

Jha-Thakur, U, T Fischer, V Onyango and P Labhasetwar, (2012). *Health within EIA in the UK and India: An International Comparative Study (Unpublished Manuscript).* Liverpool: University of Liverpool.

Kidd, S and TB Fischer (2007). Towards sustainability: is integrated appraisal a step in the right direction? *Environment and Planning C*, 25(2), 233—249.

Kørnøv, L and WAH Thissen (2000). Rationality in decision- and policy-making: implications for strategic environmental assessment. *Impact Assessment and Project Appraisal*, 18(3), 191–200.

Leroy, P (1996). De effecten van m.e.r.: een evaluatie. In *Milieu-effectrapportage, methoden, effecten en resultaat*, R Bonte, P Leroy and R Mooren (ed.), Best: Aeneas. pp. 127–137.

Ligtermoet and partners (2007). *Betrokkenheid van burgers en maatschappelijke organisaties in m.e.r. procedures.* Gouda: Ligtermoet en partners.

Marr, K (1997). *Environmental Impact Assessment in the United Kingdom and Germany.* Aldershot: Ashgate.

Marshall, R (2001). Application of mitigation and its resolution within environmental impact assessment: an industrial perspective. *Impact Assessment and Project Appraisal*, 19(3), 195–204.

Morrison-Saunders, A and J Arts (eds.) (2004). *Assessing Impact – Handbook of EIA and SEA Follow-Up*. London: Earthscan.

Mostert, E (1995). *Commissions for Environmental Impact Assessment. Their Contribution to the Effectiveness of Environmental Impact Assessment*. Delft: Proefschrift, Delft University Press.

Ortolano, L and A Shepherd (1995). Environmental Impact Assessment. In *Environmental and Social Impact Assessment*, F Vanclay and DA Bronstein (eds.). Chichester: John Wiley and Sons. pp. 3–31.

Partidário, MR (2000). Elements of an SEA framework-improving the added-value of SEA. *Environmental Impact Assessment Review*, 20(6), 647–666.

Partidário, MR and TB Fischer (2004). Follow-up in Current SEA undestanding. In A Morrison-Saunders and J Arts (eds.). *Assessing Impact – Handbook of EIA and SEA Follow-Up*. London: Earthscan. pp. 224–247.

Petts, J (1999). *Handbook of Environmental Impact Assessment*. Vols. 1 and 2. Oxford: Blackwell Science.

Piper, JM (2001). Barriers to implementation of cumulative effects assessment. *Journal of Environmental Assessment Policy and Management*, 3(4), 465–481.

Pokorný-Versteeg, WH (2003). *Milieu-effectrapportage in Nederland en de direct aangrenzende EU-lidstaten. Een rechtsvergelijkend onderzoek*, PhD dissertation, Leiden University, Kluwer, Deventer.

Richardson, T and C Cashmore (2011). Power, knowledge and environmental assessment: the World Bank's pursuit of "good governance". *Journal of Political Power*, 4(1), 105–125.

RTPI (2001). *Planning Practice Standard, Environmental Impact Assessment*. Available at www.rtpi.org.uk.

Runhaar, HAC and PPJ Driessen (2007). What makes strategic environmental assessment successful environmental assessment? *Impact Assessment and Project Appraisal*, 25(1), 2–14.

Runhaar, HAC, EJMM Arts, F van Laerhoven and PPJ Driessen (2011). Naar een toekomstbestendige m.e.r., *lessen uit 25 jaar m.e.r. in Nederland en een verkenning van kansen en bedreigingen voor de m.e.r. in de nabije toekomst*, in opdracht van: Ministerie van Infrastructuur and Milieu. Utrecht/Groningen: Universiteit Utrecht/Rijksuniversiteit Groningen (available from www.rug.nl/frw/onderzoek/duurzamewegen/25jaarmer).

Runhaar, H., F. van Laerhoven, P. Driessen and J. Arts (2013). Environmental Assessment in the Netherlands: effectively governing environmental protection? A discourse analysis. *Environmental Impact Assessment Review*, 39, 13–25.

Sadler, B (ed.) (1996). *International Study of the Effectiveness of Environmental Assessment. Final Report: Environmental Assessment in a Changing World: Evaluating Practice to Improve Performance*. Ottawa: Canadian Environmental Assessment Agency. Available at www.ceaa.gc.ca/Content/A/F/B/AFB7DF6E-7212-451B-A27E-9A56035CD913/iaia8_e.pdf.

Sadler, B (2004). On evaluating the success of EIA and SEA. In *Assessing Impact — Handbook of EIA and SEA Follow-Up*, A Morrisson-Saunders and J Arts (eds.). London: Earthscan. pp. 248–285.

Sadler, B and R Verheem (1996). *Strategic Environmental Assessment, Status, Challenges and Future Directions*. EIA series 53. Den Haag: Ministerie van VROM.

Scholten, J and M van Eck (1994). Effectiveness of environmental impact assessment in the Netherlands. In EIA Commission, *EIA Methodology in the Netherlands, Views of the Commission for EIA*. Utrecht. pp. 9–21.

Sheate, W, S Dagg, J Richardson, R Aschemann, J Palerm and U Steen (2003). Integrating the environment into strategic decision-making: conceptualizing policy SEA. *European Environment*, 13(1), 1–18.

Stoeglehner, G, AL Brown and LB Kørnøv (2009). SEA and planning: "ownership" of strategic environmental assessment by the planners is the key to its effectiveness. *Impact Assessment and Project Appraisal*, 27(2), 111–112.

Taylor, S (1984). *Making Bureaucracies Think, Environmental Impact Assessment Strategy of Administrative Reform*. Stanford CA: Stanford University Press.

Ten Heuvelhof, EF and CM Nauta (1996). *(M)erkenning, onderzoek naar de doorwerking van m.e.r.* Den Haag: Evaluatiecommissie Wet Milieubeheer.

Ten Heuvelhof, E and C Nauta (1997). The effects of environmental impact assessment in the Netherlands. *Project Appraisal*, 12(1), 25–30.

Thérivel, R and P Minas (2002). Measuring SEA effectiveness. Ensuring effective sustainability appraisal. *Impact Assessment and Project Appraisal*, 20(2), 81–91.

Therivel, R and TB Fischer (2012). Sustainability appraisal in England. *UVP Report*, 26(1), 16–21.

Van Buuren, A and S Nooteboom (2009). Evaluating strategic environmental assessment in the Netherlands: content, process and procedure as indissoluble criteria for effectiveness. *Impact Assessment and Project Appraisal*, 27(2), 145–154.

Van Doren, D, PPJ Driessen, B Schijf and HAC Runhaar (2013). Evaluating the substantive effectiveness of SEA: towards a better understanding. *Environmental Impact Assessment Review*, 38(1), 120–130.

Van Kessel, H, T Boer, B Roelofs and KA Klein Koerkamp (2003). *Evaluatie m.e.r. 2003*. In opdracht van Ministerie van VROM. Nijmegen: NovioConsult.

VROM (1994). *Handleiding Milieu-Effectrapportage*. opgesteld door JJFM van Haeren, JW Gravendeel (DHV) en D Samkalden, R van Tilburg (Twijnstra Gudde). Den Haag: Koninklijke Vermande.

Wassen, MJ, H Runhaar, A Barendregt and T Okruszko (2011). Evaluating the role of participation in model studies for environmental planning. *Environment and Planning B*, 38(2), 338–358.

Weston, J (2002). From Poole to Fulham: a changing culture in UK Environmental Impact Assessment decision making? *Journal of Environmental Planning and Management*, 45(3), 425–443.

Wathern, P (ed.) (1990). *Environmental Impact Assessment, Theory and Practice*. London: Unwin Hyman.

Wood, C (1995). *Environmental Impact Assessment: A Comparative Review*. Harlow: Longman Scientific and Technical.

Wood, C (2003). *Environmental Impact Assessment — A Comparative Review*, 2nd ed. Harlow: Pearson Education Ltd.

Wood, G, J Glasson and J Becker (2006). EIA scoping in England and Wales: practitioner approaches, perspectives and constraints. *Environmental Impact Assessment Review*, 26(3), 221–241.

CHAPTER 9

Advancing Regional Strategic Environmental Assessment in Canada's Western Arctic: Implementation Opportunities and Challenges

Courtney Fidler* and Bram F. Noble

Department of Geography and Planning
University of Saskatchewan, Canada
crfidler@gmail.com

The absence of Regional Strategic Environmental Assessment (R-SEA) in Canada's western Arctic has raised many questions concerning the country's preparedness for offshore Arctic energy development, given the constraints of project assessments in addressing long-term cumulative impacts of energy development on the marine environment and local communities. There has been much interest in R-SEA in recent years, and a growing body of research on the benefits of strategic approaches to environmental assessment, but relatively little attention has been given to implementation. This paper examines key opportunities for, and challenges to, the implementation of R-SEA in Canada's western Arctic. Results reinforce concerns that the current approach to environmental assessment in Canada's western Arctic is insufficient to address expanding offshore energy development. However, results also indicate several challenges to be addressed to advance R-SEA in the offshore environment including governance, stakeholder resistance to a futures-based approach, the timing of implementation, managing the diversity of expectations about R-SEA, and the nature and scope of alternatives assessment.

Keywords: Regional strategic environmental assessment; Beaufort Sea; offshore energy; Arctic; resource development.

*Corresponding author.

Introduction

Oil and gas exploration licenses continue to be let in the deep offshore of Canada's western Arctic, resulting in new questions around Canada's preparedness for drilling in frontier regions (Elvin and Fraser, 2012). Tension regarding the planning and management of offshore energy development is particularly evident in the hydrocarbon rich Beaufort Sea, where challenges over energy development are intensified by controversial relief well regulation; risk and benefit debates for those who inhabit the region (see Porta and Bankes, 2011); concern regarding long-term impacts on an ecologically sensitive marine environment (Burkett, 2011); and the uncertainties of Arctic climate change (Prowse *et al.*, 2009). This culmination of conditions, along with the current project-driven regulatory environment, has generated considerable discussion on the need for regional strategic environmental assessment (R-SEA) (see Doelle *et al.*, 2012; Noble *et al.*, 2013; WWF, 2005).

The Canadian Council of Ministers of the Environment (CCME) (2009) defines R-SEA as a "process designed to systematically assess the potential environmental effects, including cumulative effects, of alternative strategic initiatives, policies, plans or programs (PPPs) for a particular region." The overall objective of R-SEA is to inform the preparation of a development strategy and environmental management framework for a region (CCME, 2009) and, in doing so, identify strategies and priorities for future management and development while enhancing the efficacy of project-level environmental assessment (EA) (see Harriman Gunn and Noble, 2009). The interest in regionally-based, strategic approaches to EA stems from the limited capacity of project-based EA to address cumulative effects (Elvin, and Fraser, 2012; Gunn and Noble, 2011), the desire for Aboriginal groups to be more involved in development planning (see Porta and Bankes, 2011), and the recognised need to ensure that development decisions are set within a broader environmental planning and management framework (CCME, 2009). In some offshore jurisdictions, such as Norway's Barents Sea, Atlantic Canada, and United Kingdom, such higher-level approaches to assessment are beginning to emerge in practice (Fidler and Noble, 2012; Hasle *et al.*, 2009). Fidler and Noble (2012) report on several potential and realized benefits from international experiences with R-SEA offshore, including improvements in the efficacy and efficiency of project-based assessments, identifying ecologically sensitive marine areas where development should not take place, strengthening opportunities for cumulative effects assessment, and minimizing environmental and safety disasters (see also Doelle *et al.*, 2012).

The absence of R-SEA in the western Arctic has raised concerns about Canada's preparedness for offshore energy development (see Doelle *et al.*, 2012;

Elvin and Fraser, 2012; Porta and Bankes, 2011; Arctic Council, 2009). These concerns are exacerbated by the uncertainties associated with Arctic climate change and due to a number of recent high-profile incidents, such as the 2010 Gulf of Mexico Deep Water Horizon oil spill, that elevated global awareness of the risks of offshore energy development. Notwithstanding several regional planning, baseline collection and science-based marine initiatives ongoing in Canada's western Arctic (e.g. Integrated Oceans Management Plan, Beaufort Regional Environmental Assessment, Integrated Regional Impact Study), and recognition at the local (e.g. IGC, 2004), national (e.g. CCME, 2009) and international levels (e.g. Arctic Council, 2009) of the importance of R-SEA, there remains no framework of R-SEA in the Beaufort Sea and whether and how such a framework can or should be implemented remains unaddressed. Research by Ketilson (2011) and Noble *et al.* (2013), for example, revealed scepticism around R-SEA in Canada's Beaufort Sea, describing it as "unchartered territory" with many questions of whether and how it can deliver benefits and influence decision-making. Part of the problem is that the majority of attention on the subject has been on the need for, and benefits of, such a framework (see Bina, 2007) rather than on implementation issues to advance R-SEA from a novel concept to a practical planning, assessment and decision support process.

The purpose of this paper is to examine key opportunities for, and challenges to, the implementation of R-SEA in the Beaufort Sea of Canada's western Arctic. The current Beaufort Sea planning, assessment and decision-making environment is examined against the normative model and expectations of R-SEA. Current support for R-SEA implementation for offshore planning and management is explored and areas of disagreement amongst stakeholders identified. The intent is not to propose another layer of legislation to what is perceived by some as an already complex regulatory offshore environment (see Callow, 2012), but to provide a foundation for advancing R-SEA from concept to practice. In the following sections the conceptual basis for R-SEA is introduced and the current state of EA in the Beaufort Sea discussed. Results report on the overall readiness for R-SEA implementation in the Beaufort Sea and the needs and interests of key actors. The paper concludes with a discussion of the implications of applying R-SEA and foreseeable challenges to implementation.

Regional Strategic Environmental Assessment

Canada is recognised internationally as a pioneer in strategic environmental assessment (SEA) (see Sadler, 2005), but the application of formal SEA in Canada is restricted to federal matters under a federal Cabinet Directive (see Noble, 2009).

Table 1. Potential benefits of R-SEA.

• Facilitates the development of improved PPPs and strategic initiatives
• Provides a more regional focus for development and decision-making
• Ensures that cumulative effects assessment is captured at the appropriate tier and scale
• Contributes to regional sustainability goals
• Enables and encourages data sharing from regional and project impact monitoring programs
• Facilitates state-of-the-region environmental monitoring and reporting
• Saves time and resources by providing a means to streamline subsequent project EA
• Establishes goals, thresholds, or maximum allowable limits against which to conduct project-based performance and impact assessment

Source: CCME (2009); Harriman and Noble (2009).

In 2008, the CCME, a forum of federal, provincial and territorial governments, commissioned a research programme to develop a more regionalised framework for SEA in an attempt to extend SEA beyond the federal Cabinet Directive and to facilitate a more regional-based, planning approach to impact assessment. The intent was to develop a framework to guide the development of PPPs and strategic initiatives above the project tier, influencing regional development programs and project specific actions and decisions (see Table 1).

Developed by Harriman Gunn and Noble (2009), R-SEA was founded on SEA principles (see Vicente and Partidário, 2006; Partidário, 1996) as a means to ensure that planning and assessment for a region supports the most desired outcomes rather than the most likely ones. The goal was to advance a more regionally based EA and planning approach to address environmental and social issues at the strategic tier of decision-making (Harriman Gunn and Noble, 2009). The focus of R-SEA is on informing the development or evaluation of alternative scenarios or PPPs for a region and then assessing those alternatives based on their potential for cumulative effects and in consideration of social, economic, environmental and planning goals (CCME, 2009). Amongst its defining features are its regional scope, strategic nature, and consideration of cumulative effects (Noble and Harriman, 2009). Its methodological approach is founded on the integration of existing knowledge, experience, and theory drawn from strategic (Vicente and Partidário, 2006; Noble and Storey, 2001; Partidário, 2000), cumulative (Noble, 2008; Dubé, 2003; Bonnell and Storey, 2000) and regional EA (Noble and Harriman, 2008). As a framework for regional application, it is intended to be context-sensitive and "tailor-made to the kind of decisions at stake and the nature of the decision-making processes in place" (Partidário and Clark, 2000).

R-SEA Framework

The R-SEA framework and guiding principles are described in detail by Harriman Gunn and Noble (2009) and the CCME (2009). R-SEA consists of three interrelated phases: a pre-assessment phase; an impact assessment and evaluation phase; and a post-assessment phase, each with corresponding stages (see Fig. 1). The pre-assessment phase involves developing a reference framework, scoping key issues, and identifying drivers and patterns of change in the region and how they have influenced valued ecosystems components (VECs). This requires identifying the main issues and concerns in the region along with the core principles that will be used to guide the assessment process. Tiering opportunities with regional initiatives and EA are also identified, as are necessary partners and partnerships, and terms of reference for the R-SEA developed. Scoping serves to establish VECs, temporal and spatial boundaries for the assessment, and sets management targets, objectives or thresholds against which future conditions and alternative scenarios of cumulative change can be assessed.

The impact assessment phase focuses on developing alternative options or scenarios of development and conservation for the region, and an assessment of the potential effects on, threats to, or changes in the state of VECs under each scenario. Cumulative effects and condition changes under each scenario are identified, and the significance of those changes and effects relative to current or past baseline conditions, thresholds, or desirable levels of change examined. More than one strategic alternative may be the outcome, and the iterative nature of the process may require re-assessing alternatives based on consideration of mitigation and management needs. The objective is to establish a preferred or satisficing PPP direction.

The post-assessment phase involves making plans to mitigate or compensate for potentially unavoidable impacts associated with the identified PPP, and developing

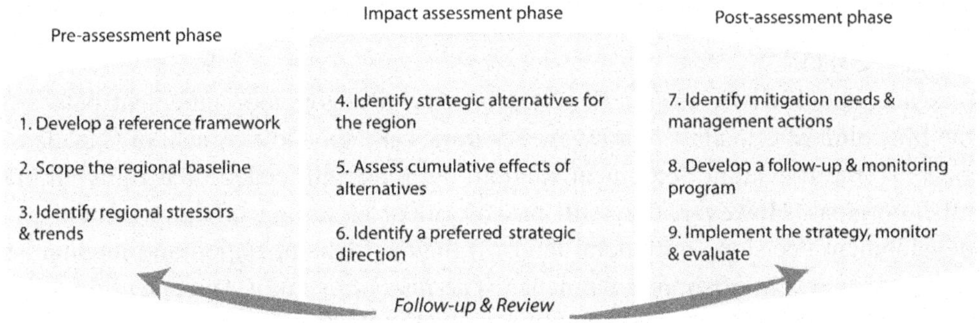

Figure 1. Methodological framework for R-SEA.

a follow-up and monitoring program for effects monitoring, goals achievement or performance evaluation, feedback on management effectiveness, and communication. This stage of the process is essential because PPPs are often formulated under uncertainty and their effectiveness is often sensitive to broader social and economic conditions (see CCME, 2009; Cherp *et al.*, 2007). Finally, in order to facilitate implementation and ensure that PPP strategies are put into action, there is a need to prescribe roles and resources for implementation and ongoing monitoring; undertake a formal public review process of the proposed strategy; and establish a regular review period to revisit the PPP, evaluate its efficacy and adjust accordingly.

Research Methods

Beaufort Sea

The Beaufort Sea Large Ocean Management Area covers over one million km^2, encompassing the marine portion of the Inuvialuit Settlement Region (Fig. 2), and has been identified as one of five priority areas for integrated ocean management by the Government of Canada (Cobb *et al.*, 2008). The area contains the Tarium Niryutait Marine Protected Area and is rich in biological diversity, including benthic fauna, birds, marine and anadromous fish (e.g. arctic cod, arctic char, salmon) terrestrial mammals (e.g. caribou, arctic fox, lynx, arctic hare) and marine mammals (e.g. bowhead whales, beluga whales, ringed seal, walrus, polar bear) (Cobb *et al.*, 2008). The marine resources of the Beaufort Sea provide sustenance and have been part of Inuvialuit fishing and hunting practices for centuries. The Beaufort Sea is also rich in hydrocarbon resources (Callow, 2012; INAC, 2010; Harrison, 2006). Prior to 2007, the majority of hydrocarbon exploration and development was restricted to the continental shelf and nearshore region. In recent years exploration licenses have been let in the deep offshore, raising concern about industry and government preparedness for drilling in frontier regions (Porta and Bankes, 2011).

Under the 1984 *Inuvialuit Final Agreement*, a negotiated agreement between the Inuvialuit of Canada's Northwest Territories and the Government of Canada to establish the Inuvialuit Settlement Region, the Inuvialuit ceded their rights in the offshore areas. However, they still provide input regarding offshore oil and gas development decisions on federal lands in their settlement region and maintain a role and interest in offshore governance. The *Inuvialuit Final Agreement* provides a framework for co-management in the Inuvialuit Settlement Region between the Inuvialuit and the federal government. However, for all intents and purposes, the

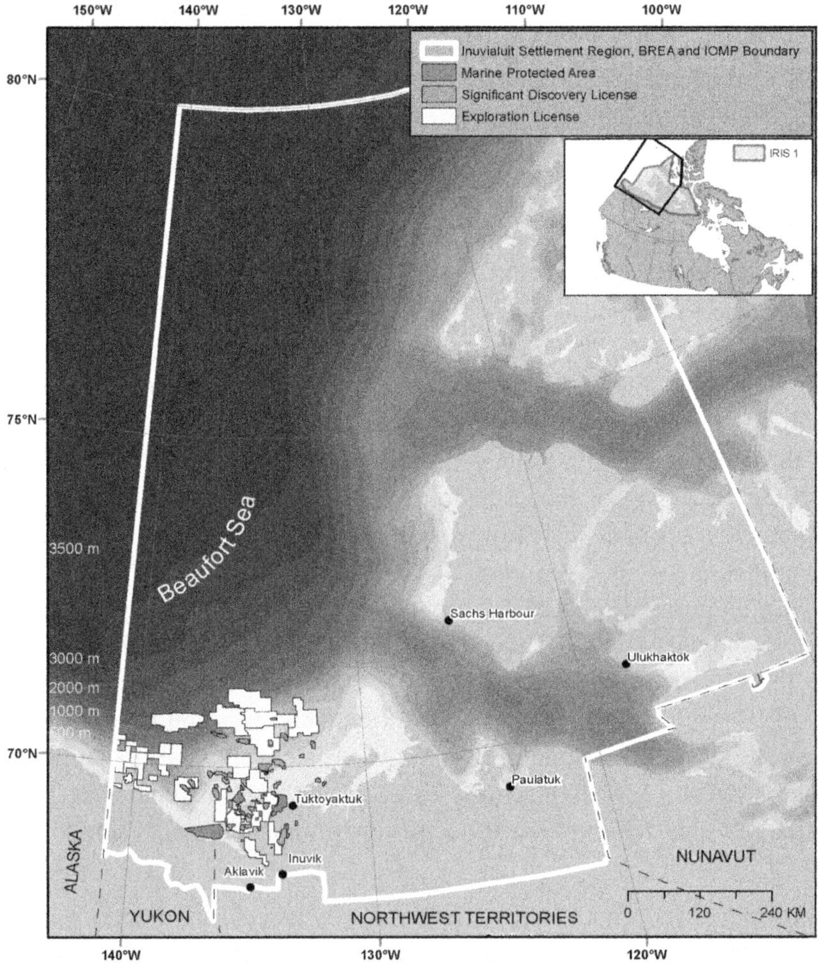

Source: Map developed by Michael St. Louis, University of Saskatchewan.

Figure 2. Inuvialuit Settlement Region of Canada's western Arctic.

Government of Canada retains ownership and authority to manage the offshore resources of the Beaufort Sea. The National Energy Board and Aboriginal Affairs and Northern Development Canada (AANDC) have independent but complementary roles in energy exploration and development in the Beaufort Sea: the National Energy Board authorises drilling, while AANDC administers rights to oil exploration. Fisheries and Oceans Canada manages development authorizations under the *Fisheries Act* (1985), and the *Canadian Environmental Assessment Act* (2012) applies to projects whenever a federal authority has a decision-making responsibility.

Project-based EA is the primary tool for planning for, assessing, and managing the impacts of offshore energy development. Challenges to EA and regulation for offshore oil and gas in the Beaufort Sea have been documented by Noble *et al.* (2013) and others (Doelle *et al.*, 2012; Ketilson, 2011; Voutier *et al.*, 2008; Erlandson and Sloan, 2001), and include the narrow scope of EA and limited reach beyond the individual development project, uncertainty and efficiency of the process, the potential for duplication of efforts in project assessment, and the lack of capacity to implement EA and regulatory programs. These limitations, combined with recent legislative and regulatory reforms to EA (Canadian Environmental Assessment Act, 2012) in Canada at the federal level that further limit the scope of EA application, including the exemption of small projects (e.g. exploration), and set maximum 24-month time frames for the review of large projects (e.g. offshore development), have generated considerable interest in regional and strategic approaches to offshore planning and assessment in the western Arctic. Notwithstanding current, independent programmes in the western Arctic focused on marine planning (e.g. Integrated Oceans Management Plan (IOMP)), baseline studies (e.g. Beaufort Regional Environmental Assessment (BREA)), and scientific understanding of climate change (e.g. Integrated Regional Impact Studies (IRIS)), there is no integrative planning and impact assessment framework to support PPP development and decisions about marine resource use, particularly with regard to deep offshore energy. What R-SEA is and what it should deliver in the western Arctic is "still far from consolidated" (Ketilson, 2011: 2). R-SEA remains largely untested and the capacity for implementation is unknown.

Data collection and analysis

Data collection was based on semi-structured interviews with Inuvialuit co-management boards and agencies and the Joint Secretariat ($n = 11$); the Inuvialuit Settlement Region Municipal Government ($n = 2$); the Gwich'in Chiefs and Gwich'in Resource Board ($n = 3$); the oil and gas industry ($n = 6$); the Government of the Northwest Territories, specifically the Water Board, Industry Tourism and Investment, and the Department of Executive ($n = 4$); the Federal Government, comprised of Aboriginal Affairs and Northern Development Canada, the Canadian Environmental Assessment Agency, Fisheries and Oceans Canada, the National Energy Board, and Natural Resources Canada ($n = 13$); private consultants ($n = 5$); environmental non-government organisations ($n = 2$); and academics and other energy interest groups ($n = 4$) researching or working in the region. Interviews with Inuvialuit and northern governments were conducted in-person in Inuvik, Northwest Territories. The remainder occurred over the telephone. Interview participants were selected based on their experience in, and knowledge of, offshore oil and gas development, planning and decision-making in the Beaufort Sea.

Participants were asked a series of semi-structured questions exploring: (i) existing challenges and opportunities with the current approach to, and provisions for, offshore oil and gas impact assessment, planning, management and decision-making in the Beaufort Sea; and (ii) critical factors, issues and concerns to be addressed through R-SEA in the Beaufort Sea, including participant views on stakeholder roles and the responsible authorities to lead and manage R-SEA. Interview results were organised using the CCME R-SEA framework described above as a general guide, coded thematically and analysed using QSR NVivo© v.9 to classify and manage qualitative information.

Results

Results are presented below based on each of the three broad phases of R-SEA: pre-assessment; assessment; and post-assessment. In cases where participants did not want their comments attributed to their department, they are referenced by the broader organisation with which they are affiliated.

Pre-assessment phase

There was a diversity of expectations on an appropriate reference framework for R-SEA, specifically what R-SEA should deliver in the Beaufort Sea. For industry, an expectation of R-SEA was to deliver greater certainty through predictability and consistency in the regulatory process. "Being able to have clear guidance on the where's and when's allows for more effective planning," explained one industry proponent. A federal government participant explained that R-SEA should be designed to "expedite the EA permitting stage by addressing problems early, and proactively solving them, rather than the current approach that is focused on mitigating negative impacts." Inuvialuit, environmental non-governmental organisations, and some government participants advocated for R-SEA as an early assessment tool that could potentially restrict development activity if the results from a preliminary analysis were unfavourable, either in terms of environmental sensitivities or data gaps. As one environmental non-governmental organisation participant explained, the current approach under EA is "how can we; with R-SEA it is 'should we'." The majority of Inuvialuit participants said that R-SEA should be a front-end assessment of conditions, when preparing leases and putting areas up for auction and tenure, to guide whether and how areas are opened for future development.

There was agreement amongst participants regarding enforceability and responsibility for R-SEA. Most said that R-SEA should be regulated so that it is enforceable and provides certainty in planning and decision-making. However, one environmental non-governmental organisation participant explained that greater certainty meant different things to different stakeholders: for some, certainty through R-SEA means greater support for establishing marine conservation areas; for industry certainty means that project approvals are expedited; and for the Inuvialuit it means that their right to access and benefit from marine resources is protected.

There was also agreement that responsibility for R-SEA implementation should be shared between the federal government and Inuvialuit, with strong industry involvement. An environmental consultant noted that creating a new body for R-SEA "would add to the perception of regulatory complexity…thus it would be better to expand and support [an existing] regulatory body through better funding, staff and technical support." A federal government participant maintained that the only way for R-SEA to be successful is through a process founded on stakeholder involvement so that "all interested and affected parties have a meaningful place at the table to influence decision-making and identify potentially conflicting values and ensuring such issues are addressed in the scoping stage." An 'industry-led' initiative was proposed by many industry and private consultant participants, with one participant explaining that "industry has financial resources so a partnership is ideal, and an umbrella partnership with industry, federal government and Inuvialuit could eliminate the 'us and them' mentality that currently exists." Such an umbrella partnership with industry was viewed with scepticism by other participants, including Inuvialuit. According to an Inuvialuit participant, the Inuvialuit view R-SEA as a means to "long term empowerment contributing increased confidence in decision-making, versus the current landscape which is plagued with contempt at the regional level as to how decisions are made at the federal level."

Participants deemed it necessary that R-SEA incorporate existing initiatives in the Beaufort Sea, namely IOMP, BREA and IRIS, and build upon already identified data and knowledge gaps, and address sociocultural and economic impacts. Most Inuvialuit participants recognised the need to "scope-in" onshore sociocultural and economic values in R-SEA for offshore development as a priority issue. Industry, government and environmental non-governmental organisations alike supported this need. The difficulty, one Inuvialuit participant explained, is that legislation does not address socioeconomic impacts the same way it does biophysical, and offshore activities have shown very little socioeconomic benefit for the Inuvialuit. When industries seek access to resources onshore they are required under the *Inuvialuit Final Agreement* to negotiate benefit agreements. Although industry is not legally required to do so offshore, some companies have voluntarily struck benefit agreements. For example, one industry participant explained that "the Inuvialuit wish to have offshore plans tailored to them as long-term development; we [industry proponent] feel this is reasonable to respect and have entered into agreements with the IRC [Inuvialuit Regional Corporation]." However, this particular company's opinion was not widely supported: other industry interviewees were opposed to reaching voluntary agreements with the Inuvialuit for offshore development, noting that regulatory framework does not require such actions.

A final issue raised concerned baseline data. Results indicated disagreement, particularly amongst those consultants involved in offshore EA practice, as to the amount and adequacy of existing baseline data. Some participants said that there is a large amount of existing information, but that there are still significant data gaps that need to be addressed. One environmental consultant noted that "EAs conducted offshore in the 80s were nearshore, primarily scientific, focused on oil spill response, and the regulatory regime was quite different then," thus highlighting the need for new data to adequately assess proposed drilling in the deep offshore. In contrast, another environmental consultant stressed that the necessary data is available, but it is not effectively managed. The participant explained "there is a tremendous amount of information that isn't being as used as effectively as it could be. R-SEA would not involve ascertaining new data, rather taking existing information and formatting it into a vehicle more readily accessible and understandable."

Impact assessment phase

The use of alternatives assessment in R-SEA, and identifying a "preferred" PPP direction, was an issue of considerable debate. The academic literature identifies alternatives as fundamental to R-SEA (see Gunn and Noble, 2009; Duinker and Greig, 2007), but there was dissent as to the usefulness of, and approach to, alternatives for R-SEA in the Beaufort Sea — specifically scenario-based approaches. There was no support for or against the use of alternatives assessment based on participant affiliation. For example, one industry participant explained that strategic planning is useful with scenarios, noting that it "doesn't have to be detailed per se, but can offer direction of how much development to permit and at what frequency." Similarly, an environmental consultant referred to the Norwegian and American regulatory regime, where cumulative impacts are examined before licensing, adding that in Canada, the practice has been to view each licensing call as a standalone exercise, and "we need to formalize the process, and look at the potential cumulative effects of licenses." A Fisheries and Oceans Canada regulator also saw the value in scenarios and explained: "if R-SEA could tell me where development needs to stop in terms of effects, that is the answer I need."

Others disagreed as to the usefulness and applicability of alternatives-based assessment and the use of scenarios. Some participants noted that identifying strategic alternatives or ways to proceed, for instance scenario planning, staged development, or restrictions toward permitting quotas, is not necessary in the Beaufort Sea given the lack of current development activity. The "pace of development will occur naturally," an environmental consultant participant expressed, "and will be determined, in part, by infrastructure capacity." Some industry participants were also sceptical, and concerned that entertaining alternative scenarios of development and conservation, including the scenario of 'no development', would not support their economic bottom line. For example, one industry participant explained that:

> "Industry is opposed to preferred scenarios because *whose* preferred scenario? Preferred scenario as an oil proponent is access to *my* resources. The Inuvialuit's preferred scenario is they get benefits with minimal impact. How does a preferred scenario unite industry, Inuvialuit and government? Preferred scenarios are polarizing, and latching onto a preferred development can end up killing a project."

An environmental non-governmental organisation participant added fodder to industry's unease, noting: "it's entirely possible that you could have policy on the energy and environment side that could restrict the ability to develop resources."

When PPPs and strategic options are developed through R-SEA, results reveal the importance of accommodating the preferences and priorities of stakeholders. Ensuring that risk and benefit debates take place, for instance, was posed by an Inuvialuit participant as an important issue in prioritizing regional PPPs and development strategies: "are there economic benefits for the Inuvialuit if development activities proceed, if so is it worth the trade-off for potential disruption of the ecosystem if there was an oil spill?" Results also showed that industry goals require careful consideration so that R-SEA does not automatically foreclose development opportunities. One participant from the territorial government explained that "industry pay huge sums for rights and without industry there is no economic development and without a regulatory environment that supports industry then we fail to support a main element of Canada's Northern Strategy, economic development." It is important when identifying a strategic alternative for offshore development that there is the potential to deliver favourable results for industry. An industry participant explained that sequencing development through a strategic planning process should be viewed in a positive fashion in that data could support more, as opposed to fewer wells.

A final issue was assessing cumulative effects. Some participants saw the potential to address cumulative effects through R-SEA, and the uncertainty that characterises the cumulative effects assessment of alternative futures as a positive attribute of R-SEA: others viewed the lack of concreteness around regional cumulative effects as a negative feature of R-SEA. An industry participant explained that "there are so many approaches to assessing cumulative effects that it is confusing, and less than useful, even harmful to industry if the assessment isn't carried out effectively with clearly established boundaries, funding and expertise." Consistent with the CCME's (2009) view of R-SEA, some participants emphasised that long-term assessment horizons are characterised by vast uncertainties about change and external drivers and can lead to corresponding uncertainty about what effects are most likely to occur in the future. Those in favour of R-SEA for cumulative effects assessment maintained that it must be stakeholder-based and a participatory process. An industry participant added that "regional scale, with multi-stakeholder participation, holds potential for cumulative effects assessment – a staple to success." Participants also identified various elements that required attention when approaching cumulative effects through R-SEA, including regional baseline data and the importance of considering the complexity of the cumulative effects issues and pathways in marine environments.

Post-assessment phase

Follow-up and monitoring emerged as a key issue in the post-assessment phase, specifically regarding R-SEA's added value to management and decision-making. An Inuvialuit participant explained the value that could come from understanding the outcomes of decision-making through follow-up, not only as it pertains to R-SEA, but also by applying performance evaluation results to implementation targets of the *Inuvialuit Final Agreement*: "follow-up monitoring would be beneficial to offer land claim beneficiaries some tangible data on the effectiveness of mitigation and management actions." A significant shortcoming identified under current EA in the Beaufort Sea was the lack of follow-up and monitoring. As one industry participant explained, "regulatory agencies do not monitor very well — they don't look at it from a larger regional perspective and it doesn't go back into a planning process." A Joint Secretariat participant agreed, noting that monitoring is "more for the purposes of meeting regulatory requirements than fulfilling a practical purpose." Congruence in monitoring methods and data collection emerged as important to R-SEA, with industry's project-specific monitoring efforts able to provide a valuable contribution to the monitoring of VECs. Consistent with the CCME (2009) it is necessary to ensure monitoring and data are quality controlled, transferable and comparable. Failure to achieve congruence may lead to problems in comparing monitoring results. As an environmental non-governmental organisation participant explained: "if you allow proponents to form their own monitoring plan you get a mishmash of data with different methodologies that makes obtaining a coherent picture challenging. Make sure the monitoring plan is well defined for the proponent so the data received is comparable and can be used to inform a regional picture."

Finally, participants identified the need for an adaptive approach to implementation. An industry participant expanded on the need for flexible and adaptable implementation:

> "Dealing with so many uncertainties we need to reserve flexibility to adjust as development unfolds. R-SEA model might be the best thing in the world, but it could be useless in affecting change if it cannot accommodate public opinion and values. When dealing with so many uncertainties with regard to how the offshore will develop, it must be modifiable."

An environmental non-governmental organisation participant added that it must demonstrate that the views of the affected public are taken seriously and integrated, where possible, in the final design of the strategic alternative, and "it has to be a living document informed by public opinion, so that the outcomes are reflected by inputs." The importance of implementing R-SEA as an iterative and adaptive process was deemed as core to R-SEA's success, particularly when it comes to VECs and having "a system built in for ensuring VECs are validated regularly, as values change over time."

Discussion: Advancing R-SEA in the Beaufort Sea

There are several issues and challenges that need to be addressed to advance R-SEA in the Beaufort Sea, yet there appears to be a foundation from which to move forward. The goal of this paper was to identify key issues and gaps that may pose challenges or opportunities for R-SEA implementation in the Beaufort Sea. What follows are a number observations concerning R-SEA implementation. Although based on Canada's western Arctic, the lessons and observations are relevant to other national and international offshore jurisdictions.

Diversity of expectations

Diversity in expectations of what R-SEA is expected to address and deliver in the Beaufort Sea is a challenge to its development and implementation, particularly when attempting to establish a reference framework. Though most expectations of what R-SEA should accomplish in the Beaufort Sea were consistent with what R-SEA can potentially deliver (see CCME, 2009), results reinforce the work of Noble *et al.* (2013), Ketilson (2011) and Vicenté and Partidário (2006) who respectively observed that R-SEA's role, particularly in energy sector PPP development and assessment, is neither well developed nor understood, and hence the nature of R-SEA remains unclear to many.

Part of the challenge may be attributed to the flexibility of R-SEA as an approach to planning and assessment. Although some have argued that structure and consistency is core to strategic approaches (Fischer, 2003), others like Partidário (2000), Nilsson and Dalkmann (2001), Fischer (2005), and the CCME (2009) maintain that SEA-based initiatives must be flexible to regional and local context. In the Beaufort Sea, flexibility proved to be a desirable characteristic of R-SEA. However, a challenge associated with flexibility was the establishment of clear purpose and scope for R-SEA and reaching agreement on the specific questions and problems to be addressed. Failure to agree on the scope of R-SEA early on in the process poses a major impediment to its adoption, implementation and ultimate effectiveness.

Poor understanding of, or disregard for, context can lead to simplified judgements about resource systems, and consequently failures in management efforts (Bina, 2008). To ensure that R-SEA can emerge as an accepted and worthwhile part of planning and decision-making processes, constraints and intent need to be agreed upon at the outset and made clear to all interested parties (see Dalal-Clayton and Sadler, 2005). This is currently not the case in the Beaufort Sea. The diversity of expectations reinforce Dalal-Clayton and Sadler's (2005) findings that R-SEA needs to be better understood if it is to be welcomed into decision-making for resource planning and development. However, notwithstanding the variety of opinions on what R-SEA should do in the Beaufort Sea, there was consensus that something needs to be done and that the current project-based approach is not sufficient.

Scepticism and resistance about a futures-based approach

Notwithstanding increasing calls for R-SEA in the Beaufort Sea, there remains scepticism and resistance, albeit by the minority, that R-SEA will not be a fair process for all stakeholders and certain procedural benefits, such as establishing regional targets and thresholds, may have adverse impacts for industry. Resistance to R-SEA implementation was rooted in one of the most fundamental components of R-SEA, that of examining alternative futures (see CCME, 2009; Jones *et al.*, 2005) and selecting a strategic course of action. Particular scepticism emerged regarding the need for scenario planning given the lack of current offshore energy activity in the region, suggesting limited understanding of the purpose of R-SEA and the role of scenarios in planning and assessment. This was not surprising. Duinker and Greig (2007) report that, although most appropriate to the task of planning for and assessing the impacts of future development, scenario-based approaches are significantly underutilised and promoted in EA practice. Scenario-based approaches are about examining "what may happen," "what is most likely to happen," and "what would we prefer to happen" concerning the development and outcomes of development in a particular region (see Rubin and Kaivo-oja, 1999). In a region characterised by much uncertainty concerning the future impacts of offshore development, as noted by interview participants, scenario-based analysis is ideally suited for grappling with such uncertainties and a shift away "from trying to estimate what is most likely to occur toward questions of what are the consequences and most appropriate responses under different circumstances" (Duinker and Greig, 2007: 209). Interestingly, the consideration of strategic alternatives early in decision-making, before irreversible development decisions are taken, was an opportunity and potential benefit of R-SEA that most Inuvialuit participants supported. However, among those who accepted the need for scenario use and evaluation in the Beaufort, there were many different opinions on what scenarios should be considered and what they would be willing to accept — ranging from seismic-free areas to energy policy on renewable versus non-renewable energy futures.

Understanding stakeholder resistance and scepticism

Resistance and scepticism regarding alternative assessments and scenario-based approaches may be a result of the perceived risk of being locked into a PPP that is seen as "less than desirable" for any individual interest. This was a particular concern raised by industry. However, more attention is needed in understanding stakeholder resistance to address the root cause of this perception. Such scepticism could stem from fear of legislation of R-SEA outputs, or the lack of flexibility for industry to plan for and control its future as it has had the liberty to do to date in the Beaufort. The level of disdain regarding the current regulatory system by industry participants is not improving. Erlandson (2009) argues that the problem with northern regulatory systems is not that they are complex, but that they don't work in relation to the business cycle for oil and gas development. Erlandson

(2009: 419) further reports that operators in the petroleum sector "can't count on standards of administrative law such as consistency, predictability and timeliness that are taken for granted elsewhere," and their perceptions of risk have risen over the last 10 years. Such stark findings illustrate deeply rooted public policy problems that have been well documented by others (see Abele, 2009; McCrank, 2008), which have lead in part to cynicism on behalf of industry on the government's ability to effectively and efficiently manage the offshore. In this way, industry seemed sceptical not of change per se, but rather the government's capacity to effectively carry out and implement an R-SEA.

Vicente and Partidário (2006) argue that such perceptions are influenced by individual cultural and social contexts, including past experience, and that understanding the root of the problem and the different perceptions of a problem lie at the core of SEA activities. Ultimately, however, if no PPP direction is identified, then R-SEA is no more than a data collection exercise with little value added to development assessing and decision-making (see CAPP, 2009) and, arguably, no different than past efforts in the Beaufort region. If there is a lack of commitment to identifying and charting a course in the western Arctic through R-SEA, then perhaps R-SEA is not the right tool. On the other hand, if there is a lack of commitment from certain stakeholders, as is the case in the Beaufort, then perhaps there is a need for better education regarding R-SEA. This could be accomplished through, for example, empirical case study learning from other offshore jurisdictions to demonstrate that R-SEA does not arrive at polarizing decisions, nor is it a command and control type process (see Knol, 2010; Hasle *et al.*, 2009). Instead, it is a means of integrating environmental and broader sustainability considerations in regional PPPs. There is a need to bring forward evidence from case applications that R-SEA can achieve its objectives without adding significant delays to regulatory and decision-making processes. Nevertheless, if industry is fundamentally opposed to R-SEA this will pose a significant challenge to implementation. This is particularly the case given the current national political climate that is driving energy investment as a result of Prime Minister Harper's commitment for Canada to be an energy superpower (see Hester, 2007), and the relaxing of federal environmental assessment laws and regulations to promote more rapid exploration, development and approval processes (see Gibson, 2012).

Good governance

R-SEA can help advance good regional governance for resource planning and assessment, as intended by the CCME (2009), but there was concern by some participants that R-SEA would simply be another layer of government. Scott (2008) maintains that SEA's contribution to good governance has been made without examination of the tensions inherent within good governance discourse, and that the SEA-good governance nexus is more an objective than a reality. In other words, R-SEA can help advance good governance, but its success in doing so is largely dependent on the extent to which stakeholders are engaged and able to participate in and influence decision-making. This kind of circumspect thinking emerged in the results, as some participants saw R-SEA adding another layer of bureaucracy with no guaranteed benefits. For some Inuvialuit participants, there was concern that R-SEA could be a process dictated at the federal level, eroding decision-making power and influence within the Inuvialuit Settlement Region. This perceived risk of loss of control was exacerbated by concerns over the possibility of industry leading R-SEA initiatives, and playing a key role in the environmental management of the region. Despite these concerns, there was broad agreement that government should play a lead role in R-SEA through a multi-stakeholder approach.

The OECD (2006: 46) reports that "SEA supports *good governance* by: encouraging stakeholder participation…[and] Increasing transparency and accountability' as well as 'clarifying institutional responsibilities."' In doing so, there is an opportunity for R-SEA to identify capacity needs and legislative and policy gaps to support effective environmental and socioeconomic PPP development and implementation. The increased transparency and understanding afforded by SEA at the regional level can be a potential catalyst for increased efforts at joined-up government through "increas[ing] pressures to overcome bureaucratic fragmentation and jurisdictional conflict", according to Stinchcombe and Gibson (2001: 354) in Scott (2008). The Inuvialuit have been active participants in multi-stakeholder government processes to improve and prepare for offshore development for decades (see CAPP, 2009). In this way, existing multi-stakeholder initiatives such as the IOMP, BREA and IRIS may help set a foundation or establish pre-conditions for R-SEA success. As laudable as these initiatives have been, however, ubiquitous governance challenges prevail (see Erlandson, 2009), which from the Inuvialuit perspective comes from the "requirements of legislation and institutions, mainly federal, that are layered on top of those of the Inuvialuit Settlement Region, and which were not designed to be coordinated with each other" – or even complementary (Cournoyea, 2009: p. 391). Fundamental governance and coordination challenges such as these may be addressed in part through R-SEA, however, given their legislative depth they will likely continue to challenge offshore planning and management efforts and be a significant challenge to R-SEA implementation.

Social impacts

Results support Norman's (2005) findings that the well-being of coastal communities should be paramount in marine resource development, and there is a need for R-SEA to focus on the onshore socioeconomic implications of offshore development. The scale and pace of future hydrocarbon development are more than a matter of deposits and reserves, and will be determined by a variety of factors including supportive infrastructure and technology, a stable political and public policy environment, and a skilled workforce (Fenge, 2009). R-SEA offshore must therefore accommodate onshore socio-economic PPP issues associated with energy development, such as community development (e.g. influx of workers), if it is to reflect broader regional sustainable development goals and objectives.

Porta and Bankes (2011) echo the concern that many Inuvialuit participants raised regarding risks and benefits: "of all the players involved in an Arctic offshore oil and gas program, Inuit communities bear the greatest risk, and with the current system benefits are not proportionate to these risks." Oil and gas investments can have significant positive local, and in the case of major projects, regional and national economic impacts. However, extractive industry investments can also give rise to a range of negative effects (Wagner and Armstrong, 2010). The potential negative effects of resource extraction proved to be a major concern for participants, especially the Inuvialuit.

R-SEA can also be a critical tool in engaging those potentially affected by development. As evidenced by local Aboriginal engagement in setting mitigation standards for the Labrador Shelf SEA, Atlantic Canada (see Fidler and Noble, 2012), even at the strategic tier local communities can influence decision outcomes. Land claim settlements have empowered Inuit communities to document traditional ecological knowledge and apply it to land and marine use planning. Traditional knowledge alongside scientific studies can help fill gaps such that, for example, biologically and culturally significant marine areas are identified for priority protection. In this way R-SEA can play an important role in supporting the objectives of the Inuvialuit Final Agreement (see IFA, 1984).

Benefit plans emerged as a contentious topic: the seabed where offshore drilling occurs is Crown land but unlike for onshore arrangements industry is not legally obliged in the offshore to establish impact and benefit agreements. Porta and Bankes (2011) maintain that the absence of a legal trigger to negotiate benefit agreements for offshore production fails to balance the risks and benefits of those operations for local communities. There is an opportunity for R-SEA to integrate socioeconomic considerations early in the planning of offshore oil and gas systems

(CCME, 2009), and, R-SEA could potentially integrate benefit agreements into the assessment process and offer decision-makers and local interests a better understanding of the broader development planning cycle. This could enable an evaluation of how such agreements contribute to the region and maximize economic benefits, prior to specific project-based negotiations, and whether they support regional goals and R-SEA objectives. The roles of these two potentially complementary processes, R-SEA and negotiated benefit agreements, in contributing toward more informed decision-making requires further examination. To date, the focus on research on benefit agreements in relation to environmental management has focused exclusively on project-level EA, namely in the mineral sector (see Noble and Fidler, 2011), and suggest that such agreements have emerged, in part, due to the inadequacies of project-based assessments in addressing cumulative effects, socioeconomic impacts and follow-up (Galbraith *et al.*, 2007). It may well be the case that such agreements need to remain outside the scope of R-SEA, but in the context of impact management and Arctic offshore planning the current system requires reform (Doelle *et al.*, 2012). Cournoyea (2009: p. 392) argues that while the Inuvialuit are adaptable people, it is the federal government's responsibility to invest in public services such as health and social services, concluding that "if the only investment in capacity is in reaction to development after the fact, it will come too late."

Post-assessment and implementation

Overall, participants had much more to say, and raised many more concerns, regarding the pre-assessment and assessment phases of R-SEA than they did about implementation and post-implementation evaluation of the PPPs that emerge from R-SEA. Although the lack of follow-up and monitoring under EA (see Morrison-Saunders and Arts, 2004; Arts *et al.*, 2001) was raised by study participants, relatively little comment was made regarding what comes after R-SEA, when a PPP or course of action is identified. This is not surprising, as Gachechiladze *et al.* (2009) report that most SEA research and practice has focused almost exclusively on pre-decision stages, in particular the development and application of SEA systems and methodological frameworks, with limited attention to the post-decision follow-up stage. With limited attention to monitoring for undesirable effects during the PPP's implementation, it will prove difficult to draw conclusions on planning assumptions and apply these to subsequent assessments (see Cherp *et al.*, 2007).

This lack of consideration to post-implementation is concerning given the noted uncertainties associated with development in the offshore region, particularly those associated with climate change. Changes in landfast ice patterns may have significant effects on offshore access and managing the risks to infrastructure due to melting permafrost (see Voutier *et al.*, 2008). Arguably, R-SEA does provide a framework for integrating climate change considerations into PPPs (see OECD, 2009). However, such emerging and external factors must be considered in R-SEA from the outset with follow-up programs post-implementation designed to inform the need for changes in previously formulated strategies (see Cherp *et al.*, 2007).

Part of the challenge, however, is that the necessary supporting institutional environments to ensure that a PPP or strategy can succeed, or that management actions are in place in order for a preferred strategy to be considered viable, remains underdeveloped both in the academic literature (Gachechiladze *et al.*, 2009) and in practice. Problematic in this regard, and consistent with Noble (2003), is that R-SEA output and decision outcomes are to a significant extent a function of the input and quality of the R-SEA process. If institutional arrangements to support the implementation of R-SEA output have not been considered during the entire R-SEA process, efforts performed in the pre-assessment and assessment phase could prove futile. The result is an R-SEA process that resembles no more than a visioning and data collection exercise but with no implementation or links to decision-making. Shannon (1998) maintains that institutional arrangements are just as important as the scientific and technical aspects of understanding and conducting assessments, as they establish a template for patterns of relationships and administrative mechanisms to manage effects. Although partnerships exist in the Beaufort Sea through various multi-stakeholder programs and initiatives, no research on the network of institutional arrangements has been conducted to establish whether existing arrangements can support R-SEA implementation. There is a documented need to better understand the ways in which institutional arrangements affect the advancement and implementation of R-SEA. Even the most well intended strategies are of little value if they are not put into action (CCME, 2009).

Conclusion

Large-scale offshore hydrocarbon development is looming in the Canadian Beaufort Sea, but there is no R-SEA framework to help plan for future energy development, establish a long-term regional vision, or to assess and effectively manage potential cumulative effects. A majority of stakeholders have advocated for R-SEA in the Beaufort (e.g. IGC, 2004; Arctic Council, 2009), and there is optimism that a foundation exists through the application of the CCME R-SEA model to guide and advance offshore planning and assessment to enable more informed regional development decisions in support of sustainability. This paper set out to identify key opportunities for, and challenges to, the implementation of R-SEA in the Beaufort Sea, and in doing so examined support for its implementation and identified areas of stakeholder disagreement. Although acute implementation challenges were identified, such as issues of governance and the nature and scope of alternatives assessment, findings reveal particular challenges to the current project EA approach and show that R-SEA offers a much needed framework to begin addressing stakeholder concerns about future offshore development in the region. The CCME framework provides a starting point to advance R-SEA in Canada's western Arctic and ensure that offshore development in the region reflects the intended and desirable outcomes of stakeholders, as opposed to the most likely ones.

Acknowledgments

This research was funded by the Social Sciences and Humanities Research Council of Canada, with additional support provided by the Northern Scientific Training Program. The authors wish to acknowledge the study participants for their time and insight, and the constructive feedback of two anonymous reviewers.

References

Abele, F (2009). Northern development: past, present and future. In Abele, F, TJ Courchene, FL Seidle and F St-Hilaire (eds.), *Northern Exposure: Peoples, Powers and Prospects in Canada's North*. Montreal, Quebec: The Institute for Research on Public Policy. pp. 19–25.

Arctic Council (2009). Protection of the arctic marine environmental working group: Arctic offshore oil and gas guidelines. Retrieved from http://www.arctic-council.org.

Arts, J, P Caldwell and A Morrison-Saunders (2001). EIA follow-up: good practice and future directions: findings from a workshop at the IAIA 2000 Conference. *Impact Assessment and Project Appraisal*, 19(3), 175–185.

Bina, O (2008). Strategic environmental assessment. In Jordan, A and A Lenschow (eds.), *Innovation in Environmental Policy? Integrating the Environment for Sustainability*. Massachusetts: Edward Elgar Publishing, pp. 134–156.

Bina, O (2007). A critical review of the dominant lines of argumentation on the need for strategic environmental assessment. *Environmental Impact Assessment Review*, 27(7), 585–606.

Bonnell, S and K Storey (2000). Addressing cumulative effects through strategic environmental assessment: a case study of small hydro development in Newfoundland,

Canada. *Journal of Environmental Assessment Policy and Management*, 2(4), 477–499.

Burkett, V (2011). Global climate change implications for coastal and offshore oil and gas development. *Energy Policy,* 39(12), 7719–7725.

Callow, L (2012). Oil and gas exploration and development activity forecast. Canadian Beaufort Sea 2012–2027. Prepared for Beaufort Regional Environmental Assessment Aboriginal Affairs and Northern Development Canada. LTLC Consulting in association with Salmo Consulting Inc.

Canadian Association of Petroleum Producers (CAPP) (2009). Research report: Beaufort regional environmental reports summary. Retrieved from http://www.capp.ca/getdoc.aspx?DocId=164181&DT=PDF.

Canadian Council of Ministers of the Environment (CCME) (2009). Regional strategic environmental assessment in Canada: Principles and guidance. Ottawa, ON.

Cherp, A, A Watt and V Vinichenko (2007). SEA and strategy formation theories: from three Ps to five Ps. *Environmental Impact Assessment Review*, 27(7), 624–644.

Cobb, D, H Fast, MH Papst, D Rosenberg, R Rutherford and JE Sareault (2008). Beaufort Sea Large Ocean Management Area: Ecosystem overview and assessment report. Central and Arctic Region Freshwater Institute, Winnipeg, MB.

Cournoyea, NJ (2009). Navigating and managing economic, environmental, and social change in the Inuvialuit Settlement Region. In Abele, F, TJ Courchene, FL Seidle, and F St-Hilaire (eds.), *Northern Exposure: Peoples, Powers and Prospects in Canada's North*. Montreal, Quebec: The Institute for Research on Public Policy, pp. 389–394.

Dalal-Clayton, B and B Sadler (2005). *Strategic Environmental Assessment: A Source-book and Reference Guide to International Experience*. London, England: International Institute for Environment and Development.

Doelle, M, N Bankes and L Porta (2012). Using strategic environmental assessment to guide oil and gas exploration decisions in the Beaufort Sea: lessons learned from Atlantic Canada. Retrieved from http://ssrn.com/abstract=2142001.

Dubé, MG (2003). Cumulative effect assessment in Canada: a regional framework for aquatic ecosystems. *Environmental Impact Assessment Review,* 23(6), 723–745.

Duinker, PN and LA Greig (2007). Scenario analysis in environmental impact assessment: improving explorations of the future. *Environmental Impact Assessment Review*, 27(3), 206–219.

Elvin, SS and GS Fraser (2012). Advancing a national strategic environmental assessment for the Canadian offshore oil and gas industry with special emphasis on cumulative effects. *Journal of Environmental Assessment and Policy Management*, 14(3), 1250015.

Erlandson, G (2009). Aligning petroleum sector activities with northern regulatory requirements. In Abele, F, TJ Courchene, FL Seidle and F St-Hilaire (eds.), *Northern Exposure: Peoples, Powers and Prospects in Canada's North*. Montreal, Quebec: The Institute for Research on Public Policy, pp. 415–423.

Erlandson, G and J Sloan (2001). Oil and gas approvals in the Northwest Territories – Inuvialuit Settlement Region: a guide to regulatory approval process for oil and natural gas exploration and production in the Inuvialuit Settlement Region. Alvarez, Sloan and Associates and Erlandson and Associates.

Fenge, T (2009). Economic development in Northern Canada: challenges and opportunties. In Abele, F, TJ Courchene, FL Seidle and F St-Hilaire (eds.), *Northern Exposure: Peoples, Powers and Prospects in Canada's North*. Montreal, Quebec: The Institute for Research on Public Policy, pp. 375–388.

Fidler, C and BF Noble (2012). Advancing strategic environmental assessment in the offshore oil and gas sector: lessons from Norway, Canada and the United Kingdom. *Environmental Impact Assessment Review*, 34, 12–21.

Fischer, TB (2005). Having an impact? Context elements for effective SEA application in transport policy, plan and programme making. *Journal of Environmental Assessment Policy and Management*, 7(3), 407–432.

Fischer, TB (2003). Strategic environmental assessment in post-modern times. *Environmental Impact Assessment Review*, 23(2), 155–170.

Gachechiladze, M, BF Noble and B Bitter (2009). Following-up in strategic environmental assessment: a case study of 20-year forest management planning in Saskatchewan, Canada. *Impact Assessment and Project Appraisal*, 27(1), 45–56.

Galbraith, L, B Bradshaw and MB Rutherford (2007). Towards a new supraregulatory approach to environmental assessment in Northern Canada. *Impact Assessment and Project Appraisal*, 25(1), 27–41.

Gibson, RB (2012). In full retreat: the Canadian government's new environmental assessment law undoes decades of progress. *Impact Assessment and Project Appraisal*, 30 (3), 179–188.

Gunn, J and BF Noble (2011). Conceptual and methodological challenges to integrating SEA and cumulative effects assessment. *Environmental Impact Assessment Review*, 31(2), 154–160.

Gunn, J and BF Noble (2009). A conceptual basis and methodological framework for regional strategic environmental assessment (R-SEA). *Impact Assessment and Project Appraisal*, 27(4), 258–270.

Harrison, C (2006). Industry perspectives on barriers, hurdles, and irritants preventing development of frontier energy in Canada's Arctic islands. *Arctic*, 59(2), 238–242.

Hasle, JR, U Kjellen and O Haugerud (2009). Decisions on oil and gas exploration in an Arctic area: case study from the Norwegian Barents Sea. *Safety Science*, 47(6), 832–842.

Hester, A (2007). Canada as the "emerging energy superpower": Testing the case. A Paper presented for the Canadian Defence and Foreign Affairs Institute. Waterloo, Ont.

Indian and Northern Affairs Canada. (INAC) (2010). Northern Oil and Gas Annual Report. Retrieved from http://www.aadnc-aandc.gc.ca/.

Inuvialuit Final Agreement (IFA). (1984). Retrieved from: http://www.inuvialuitland.com/resources/Inuvialuit_Final_Agreement.pdf.

Inuvialuit Game Council. (IGC). Letter to Honourable David Anderson, PC, MP; 21 June 2004.

Jones, C, M Baker, J Carter, S Jay, M Short and C Wood (2005). *Strategic Environmental Assessment and Land Use Planning: An International Evaluation*. London: Earthscan.

Ketilson S (2011). Regional strategic environmental assessment roles and stakes in Arctic oil and gas development. M.E.S. Thesis. Saskatoon, SK: University of Saskatchewan.

Knol, M (2010). Scientific advice in integrated ocean management: The process towards the Baronts Sea Plan. *Marine Policy*, 34, 252–260.

McCrank, N (2008). Road to improvement: the review of regulatory systems across the North. Indian and Northern Affairs Development Canada.

Morrison-Saunders, A and J Arts (2004). Introduction to EIA follow-up. In Morrison-Saunders A and J Arts (eds.), *Assessing Impact: Handbook of EIA and SEA Follow-up*. London: Earthscan, pp. 1–21.

Nilsson, M and H Dalkmann (2001). Decision making and strategic environmental assessment. *Journal of Environmental Assessment Policy and Management*, 3(3), 305–327.

Noble, BF (2008). Strategic approaches to regional cumulative effects assessment: a case study of the Great Sand Hills, Canada. *Impact Assessment and Project Appraisal*, 26(2), 78–90.

Noble, BF (2009). Promise and dismay: the state of strategic environmental assessment systems and practices in Canada. *Environmental Impact Assessment Review*, 29(1), 66–75.

Noble, BF (2003). Auditing strategic environmental assessment practice in Canada. *Journal of Environmental Assessment Policy and Management*, 5(2), 127–147.

Noble, BF and C Fidler (2011). Advancing indigenous-community corporate agreements: lessons from practice in the Canadian mining sector. *Oil, Gas and Energy Law Intelligence*, 9(4), 1–30.

Noble BF, S Ketilson, A Aitken and G Poelzer (2013). Strategic environmental assessment opportunities and risks for Arctic offshore energy planning and development. *Marine Policy*, 39(1), 296–302.

Noble, BF and J Harriman (2009). Strategic environmental assessment. In Hanna K (ed.), *Environmental Impact Assessment: Practice and Participation*. Don Mills, ON: Oxford University Press.

Noble, BF and J Harriman (2008). Regional strategic environmental assessment: Methodological guidance and good practice. Report prepared for the Canadian Council of Ministers of the Environment. Environmental assessment task group: Ottawa, ON.

Noble, BF and K Storey (2001). Towards a structured approach to strategic environmental assessment. *Journal of Environmental Assessment and Policy Management*, 3(4), 483–508.

Norman, D. (2005). A Review of the Potential Implications of an Offshore Oil and Gas Industry on Coastal Communities. UNBC Community Collaborative Studies on British Columbia Offshore Oil and Gas #5, Vancouver, BC.

Organisation for Economic Co-operation and Development (OECD) (2006). *Applying Strategic Environmental Assessment: Good Practice Guidance for Development Co-operation.* Paris.

Organisation for Economic Co-operation and Development (OECD) (2009). OECD policy guidance on integrating climate change adaptation into development co-operation. Retrieved from http://www.oecd.org/dac/43652123.pdf.

Partidário, MR (2000). Elements of an SEA framework – improving the value added of SEA. *Environmental Impact Assessment Review*, 20(6), 647–663.

Partidário, MR (1996). Strategic environmental assessment: key issues emerging from recent practice. *Environmental Impact Assessment Review*, 16(1), 31–55.

Partidário, MR and R Clark (2000), *Perspectives on Strategic Environmental Assessment.* Boca Raton, FL: CRC-Lewis.

Porta, L and N Bankes (2011). Becoming Arctic-ready: policy recommendations for reforming Canada's approach to licensing and regulating offshore oil and gas in the Arctic. The Pew Environment Group.

Prowse, TD, C Furgal, R Chouinard, H Melling, D Milburn and SL Smith (2009). Implications of climate change for economic development in northern Canada: energy, resource, and transportation sectors. *Ambio*, 38(5), 272–281.

Rubin, A and J Kaivo-oja (1999). Towards a futures-oriented sociology. *International Review of Sociology*, 9(3), 349–371.

Sadler, B (2005). Canada. In Jones C, M Baker, J Carter, S Jay, M Short and C Wood, (eds.), *Strategic Environmental Assessment and Land Use Planning: An International Evaluation.* Sterling, VA: Earthscan, pp. 44–58.

Scott, C (2008). Strategic environmental assessment (SEA) and 'good governance:' Tryst or Tension? *The International Journal of Public Participation*, 2(2), 5–24.

Shannon, M (1998). Social organizations and institutions. In Naiman R and R Bilby (eds.), *River Ecology and Management: Lessons from the Pacific Coastal Ecoregion.* New York: Springer Verlag, pp. 529–552.

Stinchcombe, K and RB Gibson (2001). Strategic environmental assessment as a means of pursuing sustainability: ten advantages and ten challenges. *Journal of Environmental Assessment Policy and Management*, 3(3), 343–372.

Vicente, G and MR Partidário (2006). SEA — enhancing communication for better environmental decisions. *Environmental Impact Assessment Review*, 26(8), 696–706.

Voutier, K, B Dixit, P Millman, J Reid and A Sparkes (2008). Sustainable energy development in Canada's Mackenzie Delta-Beaufort Sea coastal region. *Arctic*, 61(1), 103–110.

Wagner, J and K Armstrong (2010). Managing environmental and social risks in international oil and gas projects: perspectives on compliance. *Journal of World Energy Law and Business*, 3(2), 140–165.

World Wildlife Fund (WWF) (2005). *Where are all the SEAs? Project Finance, and Strategic Environmental Assessment of Major Oil and Gas Developments.* UK: WWF.

CHAPTER 10

Perspectives on Inter-Linking Impact Assessment and Green Procurement: The Case of Green Energy

Kedar Uttam*, Berit Balfors, Charlotta Faith-Ell and Ulla Mörtberg

Department of Land and Water Resources Engineering
KTH Royal Institute of Technology, Brinellvagen 28
10044 Stockholm, Sweden
**kedar@kth.se*

The construction sector consumes significant quantities of energy and contributes substantially to greenhouse gas emissions. It is imperative that the sector considers a renewable component to its energy procurement. The sector has adopted various policy instruments such as impact assessment and green procurement for improving its environmental performance. Green procurement involves the procurement of services, products and also energy that meet environmental requirements. This paper focused primarily on the conceptualisation of the inter-link between impact assessment and green procurement. Furthermore, interviews were conducted with three fundamental categories of stakeholders that involved client, contractor and energy company. Although the interview responses showed varied perspectives on the inter-link, most of the interviewees acknowledged its need and suggested certain approaches towards establishing the inter-link. The highlighted approach was to consider the use of commercially available environmental certification tools and focus on areas such as partnerships. Future research should examine how environmental certification tools could facilitate the inter-link and how partnerships need to be arranged therein.

Keywords: Impact assessment; green procurement; renewable energy; construction sector.

---

*Corresponding author.

Introduction

The construction sector consumes a significant amount of energy. Energy is used during the manufacturing and transportation of building materials, installation, and construction activities (Yan *et al.*, 2010). In addition, studies show that the energy use and the environmental impacts during operation constitute a majority of the total in the construction sector (cf: Adalberth, 2000; OECD, 2003). According to the Organisation for Economic Co-operation and Development (OECD), the construction sector's energy use accounts for 25 to 40% of final energy consumption in OECD countries (OECD, 2003). Moreover, the substantial consumption of non-renewable energy by the sector results in emissions of greenhouse gases (Sterner, 2002). Emissions are also associated with the use of energy in construction-related activities, primarily in the construction procurement supply chain (Acquaye and Duffy, 2010). The construction sector annually generates 50% of the global greenhouse gases (IWMB, 2000). Furthermore, the energy consumption and the resulting contribution to the greenhouse gases influence the environmental performance (Rikhardsson, 1998) of the construction industry. Kasius and Seth (1998) have highlighted that the industrial consumers who are concerned about their environmental performance must consider a renewable component to their energy procurement. A similar consideration might apply for the construction sector due to the volume of its energy consumption and to improve the sector's environmental performance (Tam *et al.*, 2006).

In order to improve its environmental performance, the construction sector has adopted various policy instruments. Some of the policy instruments involve environmental impact assessment (EIA), strategic environmental assessment (SEA), and green procurement. EIA is a process that intends to seek an accord between development and the environment. In essence, it examines the environmental consequences of proposed development actions (Glasson *et al.*, 2005). It can also involve planning mitigation measures and "promoting environmental enhancement of positive impacts" (McCluskey and João, 2011). Furthermore, one of the most recent trends in the planning process has been the application of SEA at the level of policies, plans and programmes (Glasson *et al.*, 2005). SEA is defined as a "process that aims to integrate environmental and sustainability considerations in strategic decision-making" (Therivel, 2004). The third policy instrument, which has started to gain attention in the construction sector, is green procurement. Green procurement involves the integration of environmental considerations into the procurement process (PWGSC, 2006) and can also consider prioritising on green energy (CEC, 2008). By choosing green energy, clients indicate their preferences for renewable energy (Markard and Truffer, 2006). The emergence of a growing demand by clients for goods and services that do no environmental damage is generating a response from project developers. EIA can be the signal to the developer of potential conflicts, and could be used in the process of negotiating green solutions (e.g. green energy), which may eliminate or offset negative environmental impacts (Glasson *et al.*, 2005). However, the implementation of such green solutions is an issue to be analysed. Several studies have indicated challenges concerning the translation of information (produced in the EIA phase) into action during the project implementation phase (Arts *et al.*, 2001; Ridgway, 2005; Faith-Ell, 2007; Slotterback, 2008).

With regard to SEA, it is important that strategic decisions influence project decisions, and such an order of planning from strategic to project level is called tiering (van Buuren and Nooteboom, 2010). Tiering intends to link environmental assessments at different planning levels in order to facilitate better scoping of the assessment, and also, if necessary, to permit the postponement of detailed issues. A tiered approach, therefore, can minimise the problem of EIA being only a "snapshot in time" (Arts *et al.*, 2005). For instance, Fischer (2006a) has suggested the application of systems-based SEA framework for transport planning. It consists of five main assessment tiers including policy-related SEA, network-plan related SEA, corridor-plan related SEA, programme-related SEA and project-related EIA. This framework is designed as a hierarchical system, wherein each tier may be allocated with certain key tasks. Furthermore, Arts *et al.* (2005) discuss the application of an instrument called "reconnaissance study" to connect the planning arenas of SEA and EIA in the Netherlands. "Reconnaissance study" explores the scope of subsequent planning, and provides a holistic and regional analysis of infrastructure and infrastructure-related issues and solutions. In terms of the Swedish practice of SEA, detailed plans are subject to a requirement for both a SEA and an EIA if their implementation may be considered to give rise to significant environmental effects. Moreover, the Swedish Environmental Protection Agency's general guidelines on SEAs of plans and programmes states that the "supporting material produced during the SEA of a plan or programme should also be used, where appropriate, in the work on the EIA for an activity within the area covered by the plan or programme". Furthermore, it is also recommended in the guidelines that the SEA process must be integrated with the planning process in order to enable the incorporation of environmental aspects into the plan or programme (SEPA, 2010). However, a study by Isaksson and Storbjörk (2012) revealed that the SEA process in the local development planning in Västerås, Sweden was initiated when it could no longer make any significant change to the plan. Their study indicated that there was a lack of integration between SEA and the planning process. Besides, they pointed out that there was an "overall lack of concretisation" of various parallel policies, plans and strategies, which disarmed the SEA of its potential to influence the outcome of the planning process (Isaksson and Storbjörk, 2012). Hence, it can be seen that "the linkages between SEA and other assessment or planning instruments is of crucial importance for SEA" (Fischer, 2006b). Moreover, following Fischer (2004), it can be said that the time dimension may be just as crucial. Given a certain time period, the potential opportunities for formalising such linkages need to be utilised. The potential opportunities in the context of formalising application of SEA at the policy level have been explained by Fischer (2004) using Kingdon's (1995) "windows of opportunity" concept. The three windows of opportunities could arise as a "problem" that needs tackling, a "policy" to tackle a perceived problem and a "political pressure" to deal with the problem. Furthermore, the problem concerning the concretisation of various parallel policies, plans and strategies can be seen as an opportunity to strengthen the linkages between SEA and other planning instruments. Elsewhere, the lack of coordination between project planning and construction phases was one of the key reasons to postulate an inter-link between EIA and green procurement (Arts and Faith-Ell, 2010; Uttam *et al.*, 2012).

This paper focuses on the conceptualisation of such an inter-link between impact assessment (EIA and SEA) and green procurement. The aim of this paper is to gain a better understanding of how the various categories of stakeholders in the Swedish construction sector view the link between impact assessment and green procurement. In so doing, this study particularly explores the stakeholders' insights and also priority areas related to the procurement of renewable energy in the construction sector.

Method

This study is based on a literature review and interviews. The review addressed previous research on impact assessment and planning, drivers and barriers for green procurement, procurement contracts, energy planning, renewable energy policy, and innovation system. The literature review aided in identifying "*sensitising concepts*" (Blumer, 1954), which facilitated the detection of some of the priority areas that could encourage the procurement of renewable energy in the construction sector. The identified priority areas involved partnerships; planning for renewable energy procurement at early stages of a development project; renewable energy technology; and policy integration. These areas are further described in the "Results" section.

Table 1. Interviewed stakeholders and their role.

Stakeholder	Stakeholder role
Municipality (one official from the City development administration)	Responsible for managing, developing and exploiting the city's land areas. It is also responsible for the implementation of plans relating to the physical environment
Construction companies (four officials at the managerial level)	Contractors that undertake construction activities
Energy supply company (one official at the managerial level)	Involved in generation, distribution and sales of electricity and heat

Furthermore, sensitising concepts also provide foci for interviews (Holloway, 1997). Thus, these concepts facilitated the interview process.

Purposive sampling (Wright, 1996) was applied to identify categories of stakeholders. The logic behind this type of sampling is to categorise and select "information-rich" cases or resource-persons. In this study, the intention was to capture the perspectives of the stakeholder categories in the Swedish construction sector on questions related to impact assessment and green procurement, and the procurement of renewable energy. Thus, the sampling process mapped three fundamental categories of stakeholders, and these involved the client, contractor and energy supplier. Accordingly, these categories signified the municipality, construction company and energy supply company (see Table 1). In the municipality, one respondent was interviewed. This respondent was employed at the Stockholm City development administration (one of the specialist administrations within the municipality), and has been involved as an environmental expert in the Stockholm Royal Seaport project[1] (SRS project). Furthermore, the four largest Swedish construction companies were selected using the data from the Swedish construction federation. The participants from the construction companies involved employees from the senior management level, and were primarily responsible for environmental concerns in their company. One interview was conducted with an employee from the energy supply company, who was a product manager from the marketing department and was also involved in the distribution of renewable energy. The interviews focused on the inter-link between impact assessment and green procurement, and the priority areas for encouraging the procurement of renewable energy.

[1] Proposed large-scale construction project in Stockholm.

Results and Analysis

In the following sections, the interview results are presented and analysed with respect to the inter-link between impact assessment and green procurement and the priority areas.

Inter-link between impact assessment and green procurement

The environmentally friendly solutions recommended during the planning phase need to be effectively communicated at the project implementation phase. Therefore, the coordination between impact assessment and the structure of the environmental tasks (during the project implementation phase) need to be improved. In order to improve such coordination, the application of green procurement is suggested. Although in certain projects, environmental requirements described in the tender documents and environmental criteria have been based on the environmental impacts identified in the EIA (Varnäs *et al.*, 2009), planning for green procurement during the impact assessment process is still not evident. Moreover, inter-linking impact assessment and green procurement could facilitate in adhering to the rule concerning the "link to the subject matter of the contract" while promoting the desired environmental and social objectives. The subject matter of the contract sets out the scope of the content of the procurement contract. In certain cases, the Court of Justice of the European Union (CJEU) has clarified that so-called "horizontal criteria" (used to promote social, environmental, and other societal objectives) must be related to the subject matter of the contract. Furthermore, horizontal criteria that satisfy the "link to the subject matter" rule might be perceived as being located at the intersection between the functional and horizontal (societal, environmental) objectives of the contract (ClientEarth, 2011). The inter-linking of impact assessment and green procurement could facilitate in locating that intersection, and thus help in strengthening the link between environmental criteria and the subject matter of the contract. The conceptualisation of the linkage between EIA and energy inclusive green procurement can be developed further by the identification of various phases in the EIA process where discussion on green procurement can be integrated. EIA involves three phases: pre-decision; post-decision; and transition (Morrison-Saunders and Bailey, 2009). The phase that can be of particular interest is the pre-decision phase, which involves designing, developing the EIA report, its review and decision-making (Uttam *et al.*, 2012). In terms of SEA, there is a widespread agreement that lessons can be learnt from other instruments and thereby benefits can be gained from their inclusion in an SEA (Fischer, 2006b). In addition, Tang

(2008) described the integrated principle[2] for SEA, which indicates that environmental assessment should have broad scope to include all of the strategically critical environmental issues, and should also be tiered to policies in relevant sectors. However, the stages in the SEA process where such policies could be tiered needs to be investigated in the future. Discussion on such stages was not a part of this interview process because the intention was mainly to understand various perceptions on the inter-link.

The responses from the interviews showed variations between the interviewees' perceptions on the inter-link.

The respondent from the municipality mentioned two documents, namely, miljöprogrammet (environmental programme) and handlingsprogram (action programme). The municipality respondent described them in the context of the SRS project. The environmental programme has set goals to undertake project's environmental responsibilities. The goals inter alia include increasing energy efficiency and renewable energy use, and selecting environmentally friendly materials and services for construction (Exploateringskontoret, 2010). The action programme for the SRS project adopts these goals to develop specific environmental and sustainability requirements for each of the construction phases in the project (Exploateringskontoret, 2011). However, the action programme for the SRS project was voluntary, and aimed at being a comprehensive document with objectives and actions for environmental and sustainable city development in the Royal Seaport. The municipality respondent's overall opinion was that the action programme had contributed more to environmental requirements than the impact assessment. Nonetheless, the respondent also mentioned that impact assessment and the action programme are linked, and that the action programme facilitates in setting environmental requirements for the developer. However, in terms of the environmental programme, one study that examined the environmental criteria in Swedish and Finnish tender documents has indicated that several contracts included merely referrals to the procuring municipalities' own environmental programmes. There was no detailed stipulation concerning the extent to which the environmental programme needs to be followed, and thus, from a contract perspective, it hardly served any purpose (Palmujoki *et al.*, 2010). Furthermore, the municipality respondent mentioned that as far as green procurement was concerned, the construction companies had to play a major role. The role played by the construction companies in green procurement was also highlighted by one of the interviewees from a construction company. However, the interviewee stated that if projects are proposed by the municipalities, the linking of green

[2] Based on the definition from International Association for Impact Assessment.

procurement with impact assessment could be highly challenging. This is because the impact assessments of such projects are often conducted by the municipalities, whereas, the planning for green procurement is largely undertaken by the construction companies. Nevertheless, it must be noted from Varnäs *et al.* (2009) that during the implementation of a particular city tunnel project in Sweden, the information from the EIA report served in determining the environmental requirements in the tender documents. Hence, these environmental requirements that were described in the tender documents facilitated the link between EIA and the construction company's management of environmental work. Therefore, it can be argued that there is an opportunity wherein EIA can influence the green procurement process implemented by the construction company.

Another interviewee from a construction company acknowledged that it is important to inter-link impact assessment and green procurement, and that it will be challenging to do so given the problems concerning the coordination between project planning and implementation phases. The interviewee considered impact assessment as a tool that is more active only during the planning phase and therefore highlighted the need to adopt a tool that is widely in use during the implementation phase. Furthermore, the interviewee referred to the use of Building Research Establishment's Environmental Assessment Method (BREEAM) for Communities in their company. BREEAM was the first commercially available environmental assessment tool for buildings, and provides the opportunity to benchmark the performance of all new and existing buildings using proven and effective methodologies. The building is awarded credit based on the level of performance against each set criterion. The BREEAM rating achieved is presented on the BREEAM certificate, and this certification is used to verify and promote the environmental credentials of the company (Grace, 2000). In addition, a certification scheme called *BREEAM for Communities* has been developed to facilitate planners and project developers to determine whether the environmental, social and economic sustainability objectives of a development project have been addressed at the preliminary and/or final planning stages of the (applicable) developer planning application process. The central purpose of *BREEAM for Communities* is to enable planning authorities to confirm whether developers seeking planning permission have met the national, regional and local sustainability targets. *BREEAM for Communities* consists of a framework, which includes environmental, social and economic sustainability objectives that are linked to core planning policy requirements in order to guide and support the delivery of required sustainability targets. Besides, it uses compliant assessment methodology that facilitates the negotiation between planners and developers from the beginning of the planning process, and helps in setting and agreeing on a series of sustainability targets within a planning policy context. Furthermore, the scheme document highlights that it can be used in conjunction with EIAs and SEAs (BRE Global, 2011). However, there are no studies until now that demonstrate such conjunction. Although, currently, the integration of *BREEAM for Communities* into the planning system is achieved through the development plan preparation process (Sharifi and Murayama, 2013), future research is needed to understand how SEA/EIA and *BREEAM for Communities* can complement each other. One of the areas in which SEA/EIA could facilitate is public participation. Sharifi and Murayama (2013) have indicated that the public can be involved in defining the sustainability targets and identifying criteria for assessment. They have noted that in tools such as *BREEAM for Communities*, participation in the process of developing objectives and criteria for assessment is currently restricted to expert groups working with the developer.

Nevertheless, the interviewee from the construction company gave an example of their company's employees, who were not so well aware of environmental issues and had recently started to work with *BREEAM for Communities*. The interviewee highlighted that when such employees would start using *BREEAM for Communities*, they will also learn how to use the results of impact assessment. According to the interviewee, this certification requires that various environmental needs including green procurement are addressed. The interviewee noted that "we ask all the time what is in this product". However, the interviewee pointed to the difficulty that when the company purchasing manager asks for green products in a signed contract, which includes several details together with the green considerations, the delivering company would return to the purchasing manager reporting that they did not find a product that met such green considerations. If the purchasing manager is very persistent, then the delivery company would be persuaded by the purchasing manager to search for it. If not, most often, the purchasing manager decides to accept the products that the delivery company wants to provide and thereby postpone certain green considerations. The interviewee argued that in postponing they have missed the opportunity to procure green, and that there should be a mechanism for a pre-assessment before making any purchase. The pre-assessment needs to consider technology, economy and environmental issues associated with the product or service. Nevertheless, in terms of green energy, the interviewee claimed that their company buys green for most of their construction sites, and that they pay extra for it. Such a claim was also made by an interviewee from another construction company. With regard to other products, this company had developed a list of demands that were connected to environmental issues. This list of demands has been incorporated into the procurement practice and the product suppliers are supposed to comply with those demands. Although some of those demands have been based on legislation, most of them were internally established by the company. The interviewee stated that their company has a systematic way to track sources when they have to deal with international purchases and suppliers outside Sweden. However, when asked whether their company used impact assessment reports to develop such a list of demands for the product supplier, the interviewee was uncertain about it. Nonetheless, the interviewee acknowledged the need to link impact assessment and green procurement, and that impact assessment can serve as an input to develop and improve the procurement process. The interviewee did not elaborate on how impact assessment can serve as an input. However, it can be noted from the study of Varnäs *et al.* (2009) that environmental requirements in the tender documents have been based on the environmental impacts identified in the EIA and described in the environmental impact statement. Moreover, the interviewee mentioned that they have not had in-depth experience with impact assessment in order to route it in their procurement process. Furthermore, the interviewee indicated that although they have considered certain recommendations made in the impact assessment report, the connection between that report and their internal list of demands may not be so clear.

The interviewee from another construction company indicated that the idea to link impact assessment and green procurement may only be partially correct. According to this interviewee, it is rather important to focus on life cycle assessment (LCA) of the energy production process and use the analysis to procure energy that is most beneficial from a product's lifecycle perspective. LCA is a tool used to "assess the environmental impacts and resources used throughout a product's life from raw-material acquisition through production use and disposal" (Finnveden *et al.*, 2003). However, Grace (2000) argued that LCA evaluates the environmental impacts of a product on the basis of material and energy input-output data, and it only partially facilitates planners. This is because only a fraction of the impacts that are associated with the functional unit are considered, not the whole scenario (Finnveden *et al.*, 2003). The interviewee who endorsed LCA also mentioned that their construction company is attempting to investigate ways for incorporating green procurement.

However, in terms of incorporating green procurement by linking it with impact assessment, one of the interviewees from a construction company was of the opinion that there are many other drivers too for green procurement. According to this interviewee, other drivers include certain documents such as the ones that were also mentioned by the respondent from the municipality, namely, environmental programme and action programme.

The interviewee from the energy supply company perceived impact assessment as a "working tool" that in certain ways contributes to the awareness of environmental impacts of the actions proposed. However, according to the interviewee, impact assessment is not the only factor that will contribute to renewable energy procurement. There are other factors such as the awareness among the customers that environmental issues are becoming more important and all (both energy suppliers and end users) must follow environmentally friendly paths in the future. The interviewee anticipated that there will be an increasing awareness among customers on the environmental impacts of their consumption patterns and that they will demand more tools, which will facilitate environmental friendly ways of living. The respondent from the energy supply company also highlighted that such awareness among the customers is the basic driving force for development in the area of green procurement, and impact assessment could facilitate in that development process by providing information on the environmental impacts of the services/products they will use in future. Moreover, it is also imperative to adopt clear communication between the client and the contractor, and this further leads to enquiry concerning the importance of contractual arrangements and partnerships for encouraging green procurement in the construction sector.

Priority areas

The following sections include descriptions of the interview responses with regard to the priority areas.

Partnerships

The reasons for economic inefficiency in the construction industry have been explained by its fragmented nature, the divorce between planning and implementation, the role of consultants, procurement methods, and adversarial relations (Naoum, 2003; Kadefors *et al.*, 2007). These reasons may also have implications on the environmental performance of the construction sector. For construction projects that are characterised by inherent complexity and uncertainty, partnering arrangements that value relationships, trust, and communication have been suggested to be appropriate (Cheung *et al.*, 2006). Partnering aims to re-cast relations between actors in projects by encouraging a collaborative approach, which is more open, less managerial and unburdened by hierarchy (Alderman and Ivory, 2007).

The interview enquired concerning the perspectives of the interviewees on partnering. They were asked to explain how important it would be to partner in order to plan the procurement of renewable energy. During the interview, the question did not focus on any specific partnership. It was general and not restricted to the contractual arrangement between the municipality and the construction company. Partnering with other actors who could facilitate green energy procurement was also considered.

All the interviewees ascertained that it is important to have partnerships for procuring green energy in the construction sector. An interviewee from a construction company stated that partnerships between a construction company and an energy company can be very useful in specific cases wherein the contractor is not an expert on the measures applicable to the project. Another interviewee from a construction company considered the end users as partners and stated that "partnering is important to have dialogue between various partners, both the end users and also constructors". This interviewee also mentioned that "when end users come with demands, there will be quick changes and so we are talking about environmental certification such as BREEAM". However, the interviewee from the energy company indicated that the end users are not involved early in the process and highlighted that "there is a movement within the minds of people or end users and they will request more improvement on energy efficiency and renewable energy". Furthermore, the interviewee from the energy company highlighted partnerships as a top priority. The interviewee cited the names of several other companies that are merged with their company for improving the efficiency of energy systems in various construction projects. Such companies involved inter alia producers of electric installation equipment and construction companies.

Planning for renewable energy procurement at initial stages of a development project

According to Williams (2010), the planning process is viewed as being central to the delivery of deploying renewable energy sources in new development in the UK. Gutermuth (2000) has stated that the municipal area development plans are significant for the transition towards renewable energy. Furthermore, what needs to be investigated is the extent to which the SEAs of such development plans discuss on the procurement of renewable energy, and how such discussions need to be communicated to the lower tiers of assessment. Therefore, in the context of this study, the interest is primarily focused on the discussion on the procurement of renewable energy within the municipal area development and project plans (and corresponding SEAs and EIAs), and also how the interviewees perceive the influence of local energy plans on development projects. The Swedish legislation requires each municipality to provide a local energy plan (SFS 1977: 439). Tyskeng's (2006) study on the Swedish local energy plans shows that these plans discuss goals concerning oil reduction and reductions and/or restrictions on carbon dioxide emissions from a regional and local perspective. The local energy plans also discuss the use of different energy resources from both regional and local standpoints, energy efficiency, and sustainable societies.

The interview questions under this priority area focused on understanding how the interviewees perceive the need for early stage planning for renewable energy procurement and whether they associate it in any way with the impact assessment process, and additionally their perspectives on the local energy plan and its influence on the development project. The interviewee from the municipality pointed out that in the context of the SRS project, the environment programme (includes discussion on renewable energy) was developed only two years ago while the actual planning for the whole project was initiated 10 years ago. This was also observed in another project called Hammarby Sjöstad that has been the forerunner of the SRS project. Based on knowledge and experiences from Hammarby Sjöstad, the City Council conceptualised the SRS project (Lundström and Lindström, 2008). Pandis Iverot and Brandt (2011) discussed that Hammarby Sjöstad's environmental programme was introduced late in the planning process. This led to a hasty formulation process of operational goals of the environmental programme that did not align with the internal goals of the project. They also pointed out that some operational goals related to energy were not achieved. This was because the Hammarby model did not facilitate the integration of technical innovations such as solar cells and biogas stoves, and therefore such technologies were used to a small extent.

Most of the interviewees from the construction companies considered planning at an early stage as a crucial requirement for procuring renewable energy in the construction sector. It seemed however that most of the interviewees related planning to the practice of procurement planning within their organisation. None of the interviewees could explain any relation between planning for renewable energy procurement and municipal area development plans or impact assessment processes. Nevertheless, statements were made such as "we need to discuss the issues much earlier", and "I think initial stages of project planning are very important to get the solution in place." One interviewee indicated that planning for procurement at an early stage is important in order to be creative and develop measures that do not exist within the purview of given standards.

Furthermore, only two interviewees expressed their views on the local energy plan and its influence on the development project. An interviewee from a construction company stated that the goals in the local energy plans have an impact on their way to work with future energy demands, and therefore, they consider it in the development process. However, the interviewee from the energy company had a general view on the importance of local energy plans, in the sense that the interviewee mentioned that those plans were important and discussed quite often in new construction projects. Nevertheless, this interviewee, while explaining about their experiences with Hammarby, indicated that difficult and overarching goals must be set because it triggers a process to learn how new technologies can be used to a greater extent, and thereby such goals have an important role in planning. Furthermore, the interviewee stated that planning at the early stages helps in generating new ideas.

Renewable energy technology

Jacobsson and Johnson (2000) highlighted that the determinants of technology choice are present in an "innovation system". Such innovation systems facilitate as well as constrain the individual actors making a decision on the technology. Innovation systems generally consist of stakeholders, markets, networks and institutions, and many other components than the relative prices of various alternatives (Bergek, 2002). Innovation is not only about a new product or technology, it could also be the introduction of an improved process, marketing method or organisational method in business practices (Bröchner, 2010). In this way, innovation as a system also leads to diffusion of technologies.

An increasing demand for renewable energy solutions and energy-efficient products calls for innovations (Hjelm, 2011). However, Tsoutsos and Stamboulis (2005) argued that despite the acknowledged wisdom on the advantages of renewable energy technologies and their exhibited performance, their diffusion is inadequate. Furthermore, they noted that the applied policies for the diffusion of renewable energy technologies have mainly focused on technology development and investment in production facilities. The results of such policies only partially meet policy objectives concerning renewables penetration and environmental impact.

Having acknowledged that the challenges with new technologies are also related to policies, this study asked about the perceptions of the interviewees on the importance of renewable energy technology in the procurement of renewable energy. The municipality respondent focused on the local generation of electricity by renewables, which has been planned as a part of the SRS project. One of the interviewees among the construction companies indicated that in order to promote renewable energy technology, it is important to start early and involve as many stakeholders as possible. The interviewee stated that if stakeholders are involved at the early stages, it would be possible to expect cooperation among the stakeholders to test and, if necessary, modify the new technology, and that "it would probably give better products". Another interviewee from the construction sector focused on partnering with the energy sector to investigate new technologies. The interviewee from the energy company identified renewable energy technology as the next priority after partnership. Most of the interviewees associated renewable energy technology with other priorities such as planning at the early stages and partnerships. These perspectives are relevant to the concept of innovation systems. Furthermore, interviewees from the energy supply company referred to the tradable green certificates as one of the ways to promote renewable energy. A tradable green certificates system has been suggested to be a superior regulatory framework for diffusing renewable energy technologies. In such a system, the purchase of green energy in the usual electricity market is complemented by certificate trading in a separate market for green certificates (Bergek and Jacobsson, 2010).

Policy integration

According to Beck and Martinot (2004), policies that intend to promote renewable energy can be grouped into four categories. They include: (a) price setting and quantity-forcing policies that mandate prices and require a fixed amount of generation to be renewable; (b) cost reduction policies that reduce investment costs through subsidies, loans and other such measures; (c) public investment and market facilitation activities that provide public funds for direct investments or for guarantees, information or training to facilitate investments; and (d) power grid access policies that gives renewable energy equal treatment for access to power grids. Furthermore, Markard and Truffer (2006) acknowledged that a combination of policies is needed in order to achieve high diffusion rates of renewable energy sources. Moreover, Johnson and Jacobsson (2001) concluded that procurement policies have a potential to influence functions such as creating the necessary networks and bridging institutions, which is required for the diffusion of renewable energy.

The interview process, thus, explored the interviewees' perspectives on the need for policy integration to procure renewable energy in the construction sector. The combination of policies was not so clear to most of the interviewees. Some admitted that they had heard it for the first time and questioned as to what it meant. However, the municipality respondent mentioned about a world-class agreement between all concerned stakeholders, which allowed creating a commitment. In the context of the SRS project, the world-class agreement facilitates a commitment that includes important components such as trust, collective risk-taking and individual/stakeholder benefits. Trust needs to be developed in order to create an innovation arena where knowledge and ideas are shared in a fair and open environment. Collective risk-taking (risks linked to each innovative project) and stakeholder benefits (economic and environmental) are important in ensuring the long-term involvement of the stakeholders (Fortum *et al.*, 2011). Although such an agreement may not be directly concerned with policy integration, its purpose to enable the creation of a committed network (among stakeholders) is relevant in the context of diffusion of renewable energy.

Furthermore, statements from two interviewees from the construction companies indicate that they regard the possibility of combining policies even though they do not prioritise it much. An interviewee mentioned that although policy integration is important, it is not that which motivates them to act in every project. The interviewee prioritised various processes such as early planning, partnering and technology development over policy integration, and indicated that the lessons learnt from such a process would help in shaping policies. Another interviewee stated that policy integration is necessary only if there is an intention to procure renewable energy in the construction sector in a well organised and regulated way. This interviewee also identified that in so doing, there will be a need to educate concerned stakeholders on policies, decide on the actions to be taken according to combined policies and finally show them the way to work accordingly. The interviewee from the energy supply company considered policy integration to be as important as early planning for renewable energy. However, the interviewee prioritised partnering and technology over policy integration.

Other priorities

An interviewee from the construction company focused on the need to create awareness among technicians working in the construction companies on environmental issues. The interviewee stated that there is a need for people within companies who may not necessarily be specialists, but should know when to ask questions regarding environmental issues even if there are no readily available solutions. Another interviewee stated that legal and end user demands for renewable energy are very important. The interviewee also mentioned that construction companies must proactively respond to such demands and show the end users that it is possible to meet those demands in an efficient way. Furthermore, the interviewee from the energy company emphasised on setting overarching goals and also taking into account many other measures such as tradable green certificates.

Discussion and Conclusion

The stakeholders interviewed in this study held different perspectives on impact assessment per se and also on the inter-link between impact assessment and green procurement. Perspectives from the respondent at the municipality indicated that their attention is rooted in documents such as the environmental programme rather than the impact assessment. The respondent was repeatedly stating that "we can go further than EIA by adopting environmental programmes". This was in the context of the SRS project. The environmental programme was intended to function as a planning tool also in the case of Hammarby Sjöstad. Although Pandis Iverot and Brandt's (2011) case study on Hammarby Sjöstad concluded that future research should focus on the formulation of operational goals in the environmental programme and the assessment process, they have nowhere stated the relevance of impact assessment (SEA or EIA) in the development project. Therefore, the questions that emerge are where exactly does impact assessment stand in these projects and whether or not impact assessment is utilised to its full potential as a planning tool. For instance, SEA can facilitate in setting sustainability objectives and also testing whether the programme or plan will help to achieve them (Glasson *et al.*, 2005). According to Wallgren *et al.* (2011), SEA may contribute in ways less apparent, by attracting the necessary resources and attention to issues traditionally ignored in certain planning situations. The respondent at the municipality, however, claimed that there is a linkage between the impact assessment and the action programme (which adopts the goals stated in the environmental programme). Nonetheless, an analysis is inevitable in understanding the contribution of impact assessment towards the development of the environmental programme, action programme, and further procurement process.

The interviewees at the construction companies perceived impact assessment as a process that they, as senior managers, have not been directly involved with and that it is managed within the subordinate organisational tiers. Most of the interviewees from the construction companies acknowledge the need to inter-link and envisage that it will benefit in improving the green procurement process. Their responses indicated various approaches towards establishing such an inter-link. The interviewees suggested that considering *BREEAM for Communities* will be useful from the standpoint of both impact assessment and green procurement, and that *BREEAM for Communities* will help put issues in a good planning order. For instance, Haapio (2012) has discussed that the criteria of environmental assessment and rating tools can be divided into seven categories. The issues highlighted in the categories are common within environmental assessment and rating tools. In *BREEAM for Communities*, the categories *Infrastructure* and *Transportation* are emphasised the most (see Table 2). Although the criteria *Well-being* is not common in such tools, it was included to signify the importance of the public role in decision-making.

The interview process indicated that there is a tendency within the Swedish construction sector to adopt environmental rating tools such as BREEAM and

Table 2. Categorising the criteria of *BREEAM for Communities*. Adapted from Haapio (2012).

Categories	Issues covered	*BREEAM for Communities* Criteria	%
Infrastructure	Design principles, communities, buildings in the area policy and governance	12	24
Location	Land use, policy and governance, affordable housing	7	14
Transportation	Public transport, pedestrian and bicycle way, private cars, parking	11	22
Resources and energy	Waste management, use of materials, conservation, non-renewables, renewables	7	14
Ecology	Nature, biodiversity, water management	9	18
Business, economy and employment	Employment, new business, telecommuting	5	10
Well-being	Quality of life, social infrastructure, urban context	0	0
Total		51	100

BREEAM for Communities, and also to adopt a Swedish version of the tool. In addition, Malmqvist *et al.* (2011) described a research programme that is aimed at suggesting an environmental rating tool for Swedish conditions and attempting to overcome some of the criticism directed towards well-known rating tools. One possible future scenario is that the reliance on such tools would grow, and thereby, policy instruments such as green procurement may have to be routed through those tools. However, Haapio (2012) has critically questioned whether the increasing interest in such tools is based on sustainable development concerns or merely profiles and benefits out of the certification. Hence, it is necessary to retain the conceptualisation of linking green procurement with impact assessment. One important reason for this is the "legally binding character" (van Buuren and Nooteboom, 2010) of impact assessment. Moreover, it has to be noted that there is an increase in the global application of impact assessment instruments to achieve a variety of policy integration goals (Cashmore *et al.*, 2009).

However, one interviewee indicated that Life Cycle Assessment (LCA) is relevant to green procurement rather than the impact assessment process. Although there are cases where the requirement of LCA-based environmental products have been presented as a criterion in the tenders (Nissinen *et al.*, 2009), the impact assessment process may be more appropriate from an organisational perspective. For instance, EIA could be seen as a means of introducing changes in the rules and priorities that guide the organisations responsible for planning decisions (Jay *et al.*, 2007). The interviewee's suggestion on Life Cycle Assessment (LCA) is, nonetheless, appropriate from a technological perspective.

The consideration of impact assessment by one of the interviewees (a product manager) as "only a working tool useful for creating awareness" can be anticipated given the nature of the profession and the company that the interviewee was located in. Although the interviewee was aware of impact assessment legislation and that it is an obligation to adopt it in their projects, the interviewee did not perceive it as a policy instrument that can help in planning mitigation measures or serve as a planning tool. Thus, the interviewee's perception was limited to impact assessment being only a legislative requirement rather than as also a tool that can influence the planning process. This indicates that it is important to explore and utilise certain functions that facilitate SEA's potential to support the greening of policy, plan, programme and project (PPPP). Fischer (2009) summarises three main functions perceived to be connected to SEA's capacity to enable the greening of PPPP. These include *information function* where SEA provides decision-makers with better information; *changing attitudes function* wherein SEA enables perceptions to change through active involvement; and *changing routines function* in which SEA changes established routines.

Overall, no significant differences exist between the interviewees' perceptions concerning priority areas such as partnerships, planning at early stages, renewable energy technology, and policy integration. However, in terms of partnerships, some interviewees focused on the active involvement of end users. Besides, the international energy agency (IEA) has recommended that Sweden should empower consumers to participate in the retail electricity market (IEA, 2013). The involvement of end users as partners can altogether be a different priority that leads towards "a more expansive agenda for research on multi-stakeholder collaboration", or in other words, "a re-invigorated agenda on partnership" (Bendell *et al.*, 2010). According to Bendell *et al.* (2010), such "a re-invigorated agenda on partnership" could include it being: (a) interdisciplinary, particularly capturing insights from political science and development studies into management studies; (b) action-oriented, so that research delves into issues that arise due to multi-stakeholder engagement; and (c) critical, so that existing practices are questioned and new and difficult questions are asked, and should explore these features in different settings rather than relating new research to existing traditions. Partnership and planning at the early stages for procurement of green energy have been regarded as the top priorities. Although many consider renewable energy technology as a significant priority, they have associated it with partnerships and planning as well.

Impact assessment is one of the policy instruments that aid the decision-making process. In other words, along with other documents and plans pertinent to a proposed development activity, impact assessment statements contribute to the approval/disapproval of the procurement of construction projects. Hence, from that point, impact assessment plays an important role in procurement. Therefore, there is the potential to extend the influence of impact assessment to where the procurement of services and products are to be made for the project, which indicates that there is a need to inter-link impact assessment and green procurement. Moreover, such an inter-link could be one approach to strengthen the relationship between environmental criteria and the subject matter of the procurement contract. The perspectives of the interviewees indicate certain directions for future investigation regarding the inter-link. Research is required to examine how the inter-link between impact assessment and green procurement could be facilitated by tools such as *BREEAM for Communities*. Furthermore, it is important to discuss how partnerships need to be arranged within the context of the inter-link.

Acknowledgements

This work was supported by the Swedish strategic research programme StandUp for Energy and the Swedish Research Council FORMAS. The authors would like to thank all the interviewees, as well as two anonymous reviewers for helpful comments on earlier versions of the manuscript.

References

Acquaye, A and A Duffy (2010). Input-output analysis of Irish construction sector greenhouse gas emissions. *Building and Environment*, 45(3), 784–791.

Adalberth, K (2000). Energy use and environmental impact of new residential buildings. Doctoral thesis, Lund University, Dept. of Building Physics.

Alderman, N and C Ivory (2007). Partnering in major contracts. Paradox and metaphor. *International Journal of Project Management*, 25(4), 386–393.

Arts, J, P Caldwell and A Morrison-Saunders (2001). Environmental impact assessment follow-up: good practice and future directions — findings from a workshop at the IAIA 2000 Conference. *Impact Assessment and Project Appraisal,* 19(3), 176–185.

Arts, J and C Faith-Ell (2010). Impact assessment in green procurement and partnering contracts. In *Proc. 30th Int. Conf. IAIA*. Geneva: Switzerland.

Arts, J and F van Lamoen (2005). Before EIA: defining the scope of infrastructure projects in The Netherlands. *Journal of Environmental Assessment Policy and Management*, 7(1), 51–80.

Beck, F and E Martinot (2004). Renewable energy policies and barriers. In *Encyclopedia of Energy*, C Cleveland (ed.). San Diego: Elsevier.

Bendell, J, E Collins and J Roper (2010). Beyond partnerism: toward a more expansive research agenda on multi-stakeholder collaboration for responsible business. *Business Strategy and the Environment*, 19(6), 351–355.

Bergek, A (2002). Shaping and exploiting technological opportunities: the case of renewable energy technology in Sweden. Doctoral thesis, Chalmers University of Technology, Dept. of Industrial Dynamics.

Bergek, A and S Jacobsson (2010). Are tradable green certificates a cost-efficient policy driving technical change or a rent-generating machine? Lessons from Sweden 2003–2008. *Energy Policy*, 38(3), 1255–1271.

Blumer, H (1954). What is wrong with social theory? *American Sociological Review*, 18, 3–10.

BRE Global (2011). *BREEAM Communities Scheme Document* [Online], Available: http://www.breeam.org/filelibrary/BREEAM%20Communities/SD5065A_-_BREEAM_Communities_Scheme_Document_Stage_2_Version_1.pdf [24 Nov 2011].

Bröchner, J (2010). Construction contractors as service innovators. *Building Research & Information*, 38(3), 235–246.

van Buuren, A and S Nooteboom (2010). The success of SEA in the Dutch planning practice: how formal assessments can contribute to collaborative governance. *Environmental Impact Assessment Review*, 30(2), 127–135.

Cashmore, M, A Bond and B Sadler (2009). Introduction: the effectiveness of impact assessment instruments. *Impact Assessment and Project Appraisal*, 27(2), 91–93.

Cheung, SO, KTW Yiu and PS Chim (2006). How relational are construction and contracts? *Journal of Professional Issues in Engineering Education and Practice*, 132(1), 48–56.

ClientEarth (2011). Clarifying the link to the subject matter for sustainable procurement criteria, in ClientEarth, Identifying Opportunities for Sustainable Public Procurement Briefing Series [Online] Available: http://www.clientearth.org/public-procurement/public-procurement-publications/identifying-opportunities-for-sustainable-public-procurement [10 July 2012].

Commission of the European Communities (CEC) (2008). COM (2008). Communication from the Commission to the European Parliament, the Council, the European Economic and Social Committee and the Committee of the Regions — Public Procurement for a Better Environment.

Exploateringskontoret (2010) Övergripande program för miljö och hållbar stadsutveckling i Norra Djurgårdsstaden [Comprehensive programme for environment and sustainable urban development in Stockholm Royal Seaport]. City of Stockholm.

Exploateringskontoret (2011). Miljö och hållbarhet vid markanvisning HANDLINGSPROGRAM Vid planering, projektering, byggande och förvaltning av bostäder, kontor och handel inom Ängsbotten [Environment and sustainability of allocated land — action programme for the planning, design, construction, and management of residential and commercial areas in Ängsbotten]. City of Stockholm.

Faith-Ell, C (2007). EIA and then what? Client and contractor responses to environmental requirements. In *Proc. 27th Int. Conf. IAIA*. Seoul: Korea.

Finnveden, G, M Nilsson, J Johansson, Å Persson, Å Moberg and T Carlsson (2003). Strategic environmental assessment methodologies — applications within the energy sector. *Environmental Impact Assessment Review*, 23(1), 91–123.

Fischer, TB (2004). Transport policy making and SEA in Liverpool, Amsterdam and Berlin — 1997 and 2002. *Environmental Impact Assessment Review*, 24(3), 319–336.

Fischer, TB (2006a). Strategic environmental assessment and transport planning: towards a generic framework for evaluating practice and developing guidance. *Impact Assessment and Project Appraisal*, 24(3), 183–197.

Fischer, TB (2006b). Conference report: "Conclusions Stream C: Linkages between SEA and other assessment or planning tools". *Journal of Environmental Assessment Policy and Management*, 8(4), 495–504.

Fischer, TB (2009). On the role(s) of (strategic) environmental assessment in "greening" decision making. Lecture notes from Copernicus Lecture, Utrecht University [Online], Available: http://www.twoeam-eu.net/role.pdf [21 December 2012].

Fortum., ABB., KTH., City of Stockholm *et al.* (2011). *Stockholm Royal Seaport — Urban Smart Grid Pre-Study*, Stockholm Royal Seaport, [Online], Available: http://www.stockholmroyalseaport.com/wp-content/uploads/2011/02/Smart-Grid-Pre-study-report1.pdf. [6 February 2013]

Glasson, J, R Therivel and A Chadwick (2005). *Introduction to Environmental Impact Assessment*. 3rd ed. New York: Routledge.

Grace, M (2000). BREEAM – A practical method for assessing the sustainability of buildings for the new millennium. In *Proc. Sustainable Building 2000*. Maastricht: The Netherlands.

Gutermuth, PG (2000). Regulatory and institutional measures by the state to enhance the deployment of renewable energies: German experiences. *Solar Energy*, 69(3), 205–213.

Haapio, A (2012). Towards sustainable urban communities. *Environmental Impact Assessment Review*, 32(1), 165–169.

Hjelm, O (2011). The SIMPLE methodology for supporting innovations in the field of renewable energy and energy efficiency. In *Proc. World Renewable Energy Congress 2011*. Linköping: Sweden.

Holloway, I (1997). *Basic Concepts for Qualitative Research*. London: Wiley-Blackwell.

International Energy Agency IEA (2013). *Energy Policies of IEA Countries Sweden 2013 Review*. Paris: International Energy Agency.

Isaksson, K and S Storbjörk (2012). Strategy making and power in environmental assessments. Lessons from the establishment of an out-of-town shopping centre in Västerås, Sweden. *Environmental Impact Assessment Review*, 34, 65–73.

IWMB (Integrated Waste Management Board) (2000). *Designing with Vision. A technical manual for material choices in sustainable construction*. California: Integrated Waste Management Board, California Environmental Protection Agency.

Jacobsson, S and A Johnson (2000). The diffusion of renewable energy technology: an analytical framework and key issues for research. *Energy Policy*, 28(9), 625–640.

Jay, S, C Jones, P Slinn and C Wood (2007). Environmental impact assessment: retrospect and prospect. *Environmental Impact Assessment Review*, 27(4), 287–300.

Johnson, A and S Jacobsson (2001). Inducement and blocking mechanisms in the development of a new industry: the case of renewable energy technology in Sweden. In *Technology and the Market: Demand, Users and Innovation*, R Coombs, K Green, V Walsh and A Richards (eds.), 89–111. Cheltenham: Edward Elgar Publishing, Inc.

Kadefors, A, E Björlingson and A Karlsson (2007). Procuring service innovations: contractor selection for partnering projects. *International Journal of Project Management*, 25(4), 375–385.

Kasius, A and C Seth (1998). Environmental performance through energy procurement in competitive markets. *Corporate Environmental Strategy*, 5(2), 75–80.

Kingdon, J (1995). *Agendas, Alternatives, and Public Policies*. New York: Longman.

Lundström, MJ and P Lindström, (eds.) (2008). *Sustainability by Sweden — Perspectives on Urban Governance*. Sweden: Boverket.

Malmqvist, T, M Glaumann, Å Svenfelt, PO Carlson, M Erlandsson, J Andersson, H Wintzell, G Finnveden, T Lindholm and TG Malmström (2011). A Swedish environmental rating tool for buildings. *Energy*, 36(4), 1893–1899.

Markard, J and B Truffer (2006). The promotional impacts of green power products on renewable energy sources: direct and indirect eco-effects. *Energy Policy*, 34(3), 306–321.

McCluskey, D and E João (2011). The promotion of environmental enhancement in strategic environmental assessment. *Environmental Impact Assessment Review*, 31(3), 344–351.

Morrison-Saunders, A and M Bailey (2009). Appraising the role of relationships between regulators and consultants for effective EIA. *Environmental Impact Assessment Review*, 29(5), 284–294.

Naoum, S (2003). An overview into the concept of partnering. *International Journal of Project Management*, 21(1), 71–76.

Nissinen, A, K Parikka-Alhola and H Rita (2009). Environmental criteria in the public purchases above the EU threshold values by three Nordic countries: 2003 and 2005. *Ecological Economics*, 68(6), 1838–1849.

Organisation for Economic Co-operation and Development (OECD) (2003). *Environmentally Sustainable Buildings: Challenges and Policies*. Paris: OECD Publications Service.

Palmujoki, A, K Parikka-Alhola and A Ekroos (2010). Green public procurement: analysis on the use of environmental criteria in contracts. *Review of European Community & International Environmental Law*, 19(2), 250–262.

Pandis Iverot, S and N Brandt (2011). The development of a sustainable urban district in Hammarby Sjöstad, Stockholm, Sweden? *Environment, Development and Sustainability*, 13(6), 1043–1064.

Public Works and Government Services Canada (PWGSC) (2006). *Policy on Green Procurement*. Quebec: PWGSC.

Ridgway, B (2005). Environmental management system provides tools for delivering on environmental impact assessment commitments. *Impact Assessment and Project Appraisal*, 23(4), 325–331.

Rikhardsson, PM (1998). Information systems for corporate environmental management accounting and performance measurement. *Greener Management International*, 21, 51–70.

SFS 1977:439 (1977). Lag om kommunal energiplanering [Municipal energy planning act]. Näringsdepartementet.

Sharifi, A and A Murayama (2013). A critical review of seven selected neighborhood sustainability assessment tools. *Environmental Impact Assessment Review*, 38(0), 73–87.

Slotterback, CS (2008). Evaluating the implementation of environmental review mitigation in local planning and development processes. *Environmental Impact Assessment Review*, 28(8), 546–561.

Sterner, E (2002). "Green Procurement" of buildings: a study of Swedish clients' considerations. *Construction Management and Economics*, 20(1), 21–30.

Swedish Environmental Protection Agency SEPA (2010). Practical guidelines on strategic environmental assessment of plans and programmes. Stockholm: Naturvårdsverket.

Tam, VWY, CM Tam, SX Zeng and KK Chan (2006). Environmental performance measurement indicators in construction. *Building and Environment*, 41(2), 164–173.

Tang, Z (2008). Integrating the principles of strategic environmental assessment into local comprehensive land use planning. *Journal of Environmental Assessment Policy and Management*, 10(2), 143–171.

Therivel, R (2004). *Strategic Environmental Assessment in Action*. London: Earthscan.

Tsoutsos, TD and YA Stamboulis (2005). The sustainable diffusion of renewable energy technologies as an example of an innovation-focused policy. *Technovation*, 25(7), 753–761.

Tyskeng, S (2006). Environmental assessments of projects and local plans in the energy and waste sectors in Sweden — practice and potential for improvement. Doctoral thesis, Linköping University, Dept. of Mechanical Engineering.

Uttam, K, C Faith-Ell and B Balfors (2012). EIA and green procurement: opportunities for strengthening their coordination. *Environmental Impact Assessment Review*, 33(1), 73–79.

Varnäs, A, C Faith-Ell and B Balfors (2009). Practice Report: linking environmental impact assessment, environmental management systems and green procurement in construction projects: lessons from the City tunnel project in Malmo, Sweden. *Impact Assessment and Project Appraisal*, 27(1), 69–76.

Wallgren, O, M Nilsson, D Jonsson and H Wiklund (2011). Confronting SEA with real planning: the case of follow-up in regional plans and programmes in Sweden. *Journal of Environmental Assessment Policy and Management*, 13(2), 229–250.

Williams, J (2010). The deployment of decentralised energy systems as part of the housing growth programme in the UK. *Energy Policy*, 38(12), 7604–7613.

Wright, LT (1996). Exploring the in-depth interview as a qualitative research technique with American and Japanese firms. *Marketing Intelligence and Planning*, 14(6), 59–64.

Yan, H, S Qiping, LCH Fan, Y Wang and L Zhang (2010). Greenhouse gas emissions in building construction: a case study of One Peking in Hong Kong. *Building and Environment*, 45(4), 949–955.

CHAPTER 11

Knowledge Management in Environmental Impact Assessment Agencies: A Study in Québec, Canada

Luis E. Sánchez* and Pierre André†

**Department of Mining and Petroleum Engineering*
University of São Paulo
Av. Prof. Mello Moraes
2373 05508-900 São Paulo, Brazil
lsanchez@usp.br

†Department of Geography
University of Montreal
Montreal, QC, Canada

Environmental impact assessment (EIA) is a knowledge-intensive activity that benefits from a highly structured approach to knowledge management (KM). In a survey of KM initiatives in two Québec government agencies, the Environmental Assessment Department and the Environment Public Hearings Bureau, knowledge repositories were mapped and officers were invited to reply to a questionnaire enquiring about the knowledge repositories' usefulness. Their perception about knowledge creation within each agency was assessed. Three drivers were identified that have steered the implementation of KM initiatives: (i) successive managers' understandings that EIA does create knowledge; (ii) a concern with consistency and reproducibility of recommendations; and (iii) improving agencies' efficiency, alongside one additional incentive: curbing the deleterious effects of staff turnover. So far an unmet challenge is making sense of a great deal of data and information obtained in the follow-up phase of the EIA process and transforming it into knowledge used to improve both efficiency and effectiveness of an agency's work.

Keywords: Environmental impact assessment; knowledge management; organizational learning; Canada; environmental agencies; efficiency.

*Corresponding author.

Introduction

Planning and management decisions of human activities on reducing their harmful effects on the environment depend on knowledge about natural systems and society's interactions. The environmental impact assessment (EIA) process is universally used (Morgan, 2012) and promoted by several intergovernmental initiatives (Sánchez and Croal, 2012) for informing such decision-making.

EIA is often implemented or managed by a government department and typically involves different parties, especially project proponents, consultants, other government departments and the public. An EIA agency can be defined as a government authority that manages the EIA process in a particular jurisdiction. These agencies accumulate information and knowledge and can learn from experience. If properly managed, knowledge created and stored can be used to improve an agency's efficiency and to enhance EIA effectiveness. Part of the organizational knowledge is kept within the professional's mind while another part is codified and stored in repositories, making up an organization's memory.

EIA is a knowledge-intensive activity that would benefit from a highly structured approach to knowledge management (KM). Discussing how EIA agencies can profit from KM initiatives is the purpose of this paper. After a contextualization and literature review, objectives and methods are stated, followed by an explanation about EIA in Québec and the agencies involved. Results are featured and discussed in two separate sections before conclusions are drawn.

Learning Through Environmental Assessment and the Role of Knowledge Management

Framing EIA as a learning process is a recent development in the professional literature. The subject has been approached under two perspectives: social learning and organizational learning. Research on social learning is largely focused on public participation, whereas more recent interest in organizational learning (OL) investigates how key actors — project proponents, environmental consultant teams and EIA agencies — can learn to improve their competencies, efficiency or effectiveness.

The potential of EIA to foster social learning through public participation and consultation was discussed by a number of authors. Saarikoski (2000) viewed EIA as a collective process where affected citizens, interest groups, authorities and experts can exchange their views and their knowledge about the impacts of a project, provided that joint fact finding and other participatory approaches are used. Sinclair *et al.* (2008) summarized 10 years of research on public participation in EIA in Canada, recognizing ample learning opportunities and considering both social and individual learning outcomes from the EIA process when "systems of participation create opportunities for meaningful [...] involvement" (p. 417). Preceding research in Switzerland by Webler *et al.* (1995) had documented two dimensions of social learning in EIA: cognitive enhancement (acquisition of knowledge) and moral development. These are encouraged or facilitated by the participatory process. Social learning "translate[s] uncoordinated individual actions into collective actions that support and reflect collective needs and understandings" (p. 460).

Finally, the organizational learning literature usually considers three dimensions or levels of learning within an organization: individual; group; and organizational (Fiol and Lyles, 1985; Kim, 1993). Individual learning was studied by Jha-Takur *et al.* (2009), who considered the "learning dimensions" of strategic environmental assessment (SEA) by undertaking case studies of land-use planning in three medium-sized cities in three European countries and seeking to capture "the extent to which individual and organizational learning was occurring as a result of SEA" (p. 138). They noted an important role of learning for fostering long-term "individual, professional and organizational norms and practices" towards sustainability (p. 134) and concluded that learning outcomes (such as knowledge acquisition, improved comprehension, enhanced application and analysis abilities) are not only contextual-dependent, but that different individual learning styles should be taken into account in SEA practices to fulfil the potential of learning — and hence the transformative — potential of the process.

Bond *et al.* (2010) explored group learning and how practitioners organize their practical routines and the procedures used by an EIA team "in the acquisition, validation and integration of knowledge" (p. 8). They observed that both formal (explicit) and informal (implicit) knowledge played important roles in a case study they conducted in Brazil. They found that, although a "typical EIA" is conducive to reductionism and prevents team learning, informal knowledge may improve learning within the team and, thus, lead to more effective EIA. The authors conclude by stating that "the understanding of Knowledge Management (KM) in EIA" is "still little explored in impact assessment research" (p. 12).

Early studies about organizational learning in the context of project EIA focused on learning by project-proponent organizations in the USA (Brendecke and Ortolano, 1981). Chávez and Bernal (2008), who considered both social and organization learning arising from a participatory EIA process in Mexico, when looking at the project proponent didn't find clear evidence of learning, but noted that "a firm whose projects have a large impact on different sets of players needs to learn" (p. 166). As OL implies changing beliefs, attitudes or behaviour, as well as "improving action through better knowledge and understanding" (Fiol and Lyles, 1985), the strategic role of learning for proponents cannot be downplayed (see also Fischer *et al.*, 2009).

Organizational learning and Knowledge Management (KM) in EIA agencies was studied by Fitzpatrick (2006) in Canada and Sánchez and Morrison-Saunders (2011) in Australia. The first study mapped the organizational structures (information sharing and interpretation and organizational memory) that facilitate learning and their outcomes in a set of government and nongovernment organizations involved in two EIA cases in Canada. Learning outcomes included strengthening internal processes and organizational practices and insights on how to use the process to further organizational objectives. In the second study, KM initiatives taken over the years by an EIA agency were identified, knowledge repositories were mapped and their usefulness and effectiveness evaluated by the agency's personnel. It was found that those initiatives were conducive to manage the agency's ability to learn, store and use knowledge in support of its core activities and that an EIA system based around a centralized agency increases learning opportunities.

Finally, a study of KM initiatives in Brazilian consultancy firms that prepare EIAs was conducted by Costanzo and Sánchez (2014). They found that KM practices depend on how diversified these firms are, both, in terms of markets and locations.

The research described here focused on OL by a key actor: government organizations in charge of managing the EIA process in a particular jurisdiction. Previous research focusing on this kind of organizations under an OL perspective include Fitzpatrick (2006) and Sánchez and Morrison-Saunders (2011). The main task of such organizations (hereafter called agencies) is typically to review an environmental impact statement or study (EIS) in order to derive recommendations or to take decisions about the conditions of approval of the projects assessed.[1] Project proponents and other government departments usually see these agencies as merely processing information contained in the EIS and related documents. In fact, these agencies actually analyze information provided by proponents, considering relevant government policies and legislation, and taking account of the views of the public. However, in undertaking their activities and making sense of

[1] In a number of jurisdictions, EIA agencies are entitled to make project approval decisions whereas in other jurisdictions they make recommendations to a decision-making authority.

the inputs to the EIA process, officers and managers in these agencies accumulate information and knowledge and can learn from experience. In performing their tasks, they rely on both their personal knowledge and the knowledge accumulated in their organization. More importantly, they also create new knowledge that can be fed back to the agency, potentially improving its efficiency, and used to the assessment of new projects, thus potentially improving the effectiveness of the process. Knowledge created can be used both internally (to review further proposals) and externally (by consultants assessing other projects, other government organizations and by the public).

As in any organization, knowledge created by an EIA agency is subject to decay if not properly captured, stored (in repositories such as files, databases and reports) and maintained in order to be retrieved and used to improve practice. Hence, managing knowledge becomes central for any knowledge-intensive organization.

Knowledge Management (KM) provides a framework for organizational learning and for structuring the knowledge-intensive work of environmental impact assessment. An EIA agency, like many organizations, cannot solely rely on the knowledge embedded in individuals. Usually defined as "activities and processes for leveraging knowledge to enhance competitiveness through better use and creation of individual and collective knowledge resources" (CEN, 2004), or through similar wording, KM in organizations enables individual, group and organizational learning. In the EIA process, it is worth studying how project proponents, government agencies and stakeholder organizations approach KM. In our research, we focused only on centralized EIA agencies in one jurisdiction, namely Québec. Although competitiveness is not relevant for a government agency, efficiency can also be fostered by KM approaches (Arora, 2011).

Objectives and Methods

The research reported here is aimed at identifying key elements of knowledge management in EIA agencies. The objectives are detailed as investigating how knowledge in EIA is created, stored and used in Québec EIA government agencies, and to what extent KM is integrated into these agencies' procedures.

EIA practice is always context-specific, i.e. it is guided by regulations and legal rules as well as administrative structure and culture. Therefore, one jurisdiction featuring long established and stable EIA experience was chosen to conduct a case study (the Province of Québec, Canada). The research was designed, having a possible comparison in mind with results of similar research recently carried on in Western Australia, a jurisdiction which also features a long established and centralized EIA agency (Sánchez and Morrison-Saunders, 2011).

Research was conducted through document review, participant observation, interviews and questionnaires. Two government organizations were studied for reasons explained in the next section. Documents of interest to be reviewed included guidance, annual reports, EIS review reports and public consultation reports. Participant observation comes from one of the authors, who has been involved in studying Québec EIA practices for more than 20 years. Interviews were conducted with the current heads of the two agencies and one former head. Questionnaires were applied to technical staff at both agencies.

Document review served mainly to identify knowledge repositories and to explore their contents. The interviews were open and semi-structured. Their purpose was: (i) to validate the identification of knowledge repositories and to understand key procedures at each organization; (ii) to identify past and present initiatives pertaining or related to Knowledge Management (KM); and (iii) to investigate the possible reasons or drivers of such initiatives.

The questionnaires aimed at: (i) obtaining data about access to existing repositories and their perceived usefulness; and (ii) gauging staff perception on knowledge creation. Both interviews and questionnaires were virtually identical for both agencies and structured around the following topics: (1) knowledge sources and repositories; (2) usefulness of and access to repositories; (3) knowledge creation and learning environment; (4) identifying gaps and improvements; and (5) context (respondent contextual data about professional experience within the agency). Respondents remained anonymous. This structure is identical to the questionnaire used by Sánchez and Morrison-Saunders (2011) in a similar enquiry in Australia, thus enabling possible comparisons.

Access to staff was gained through the directors of both agencies, who explained the research purposes and who invited officers to participate by responding to the questionnaires. Respondents returned the questionnaires directly to the first author and remained anonymous.

The Research Context: EIA in Québec

A brief outline of the EIA process in Québec will subsequently be presented. For the sake of international comparison, the correspondence of Québec's process with a generic EIA process will be underlined.

The Québec EIA process[2] is more than 30 years' old. The Environmental Quality Act (L.R.Q., c. Q-2) of 1972 was amended in 1978 to include environmental impact assessment as a requirement for the approval of major projects and to create both the Ministry of the Environment[3] and the Québec Environment Public Hearings Bureau (*Bureau d'audiences publiques sur l'environnement* — BAPE).

Projects that seem likely to "result in an emission, deposit, issuance or discharge of contaminants into the environment or a change in the quality of the environment" (article 22) need an environmental authorization from the Minister of the Environment prior to any construction activity. Certain projects with potentially significant impacts are also subject to the environmental impact assessment and review process (division IV.1). For such projects, an environmental authorization can only be requested after government approval informed by an EIS, an internal review report and an inquiry and public hearing report.

On December 30, 1980, the Regulation on environmental impact assessment and review (R.R.Q., c. Q-2, r. 23) came into effect. This Regulation establishes a procedure subjecting listed projects to the EIA process (art. 2), defining the parameters for EIS (art. 3), and installing a public information and consultation process (art. 6 to 16). Thus, the public gained the right to be informed and to express its opinions by way of public consultations.

An EIA unit within the Ministry of the Environment was created in 1980 to implement the EIA procedure, the *Direction des évaluations environnementales* (DEE). Most tasks of the EIA process are performed by this unit, except for public consultation, which is led by the BAPE. Tasks include the management of the EIA process itself, the inter-ministerial consultation on files under study, the request to the proponent of a project for information or research necessary to fully evaluate the impact of the proposal, the internal review of the EIS, and the preparation of the decision for the Minister to be submitted to the Government.

---

[2] We limited the study to the Southern Québec EIA procedure (*Québec méridional*) where the BAPE holds authority. In the Northern Québec EIA is framed by the James Bay and Northern Québec Agreement, jointly signed on November 11, 1975, by the governments of Quebec and Canada and the Cree and Inuit communities, largely to establish the terms for assessing the big hydroelectric projects of the James Bay region. This agreement originated from two different EA procedures, one south of the 55th parallel, jointly managed by the Cree, and the other north of the 55th parallel, jointly managed by the Inuit. A number of projects may also be subject to the requirements of the Canadian Environmental Assessment Act, in which case a Federal-provincial agreement provides for harmonization and avoids duplicating procedures.

[3] The current name of this Department is Ministry of Sustainable Development, Environment, Fauna and Parks (MDDEFP).

The BAPE, created in 1979, is a consultative body under the responsibility of the Minister of the Environment. "The function of the Bureau is to inquire into any question relating to the quality of the environment submitted to it by the Minister and to make to him or her a report of its findings and of its analysis thereof. It must hold public hearings whenever required by the Minister." (L.R.Q., c. Q-2, art. 6.3). The Bureau is independent from the Ministry of the Environment. In 1981 the Rules of procedure relating to the conduct of public hearings (R.R.Q., c. Q-2, r. 45) were adopted.

A distinctive feature of the Québec EIA system is the public consultation procedure. After having received the EIS, the Minister asks the BAPE to make it public, and to initiate the 45-days period of information and consultation. During this period, any individual, organism or municipality may address a written and justified request for a public hearing to the Minister. If the Minister consents, he or she will mandate the Bureau to conduct an enquiry and a public hearing. BAPE produces a report containing findings, opinions and advises to the Minister. Furthermore, DEE prepares an Environmental review report which is also submitted to the Minister.

Figure 1 summarizes the main steps in the Southern Québec EIA procedure, providing a relationship with the steps of a standard international EIA process (André *et al.*, 2010).[4] Screening is dealt with through a positive list enacted into the Regulation, including, among others, works in the aquatic environment, ports and wharfs, mines, industrial facilities, hazardous material treatment and disposal sites, power production facilities and power transmission lines, roads and highways, railroads and railway stations, airports, aerial pesticide spraying and waste disposal sites. In most cases, there is a threshold above which a project is subjected to the procedure.

Québec's approach to scoping is to provide generic guidance for preparing an EIS according to the specific type of project. This way to proceed has been put in place during the 1990s to improve "the efficiency, the predictability and the equity of the process", following the EIA process review (Québec, 1988, 1992). In the past, DEE prepared tailor-made terms of reference for each project individually. However, there was a perception among project proponents that this could potentially result in unfair treatment of similar projects, triggering the current class *modus operandi*. Hence, an officer within DEE checks if the generic guideline is appropriate for the submitted proposal and prescribes an existing guideline (*directive*).

[4] All the documents we refer to are available on the MDDEFP or BAPE websites: www.mddep.gouv.qc.ca or www.bape.gouv.qc.ca. Last consulted on March 15, 2013.

Inputs	*Steps of the EIA process*	*Internal procedures*	*Outputs*
	Screening (1)		
Project notice (*avis de projet*) →	**Scoping** (2)	checking adequacy of generic guidelines or preparing specific guidelines, as needed →	Guideline for EIS preparation (*Directive*)
Environmental impact Study (*étude d'impact*) →	**Submission of environmental impact study**	consulting with other ministries and with other departments within the Ministry of the Environment and coordinating the review of EIS or parts of it →	Comments collected and synthesized by DEE staff Questions addressed to the proponent
Proponent responses →	**EIS review**	reviewing adequacy of responses by consulting with other departments or ministries →	Receivability Notice (*avis de* recevabilité)
Public comments →	**Public review** (3) **Part 1: information**	BAPE officers put together public remarks →	Public information reports (*Comptes rendus des séances d'information*)
Public comments, proponent responses, government officials' explanations →	**Public review Part 2: hearing**	Panel members chair hearings, BAPE officers help with document review →	BAPE report
Outputs from previous steps →	**EIS review (conclusion)**	DEE officers prepare final report	Environmental analysis report (*Rapport d'analyse environnementale*)
	Decision-making		Ministerial recommendations to cabinet
Project drawings and specifications			Certificate of authorization (4)
Monitoring and follow-up reports →	**Follow-up**	Reports are filed and can be reviewed	

Notes.

(1) Screening is based on a positive list of projects.

(2) In the past, scoping was conducted on a case-by-case basis; currently, there are generic terms of reference (guidelines) by type of project; if the DEE reviews a new type of project, a tailor-made guideline is prepared; if several projects of a new type are submitted, a new generic guideline is produced; the most recent generic guidelines were prepared for wind farms.

(3) Public consultation is conducted by the BAPE — *Bureau d'audiences publiques sur l'environnement*, an independent advisory body to the Minister of the Environment and totally separated from the DEE.

(4) Equivalent to a construction permit, a formal authorization to initiate project implementation.

Figure 1. The steps, inputs and outputs of the EIA process in Québec.

For reviewing EIA documents, officers at DEE act as coordinators or organizers of partial reviews undertaken by different specialized services in other departments within the Ministry and with other ministries, e.g. the Water Department (*Direction des politiques de l'eau*), the Soil and Land Protection Department (*Direction des politiques en milieu terrestre*) and the Ministries of Health and Natural Resources. A network of government experts, each providing advice on his or her domain, is mobilized by a project officer (*chargé de projet*) at the DEE. Once an environmental impact study is filed, DEE officers perform a first review to check the adequacy of the document before its release for public consultation. Some exchanges (questions – answers) between the project officer and the proponent may occur before the official deposit of the EIS to the Minister and the initiation of a public information and consultation phase. The DEE has to produce a receivability notice to qualify the project for the public consultation procedure. Upon completion of public consultation, DEE concludes an Environmental Analysis Report (*Rapport d'analyse environnementale*) which is made publicly available and submitted to the Minister of the Environment.

In case of project approval, a certificate of authorization is issued with terms and conditions for project construction and operation. A project proponent is usually required to prepare monitoring and follow-up reports on specific issues, which are filed with the regional offices of the Ministry.

The roles of DEE and BAPE being so closely interlinked, both agencies were considered for the purpose of the research.

Results

This section reports the study findings. Results are presented in six parts: (1) a profile of the staff of each agency; (2) a description of their knowledge repositories; (3) a synthesis of KM-related initiatives; (4) officers' understanding of the use of stored organizational knowledge; (5) their perception of the usefulness and accessibility of the repositories; and (6) their perceptions about knowledge creation. In each subsection, results are presented separately for DEE and BAPE.

Results pertaining to the first three subsections were mostly obtained from document review and interviews. Results reported in the last three subsections came from the questionnaires to which technical staff at both organizations voluntarily responded. The rate of return of questionnaires at the DEE was 49% (21 out of 43 officers and coordinators). At the BAPE, the rate was 83% (10 out of 12 officers).

Staff profile

Each organization has independent and separate technical and clerical staff. A brief overview is provided in this section as an introduction to reporting the findings.

As of 2011, the DEE had 37 project officers (*chargés de projet*) and 9 coordinators (according to types of projects such as dams and roads), organized in three units — terrestrial, aquatic and industrial environments — each led by a head of service under the supervision of one director. The project officers and the coordinators are the staff directly involved in the EIA process, making up 43 users of knowledge repositories. Technical staff is composed of professionals featuring a range of backgrounds and experience. Staff turnover is currently not considered as a problem, with officers remaining for many consecutive years. Newly admitted staff, however, is perceived as featuring higher potential of changing jobs after short periods of time at DEE. In 2011, 7% of the staff had been working at DEE for less than two years, 33% between two and five years, 21% between five and ten years, 16% between 10 and 20 years and 23% for more than 20 years. In terms of qualification, 23% hold a bachelor's degree, 70% a master's degree and the remaining 7% a Ph.D. The team features a multidisciplinary background.

BAPE has a group of 12 project officers (*analystes*) who work with and support the work of the members of enquiry panels. Six (50%) such officers have been working at BAPE for less than two years, one (8%) between two and five years, two (17%) between five and 10 years and three (25%) had more than 10 years of working experience at BAPE. One officer had a bachelor degree only, while 10 had a master's degree and 1 features a doctorate. The staff is multidisciplinary.

Mapping knowledge repositories

The identification of knowledge repositories used for performing environmental analysis was conducted by a careful scan of the organizations' websites and by interviews with senior staff.

Table 1 shows the 16 knowledge repositories identified within DEE. Some are available only for internal use, while others were prepared for the benefit of DEE's stakeholders, i.e. project proponents, environmental consultants and the public, but are also a reference for internal use. Table 2 provides a synthetic explanation of the contents of each type of repository.

A number of repositories, in particular those in a category labelled "Other Technical Guides", were prepared by other divisions within the Ministry of the Environment and are also used in analysing environmental authorizations of those

Table 1. DEE internal knowledge repositories.

	Availability		Form of access		
Type/Name	Internal use only	Publicly available	Intranet	Files/ Printed	Personal contact
1. Sectoral Guidelines		■	■	■	
2. EIA Technical Guides		■	■	■	
3. Other Technical Guides		■	■		
4. Policy guidance		■	■		
5. Officer's Manual and templates	■		■		
6. Technical sheets	■		■		
7. BAPE Reports		■	■	■	
8. Public information period reports		■	■	■	■
9. SAGO — (Operations management system)	■		■		
10. Files	■			■	
11. Environmental Analysis Reports		■	■	■	
12. Government approval documents		■	■	■	
13. Follow-up reports	■		■	■	
14. Staff meetings	■				■
15. Informal talks	■				■
16. Senior staff	■				■

Sources: Compiled from the DEE website (www.mddep.gouv.qc.ca/evaluations/inter.htm) and interviews.

projects not submitted to the EIA procedures, for compliance inspections and other applications. There are two kinds of Technical Guides, those developed by DEE and those developed by other departments or ministries. The former are applied only to projects submitted to the EIA process and the latter serve several purposes of environmental and resource management, especially the delivery of project authorizations (*certificats d'autorisation*). These "Other Technical Guides", however, are highly relevant to EIA, as officers in other departments within the Ministry and with other ministries constantly refer to them when reviewing specific parts of an EIS.

BAPE has its own knowledge repositories, not all of them being publicly available (Table 3). A short description is provided for in Table 4. Recently developed tools include the Systematic Review Approach and the Sustainable Development (SD) Analysis Framework, both in use since 2008. By comparing Tables 2 and 4, a number of common repositories appear, such as the EIA Review Reports prepared by DEE and the BAPE Reports, which are also the main outputs of each agency activities.

Table 2. DEE knowledge repositories and their main features.

Type/Name	Description
1. Sectoral Guidelines (*Directives types*)	Developed for the types of projects most frequently processed by the Ministry of the Environment and established following consultations with other departments and with environmental, professional and interest groups, they define the elements of an EIS: 16 such guides were available as of 2011, all updated in 2010.
2. EIA Technical Guides	Developed by DEE to assist a project proponent in producing a better EIS, there are currently three such guides: a general guide to preparing EIS; a guide to technological risk analysis (first version 2000, revised 2002); and a guide to follow-up (2002, revised 2005).
3. Other Technical Guides	Documents developed by other sectors within the Ministry of the Environment or other ministries: such guides are listed in the DEE webpage.
4. Policy guidance	Government guidance on natural resources and environmental protection.
5. Officer's Manual and templates (*Cahier de procédures*)	Description of 16 internal procedures from receiving a proposal to terminating a file. Each procedure describes the actions and tasks, the responsible person for performing the task and provides templates for letters and other documents. It also advises on approximate timelines for performing major tasks. The procedures are often updated, as needed.
6. Technical sheets (*Fiches techniques*)	A collection of 39 summary thematic sheets describing the main issues, from legal aspects to practical tips on how to deal with the issues. Each sheet is ascribed to one project officer for preparation and updating.
7. BAPE Reports (*Rapports du BAPE*)	Comprehensive reports prepared by the BAPE as a result of inquiry and public hearings: usually contains opinions and advices to the Minister.
8. Public information and consultation period reports	A summary of views and concerns raised by the public during the official public information and consultation period.
9. SAGO Operations management system	A database of assessments featuring information on proponents, timelines, transmission of documents, etc.: it contains a record of each assessment.
10. Files	Physical files containing all documents relative to every proposal received by the DEE: active files are maintained at the office for quick retrieval.

Table 2. (*Continued*)

Type/Name	Description
11. Environmental Analysis Reports	The main output of the review process; feature recommendations of the DEE; reports are available on the Internet and numbered since 1 January 1994; 594 such reports were delivered as of December 2012.
12. Government approval documents (*Décrets et certificats d'autorisation*)	The *décrets* are the result of the assessment process, featuring the final decision taken by the Cabinet, usually listing terms and conditions to be fulfilled by the proponent when requiring a permit (*certificat d'autorisation*).
13. Follow-up reports (*rapports de suivi*)	Proponent-prepared reports to comply with terms and conditions of the government authorization, most often containing monitoring and follow-up data.
14. Staff meetings	Formal meetings to deal with ongoing assessments; prior to finishing an environmental analysis report; an internal presentation to coordinators is made by the project officer and open discussion follows.
15. Informal talks	By definition, any exchange of views or information between staff members: this category also includes social networks both inside and outside the organization.
16. Senior staff	Advice sought with experienced officers.

Sources: Compiled from the Ministry's website (www.mddep.gouv.qc.ca/evaluations/inter.htm), internal documents and interviews.

Initiatives related to knowledge management and organizational learning

Notwithstanding that both organizations have not implemented any formal Knowledge Management (KM) program, several internal management initiatives that have relevance as enablers of KM (Agimo, 2004) have been adopted. We use the term "initiatives" as an umbrella term encompassing any action towards managing knowledge.

At DEE, the interviewees mentioned that government officers are usually concerned about the "coherence of their advice". Hence, a number of tools were developed to build consistency and reproducibility in the officers' work, while contributing for efficiency in performing some tasks. The *cahier de procédures* (procedure manual) is an example of the latter, while the former is exemplified by the *fiches techniques* (technical sheets) (Table 2). Both are periodically updated, which is essential if the tools are to be continuously used. The questionnaires did not register any comment related to a repository being out-of-date.

Table 3. BAPE internal knowledge repositories.

	Availability		Form of access		
Type/Name	Internal use only	Publicly available	Intranet	Files/ Printed	Personal contact
1. BAPE's findings and opinions	■		■	■	
2. Files of previous enquiries	■		■	■	
3. BAPE reports		■	■	■	
4. Sustainable Development Analysis Framework		■	■		
5. Synoptic Sustainable Development tables from previous enquiries	■		■		
6. Current priority issues	■		■	■	
7. Digest of panel questions	■			■	
8. Systematic review approach	■		■		
9. Environmental Analysis Reports		■	■	■	
10. Government approval documents		■	■	■	
11. Internal meetings	■				■
12. Informal talks	■				■
13. Senior staff	■				■
14. Panel members' experience	■				■

Sources: Compiled from the BAPE website (www.bape.gouv.qc.ca) and interviews.

The fact that scoping is no longer made on a case-by-case basis is perhaps another sign of the abovementioned concern with consistency and reproducibility. However, in this case, proponents' pressure to streamline the procedures and calls for quicker response from the Ministry were mentioned in interviews as major drivers for generic scoping guidance.

Guidance for preparing EIA documents is provided both by the DEE and by other departments. This is essentially substantive guidance, i.e. documents explaining "how to" fulfil a requirement such as collecting, presenting or analysing data to be presented in the EIS. Examples are: (1) the "Guide for modelling atmospheric dispersion" (Leduc, 2005), which details acceptable models, modelling parameters, type of meteorological data and other practical issues in air quality prediction; and (2) the publication "Identification and delimitation of aquatic, wetland and riparian ecosystems" (Québec, 2006), which provides guidance for field and map delimitation of these ecosystems.

As staff turnover is not currently perceived as an issue, there is no specific mentoring program for newcomers. However, a few years ago a video interview was recorded with one retiring officer on his main lessons learned, aimed at

Table 4. BAPE knowledge repositories and their main features.

Type/Name	Description
1. BAPE's findings and opinions	A searchable database containing opinions and findings from previous public enquiry and hearing reports, allowing search by report, type of project, subject or keyword. The database compiles all reports since 2000 and the most important reports prior to that date. It started in 2004 and is continuously updated upon completion of every enquiry.
2. Files of previous enquiries	Provides access to all documents produced by all stakeholders during previous enquiries since 1978. Access can be made through Intranet or a library of printed documents.
3. BAPE reports	The collection of reports prepared and published since 1978. As of December 2012, 292 reports were published and are available on the Internet and can be searched by date, number or subject.
4. Sustainable Development Analysis Framework	Guidance for considering the 16 sustainability principles adopted by Québec's Sustainable Development Act in project review: it includes a synoptic table to assess the project against the principles; available since May 2008.
5. Synoptic Sustainable Development tables from previous enquiries	Synoptic tables and associated documents resulting from the analysis done by previous commissions.
6. Current priority issues	A set of sheets providing a synthetic description of three key issues currently considered by the BAPE as top priority, i.e. forest conservation, wetlands and ecological footprint of buildings.
7. Digest of panel questions	A relation of topics and issues rose during public enquiries, organized by type of project.
8. Systematic review approach	Procedural and technical internal guidance for officers aiming at ensuring consistency in project review: adopted in early 2008 and last reviewed in 2009.
9. Environmental Analysis Reports	The final review reports prepared by the *Direction des évaluations environnementales*. Although not prepared by BAPE, the reports may contain views and analysis useful for reviewing similar projects.
10. Government approval documents	The result of the assessment process, featuring the final decision taken by the Cabinet (*Décret*).
11. Internal meetings	Work meetings of officers within the *Service de l'Expertise Environnementale*.

Table 4. (*Continued*)

Type/Name	Description
12. Informal talks	By definition, any exchange of views or information between staff members: this category also includes social networks both inside and outside the organization.
13. Senior staff	Experienced members within the *Service de l'expertise environnementale* and BAPE members.
14. Panel members' experience	BAPE members and external commissioners assigned to each enquiry are selected on the basis of their knowledge about particular issues.

Sources: Compiled from the BAPE website (www.bape.gouv.qc.ca) and interviews.

avoiding the phenomenon described as "discontinuity of knowledge", when a worker moves without transferring tacit knowledge (Frappaolo, 2006, p. 90). Staff training, on the other hand, is routinely undertaken according to an annual schedule. A noticeable initiative is the carrying out of an after-action review as a means of sharing learning among analysts: after an EIS is reviewed and the environmental analysis report is concluded, the team leader briefs other officers and a group discussion follows, aiming at understanding key issues and lessons learned; each session typically takes 1–2 hours.

At BAPE, a concern with consistency and reproducibility was mentioned as a primary driver for initiatives to manage information and knowledge. BAPE activities are guided by a five-year strategic plan. The current 2008–2013 plan deals with three major issues: (1) significantly contributing to sustainable development; (2) maintaining services adapted to the clients' needs; and (3) seeking optimum management of resources (BAPE, 2009). Under the first topic, propelled by the 2006 Sustainable Development Act (L.R.Q., c. D-8.1.1)[5] and the corresponding BAPE's action plan (BAPE, 2008), the strategic plan contains actions aimed at "accessing knowledge". Under the second issue, internal guidance has been developed and updated for a number of years, aiming at "ensuring better coherence in internal processes and the *perpetuity of the organizational memory*" (BAPE, 2010, p. 14, emphasis added).

Both organizations care for the public availability and easy accessibility — through Internet — of their main outputs: the environmental analysis report, for

[5]To access to the Québec's SD Act: www2.publicationsduquebec.gouv.qc.ca/dynamicSearch/telecharge.php?type=2&file=/D_8_1_1/D8_1_1.html.

DEE, and the inquiry and public hearing report, for BAPE. The DEE website also hosts all records of government decisions based on the EIA process. These documents are also important knowledge repositories.

Use of knowledge stored in organizational memory

Findings related to access and use of collective knowledge for performing the EIA tasks at both organizations were obtained from the questionnaires. Respondents were asked "Which of the following internal knowledge sources do you use or consult when performing your ordinary tasks?" At DEE, out of the 16 knowledge repositories (Table 1), almost all respondents have said to most often access the sources which provide personal interaction, i.e. informal talks. Among other repositories, the most frequently used are manuals and templates (81%), government approval documents (71%), previous environmental analysis reports (67%) and BAPE reports (61%). The least accessed repositories are the document management system, which is frequently accessed by 5% of respondents only and never accessed by 71% of respondents, while follow-up reports and staff meetings are frequently accessed by only 14% of respondents.

At BAPE, a similar pattern emerged from the responses. Informal talks and senior staff appear as the most frequent knowledge repositories for performing EIA-related tasks, featuring, respectively, 100% and 70% of respondents declaring they use them more than twice for each review. Previous BAPE reports are frequently used by 67% of respondents, while internal meetings were mentioned by 60% of respondents and the SD Analysis Framework (repository 4) by 56%. The least accessed repositories were the Synoptic SD tables prepared for previous enquiries and the digest of panel questions.

A second question enquired whether any external database or information sources were used when performing regular tasks. External repositories are accessed by officers at both organizations. Generic Internet was mentioned by 76% at DEE and by 70% at BAPE. The second most accessed external repository is scientific literature, used by 33% of respondents at DEE and 50% at BAPE. Personal professional networks are used by 29% of respondents at DEE and 50% at BAPE.

In order to inquire whether any particular task in the EIA process required more frequent access to the available repositories, the main tasks at each agency were mapped. The lists were based on document and literature review, and submitted to interviewees for validation. Then a question was formulated to gauge frequency of access to the repositories to perform each task. Nine tasks were mapped at DEE and ten at BAPE and the results are shown on Tables 5 and 6. The lists are different for each organization, as they have different duties. Taken collectively, all

Table 5. Frequency of internal knowledge repository usage with respect to EIA tasks at the DEE.

	Frequency				
Task	Never	Once	Twice	More than twice	Total (n)
Reviewing project notice	3	12	2	4	21
Preparing EIS guideline	6	12	3	0	21
Consulting with other ministries and departments	3	7	4	7	21
Synthesizing opinions and comments and preparing receivability notice	2	3	2	12	19
Formulating questions to the proponent and reviewing responses	0	3	4	14	21
Preparing for public hearings	0	2	2	15	19
Preparing the environmental analysis report	0	3	2	16	21
Preparing draft Ministerial recommendations	0	5	3	13	21
Reviewing follow-up reports	6	8	2	5	21
Please indicate any other task	0	0	0	0	0

repositories are used by analysts at both organizations. At DEE, two tasks require the least access: preparing guidelines or terms of reference for an EIS, for which officers most often need to consult only once, and reviewing follow-up reports. At BAPE, preparing questions to proponent and resource persons is the task that obtained most "twice" and "more than twice" hits.

Usefulness and accessibility of knowledge stored in organizational memory

To gauge how accessible (in terms of time and effort) and useful the repositories are in each organization, a number of statements were proposed in the questionnaires and the respondents were asked to declare their agreement with statements.

Eight statements were formulated in the DEE questionnaire (Table 7). The level of agreement among the respondents was high. All respondents fully agreed that knowledge sources at DEE are easily accessible. They recognize that expertise provided by other departments within the Ministry and by other ministries is the most important knowledge source for performing DEE EIA tasks. When asked if the reports they prepare deal with such specific issues they hardly ever find to be any useful knowledge in DEE internal sources, a majority disagreed. This result confirms the importance of Knowledge Management (KM) initiatives for EIA agencies. Two-thirds of respondents see the input provided by BAPE as useful for preparing

Table 6. Frequency of internal knowledge repository usage with respect to EIA tasks at the BAPE.

	Frequency				
Task	Never	Once	Twice	More than twice	Total (n)
Reading the files	1	1	1	7	10
Preparing questions to proponent and resource persons	0	0	3	7	10
Framing sustainable development analysis	0	2	3	5	10
Preparing draft report	0	3	4	3	10
Preparing for the first part of public hearings	0	3	3	3	9
Follow-up of proponent and resource persons responses	1	2	3	4	10
Reading records of public enquiries	2	3	1	4	10
Preparing for the second part of public hearings	2	4	2	2	10
Preparing an advanced draft report	0	2	1	7	10
Reviewing drafts and assisting the preparation of the final report	0	1	2	6	9
Please indicate any other task	0	0	0	0	0

the environmental analysis report and drafting the approval documents, and sustain that follow-up reports are useful for drafting environmental approval documents. Only one statement (Item D) showed a split opinion: while about one-third of the respondents show a preference for obtaining information from other people rather than from documents, one-fourth has the opposite opinion.

Six statements[6] relevant to their tasks were formulated to BAPE analysts (Table 8). The respondents recognized unanimously the relevance of public participation and the usefulness of the BAPE reports for sound environmental decision-making. They acknowledge also the accessibility of both databases and knowledge sources. They build their practice on previous Commission experiences. When asked if they are dealing with such specific issues they hardly ever find any useful knowledge in BAPE internal sources, the divergence of opinion is higher: only 30% disagreed as compared to 55% at DEE. Additionally, while 85% of DEE respondents declared their agreement or disagreement, only 50% at BAPE did, the other half feeling unable to provide a straightforward response. We can only speculate that this difference may be related to a possible greater variability in

[6] Four statements are exactly the same as those proposed to DEE staff, one has a different wording to reflect BAPE's tasks, and one is specific to the public participation tasks performed by BAPE.

Table 7. Statements about internal knowledge repositories at DEE.

Statement	Disagree	Agree
A. Databases and knowledge sources at DEE are easily accessible in terms of time and effort in retrieving information	0	21 (100%)
B. The most important knowledge source for performing your tasks is the expertise provided by other departments within the Ministry and by other ministries	2 (10%)	14 (67%)
C. Sectoral Guidance usefully condense knowledge accumulated at the DEE	0	14 (67%)
D. In my work, I have a preference for obtaining information from other people rather than from documents	7 (33%)	5 (24%)
E. The reports I prepare deal with such specific issues that I hardly ever find any useful knowledge in DEE internal sources	11 (55%)	6 (30%)
F. The most accurate and reliable way of doing an internal task is to check how it was done last time	4 (19%)	10 (48%)
G. BAPE's opinions and recommendations are useful for drafting environmental approval documents	3 (14%)	10 (62%)
H. Follow-up reports of similar projects are not very useful for reviewing a submitted project	14 (67%)	2 (10%)

Note: Results are presented in two broad groups; in the questionnaire, respondents were asked to mark "strongly disagree", "partially disagree", "unable to judge", "partially agree", and "strongly agree"; sums can be less than 100% where respondents indicated that they were "unable to judge". One respondent chose not to rate two statements.

Table 8. Statements about internal knowledge repositories at BAPE.

Statement	Disagree	Agree
A. Databases and knowledge sources at BAPE are easily accessible in terms of time and effort in retrieving information	0	9 (90%)
B. Public participation often adds new knowledge to project environmental review	0	10 (100%)
C. In my work, I have a preference for obtaining information from other people rather than from documents	4 (40%)	1 (10%)
D. The reports I prepare deal with such specific issues that I hardly ever find any useful knowledge in BAPE internal sources	3 (30%)	2 (20%)
E. The most accurate and reliable way of doing an internal task is to check how it was done last time	2 (20%)	4 (40%)
F. BAPE opinions and advice are useful for environmental approval documents	1 (10%)	6 (60%)

Note: Results are presented in two broad groups; in the questionnaire, respondents were asked to mark "strongly disagree", "partially disagree", "unable to judge", "partially agree", and "strongly agree"; sums can be less than 100% where respondents indicated that they were "unable to judge".

public reactions to projects during consultations, as compared to a more technical review performed by DEE.

There is consistency between the perceived effectiveness and the frequency of access to knowledge repositories. Effectiveness was defined as "the ease of access — in terms of time and effort — to DEE or BAPE collective experience and knowledge". The question aimed at checking whether the repositories, albeit containing potentially useful knowledge, could require such a level of effort (e.g. it could be excessively time-consuming to access) as to discourage its use. Ranking utilised the following five-level scale: "effective-very effective-slightly effective-ineffective-very ineffective". At DEE, 11 out of the 16 repositories were evaluated as effective or very effective by more than 50% of respondents. Senior staff was ranked first (100%), followed by informal talks, previous environmental analysis reports, government approval documents, officer's manual and templates and technical sheets (all at 95%).

At BAPE, 12 out of 14 repositories are considered to be effective or very effective by a majority of respondents. Similarly to the responses given by the DEE staff, senior staff and informal talks are perceived by all respondents as effective or very effective knowledge repositories. Second in ranking are previous BAPE reports (90%), followed by files of previous enquiries, the systematic review approach, DEE's environmental analysis reports and internal meetings, all ranked as effective or very effective by 80% of respondents. Tools recently developed (the collection of opinions and findings and the SD Analysis Framework) were ranked as less effective (50% of effective or very effective responses) than the "traditional" sources.

Knowledge creation

Staff's views on statements that their work creates knowledge (Tables 9 and 10) were sought. Respondents unambiguously reported that they experienced meaningful opportunities to increase their personal knowledge and that sharing information was part of their normal work activity, as expected from individuals in multidisciplinary teams. There is also strong agreement that the new knowledge generated can be used by the respective organizations in the future. Less unanimous was the appreciation about personal learning being actually fed back to the organizations in order to improve future assessments: 50% at BAPE and 71% at DEE.

In terms of learning from follow-up to improve ongoing and future assessments, DEE staff is split: 38% believe that knowledge created by follow-up can effectively be used to improve the review of new proposals, while 28% don't

Table 9. Statements about internal knowledge creation at DEE.

Statement	Disagree	Agree
A. Working at DEE provides meaningful opportunities to increase my personal knowledge	0	20 (95%)
B. Sharing information is part of my normal work activity	0	21 (100%)
C. At DEE, mistakes are viewed as learning opportunities	8 (38%)	9 (43%)
D. In my work I generate new knowledge that can be used by DEE in the future	0	17 (85%)
E. Significant parts of what I learn during my work is actually fed back into our future assessments	1 (5%)	15 (71%)
F. What DEE learnt when evaluating EIA projects is also used for evaluating projects not submitted to the EIA procedure	5 (24%)	7 (33%)
G. Follow-up reports create knowledge effectively used to improve the review of new proposals	6 (29%)	8 (38%)
H. Preparation of well-informed reports significantly contribute to decisions that protect the environment for present and future generations	1 (5%)	17 (81%)

Notes:

(1) Three respondents did not rank statement [F].

(2) Results are presented in two broad groups: in the questionnaire, respondents were asked to mark "strongly disagree", "partially disagree", "neither agree nor disagree", "partially agree", and "strongly agree". Totals may be lower than 100% due to responses marked as "neither ...".

Table 10. Statements about internal knowledge creation at BAPE.

Statement	Disagree	Agree
A. Working at the BAPE provides meaningful opportunities to increase my personal knowledge	0	10 (100%)
B. Sharing information is part of my normal work activity	0	10 (100%)
C. At DEE, mistakes are viewed as learning opportunities	2 (20%)	5 (50%)
D. In my work I generate new knowledge that can be used by BAPE in the future	0	9 (90%)
E. Significant parts of what I learn during my work is actually fed back into our future assessments	2 (20%)	5 (50%)
F. Follow-up reports prepared by proponents create knowledge effectively used to improve the review of new proposals by BAPE panels	2 (20%)	7 (70%)
G. Preparation of well-informed reports significantly contribute to decisions that protect the environment for present and future generations	0	8 (80%)

Notes:

(1) One respondent did not rank statements [C] and [F].

(2) Results are presented in two broad groups: in the questionnaire, respondents were asked to mark "strongly disagree", "partially disagree", "neither agree nor disagree", "partially agree", and "strongly agree". Totals may be lower than 100% due to responses marked as "neither...".

agree. The results are different at BAPE, where 78% of respondents feel positive about the use of knowledge created during follow-up.

Finally, the mood is generally positive (>80%) about the contribution of well-informed reports to environmental protection.

Discussion

The results presented in the previous section show that both agencies have been promoting KM in support of their core activities. In this section, the main findings will be reviewed and discussed.

Knowledge Management (KM) initiatives

Both organizations have independently implemented initiatives conducive to KM, but didn't adopt a formal approach to KM, nor were their initiatives linked to any government-wide policy. These initiatives include both technology-based and human resources approaches (Handzic, 2005) and pertain mostly to the first two out of the following objectives of KM projects identified by Davenport *et al.* (1988): (1) create knowledge repositories; (2) improve knowledge access; (3) enhance the knowledge environment; and (4) manage knowledge as an asset. Establishing repositories and facilitating access to knowledge are patent objectives for both organizations. Concerns with enhancing a knowledge environment, i.e. "establishing an environment conducive to more effective knowledge creation, transfer, and use" (p. 47) were implicit in a number of DEE initiatives, such as conducing after-action reviews and updating the Technical Sheets. They were also mentioned in BAPE's strategic plan. Managing knowledge as an asset, on the other hand, is an objective more applicable to private firms than to government agencies, as it supposes accounting for intellectual capital such as patents.

Three drivers and one additional incentive were identified as having acted for the implementation of KM. Firstly, such initiatives appear to originate in the successive managers' perceptions that EIA does create knowledge and it is therefore important that such knowledge be retained in the organization, not only at individual level.

A second and possibly the major driver for KM-related initiatives is a concern with consistency and reproducibility of advice provided. Evidence from both interviews at DEE and from one interview as well as published documents at BAPE mentioned such interest as a major reason for implementing the initiatives. The statement "reduce the variability among reports prepared by different commissions", mentioned by one BAPE interviewee and the aim of achieving "coherence in internal processes" stated in an annual report (BAPE, 2010, p. 14) clearly support this understanding.

Thirdly, Knowledge Management (KM) initiatives are imposed by time and resource constraints. They allow reviewing teams to deal with an apparently universal concern for EIA agencies, keeping up with timelines and providing quick responses (Middle and Middle, 2010). Worries about the efficiency, the predictability and the equity of the EIA process were voiced by stakeholders during public debates on the effectiveness of the Québec procedures in both 1988 and 1992.

Staff turnover may be an additional reason contributing to justify KM initiatives, as "buffering an organization from the disruptive effect of staff turnover" (Argote *et al.*, 2003). It is a major driver for KM in the literature. In Québec, as well as in many other places, EIA procedures have been put in place in the 1970s and 1980s. Therefore, officers that worked in applying the then innovative procedures are likely to be retired after 25 or 30 years of services. Within the context of reduction of the number of employees within the Québec government, BAPE has made a five-year retirement plan in 2009–10 (BAPE, 2010). In addition, movements of employees are not just related to retirement. In 2011–12, BAPE reported a voluntary departure rate for all type of employment (not only analysts) of 22,9%, including two retirements (BAPE, 2012).

BAPE and, to a lesser extent, the DEE work in a dynamic socio-political system where the environmental approval process is often commented on in the media, particularly for controversial projects. The media have specific rules to follow during public hearings and locations where to stand. Media are regularly used as a communication tool both by the BAPE and the different stakeholders of a project. All documents, from project notices to the BAPE reports, as well as the environmental analysis reports and the decrees, are available to the public and posted on BAPE's or the Ministry of the Environment's website. This situation adds a public concern component about the fairness of the government agencies involved, potentially acting as the instigation for setting up both procedural and substantive rules.

Knowledge repositories and their usefulness

Both organizations built and maintain knowledge repositories tailored to their own needs. Some repositories are intended primarily at external stakeholders — thus fulfilling multiple functions — whereas others were conceived as tools to facilitate the performance of internal tasks. The repositories were devised and developed down the years, not as a result of a one-time initiative.

Responses to the questionnaires showed that analysts access all categories of repositories when performing their tasks. At both organizations, the most frequently accessed and effective repositories were informal talks and senior staff: face-to-face interactions seem to be the preferred way of accessing knowledge. The least accessed are the operations management system and formal meetings for DEE officers, and the SD analysis framework and the digest of panel questions for BAPE analysts.

Informal knowledge exchange is a characteristic of consultancy firms (Rueschemeyer, 1983) that can also be found in government organizations. Accordingly, Taminiau *et al.* (2009) consider that informal exchange features a fruitful path to innovation in such firms. Hansen *et al.* (1999) observe that social networks are more effective ways of knowledge sharing the more intellectual the work to be performed is. Such preference for personal interactions has also been found among project officers in the Western Australia EIA agency (Sánchez and Morrison-Saunders, 2011) and in environmental consultancies firms preparing EISs in Brazil (Costanzo and Sánchez, in press).

This is not to say that other repositories are less important or less useful. Respondents at both BAPE and DEE do value other repositories. At DEE, the manual, previous reports and government approval documents, alongside the technical sheets are considered as the most effective while at BAPE, previous reports and files and the systematic review framework are considered to be the most effective.

The use by the DEE of after-action reviews as a learning mechanism is noteworthy. Defined as "a facilitated discussion, conducted immediately after a project or major activity and using a semi-structured format" that "draws on the participants experience and perceptions to help identify lessons" (Standards Australia, 2005), its systematic use had not been found in the EIA agency in Western Australia nor was it detected in the survey of Knowledge Management (KM) practices by environmental consultancies in Brazil (Costanzo and Sánchez, in press). However, elsewhere in Canada, Fitzpatrick (2006, p. 171) documented "post hoc analysis" as a task for "debriefing and documenting lessons from specific experiences" in the EIA process. At the DEE, when a review is approaching its conclusion, it is established practice to have the project officer organizing an internal workshop in which the findings are presented and discussed with other officers.

KM through codification is perceived as limited by DEE and BAPE. This result is similar to what has been found in the Australian study. Although officers are benefiting from practices such as procedural guidelines or replicable successful solutions, they acknowledge that professional judgement plays a central role in the tasks performed by officers in an EIA agency and tailor-made solutions are often needed.

Looking ahead

Despite achievements in Knowledge Management (KM) practices, huge learning opportunities lie unexplored. It was found that project performance and compliance data generated in follow-up is not systematically processed into meaningful knowledge for future assessments (Sánchez, 2012). Respondents at DEE were split when asked if follow-up reports create knowledge that is effectively integrated into current practice. During the interviews, it was mentioned that the need to meet timelines for ongoing assessments implies that low priority is assigned to review follow-up reports. As Québec's legislation does not make publicly available the follow-up and monitoring reports prepared by project proponents, it is difficult for the professionals and for the academic and research community to tap into those information resources to generate new knowledge: only the government is in a position to explore those data and transform them into meaningful knowledge in support of future assessments.[7]

A second focus of ongoing and future development is the enlargement of scope of BAPE appraisals to incorporate sustainability assessment, as guided by the Québec Sustainable Development Act. While DEE reviews are exclusively performed under the Environmental Quality Act, BAPE reviews are now based on both laws. This is aligned with further developments on Sustainability Appraisal in Canada and in a few other countries (Gibson and Hassan, 2005) and the international evolution of impact assessment (Sánchez and Croal, 2012).

Comparisons

To a large extent, EIA practice is framed by the legislation and regulations of each jurisdiction. On the other hand, everywhere the EIA process primarily aims at informing decision-making about projects potentially having significant environmental effects. Therefore, many common features exist between procedures adopted in different countries. In addition, in those jurisdictions where a centralized EIA agency manages the whole process, several similar tasks are performed and there is room for interorganizational experience sharing and comparisons. Considering the similarly aimed research performed in Western Australia, a

[7] It should be noted that even if the follow-up and monitoring reports are not publicly available, results may be divulgated.

Table 11. Types of knowledge management initiatives adopted by three EIA agencies.

Initiative	DEE	BAPE	OEPA[a]
(1) Maintaining a library of previous assessments	XXX	XXX	XXX
(2) Staff, social networks and communities of practice	XXX	XXX	XXX
(3) Provision of substantive guidance (know-how)	XX	X	XXX
(4) Provision of internal procedural guidance (know-what)	XX	XX	XX
(5) Treatment of spatial information (know-where)	o	o	X
(6) After-action reviews (including follow-up) (know-why)	X	o	o

Legend:
X used, but only for a few tasks, could be used more extensively.
XX used for many tasks performed.
XXX intensive use, adopted for most tasks performed.
o not used.
[a] the Western Australia EIA agency.

simplistic comparison is presented in Table 11. The level of adoption of six groups of Knowledge Management (KM) initiatives is indicated.

Initiatives pertaining to KM were grouped into six categories. Types (1) and (2) are generally mentioned in the KM literature and the four types of knowledge relevant to EIA, know-how, know-what, know-where and know-why were taken from Sánchez (2012, see also Fischer *et al.*, 2009).

All three organizations maintain a library of previous assessments available to both external users and their analysts, who value this repository as useful and effective. Face-to-face communications are rated as the most effective repository by officers at the three organizations.

Conclusions

Both DEE and BAPE have been adopting initiatives towards managing knowledge: repositories have been established, access to knowledge has been pursued and officers believe that they contribute to knowledge creation that is actually useful in their work. These initiatives have been driven by: (i) a shared understanding by successive managers that EIA does create knowledge; (ii) a concern with consistency and reproducibility of recommendations resulting from EIS review and public consultation; and (iii) a concern with improving work efficiency. In addition, preparing for inhibiting the negative effects of staff turnover is another contributor to such initiatives. The experience of Québec EIA agencies is similar to Western Australia's in the sense that even in the absence of a formal KM strategy, it is possible to adopt a number of initiatives conducive to managing the organization's ability to learn, store and use collective knowledge in support of its core activities.

Internal knowledge users — the analysts — often access existing repositories when performing their tasks and perceive most repositories as effective. However, face-to-face interactions are seen as the most effective way of accessing knowledge.

The key elements of Knowledge Management (KM) adopted by DEE that could be relevant for international practice are: (i) the provision (to proponents, consultants and project officers alike) of substantive guidance on methods, approaches and assessment criteria; and (ii) the mobilization of specialized knowledge within the Ministry of Environment and relevant departments in other ministries to prepare the environmental review reports.

At the BAPE, the key element towards managing organization knowledge is the compilation — in an accessible and continuously updated database — of previous findings, opinions and recommendations from enquiries and reports in order to facilitate access and provide consistency and reproducibility to ongoing enquiries.

In terms of the main functions of KM (knowledge creation, retention, and transfer or share) both DEE and BAPE appear to be successful in retention and transfer to internal knowledge users. Less spectacular results are featured for knowledge creation. As a matter of fact, there is as yet an underexplored learning potential in the follow-up activities after project approval. Québec legislation does not make publicly available the follow-up and monitoring reports prepared by project proponents. Hence, it is very difficult for the academic and research community to tap into those information sources to generate new knowledge. Currently, only the government is in a position to explore those data and transform them into meaningful knowledge in support of future assessments.

EIA agencies play a key role in capturing knowledge created in the EIA process. Adoption of KM initiatives such as those documented and reviewed in this research provide a much needed first step that could serve as inspiration for similar organizations in both developed and developing countries. So far an unmet challenge is making sense of a great deal of data and information obtained in the follow-up phase of the EIA process and transform it into knowledge to be used to improve both efficiency and effectiveness of an agency's work.

This research documented that 80% of officers at the EIA agencies believe that the preparation of well-informed reports significantly contribute to decisions that protect the environment for present and future generations. Future research could explore the possible contribution that a structured approach to KM in EIA agencies could make to positively influence decision-making and EIA effectiveness.

Acknowledgements

Research was supported by the International Council of Canadian Studies. The authors are indebted to Mrs. Marie-Josée Lizotte, Director, and Mr. Robert Joly, former Director of *Direction des évaluations environnementales*, and Mr. Pierre Renaud, President at the time the research was conducted, and Sylvie Mondor, Director of environmental expertise and sustainable development, *Bureau d'audiences publiques sur l'environnement*, for their advice and insights.

References

Agimo, Australian Government Information Management Office (2004). *Knowledge Management Best Practice Checklist*. Canberra: Agimo.

André, P, CE Delisle and JP Revéret (2010). *L'évaluation des Impacts sur l'environnement. Processus, Acteurs et Pratique pour un Développement Durable*. Montreal: Presses Internationales Polytechnique.

Argote, L, B McEvily and R Reagans (2003). Managing knowledge in organizations: an integrative framework and review of emerging themes. *Management Science*, 49, 571–582.

Arora, E (2011). Knowledge management in public sector. *Researchers World*, II, 165–171.

BAPE, Bureau d'Audiences Publiques sur l'Environnement (2008). Plan d'action de développement durable 2008–2013. Québec.

BAPE, Bureau d'Audiences Publiques sur l'Environnement (2009). Plan stratégique 2008–2013. Ensemble pour un développement durable. Québec.

BAPE, Bureau d'Audiences Publiques sur l'Environnement (2010). Rapport annuel de gestion 2009–2010. Québec.

BAPE, Bureau d'Audiences Publiques sur l'Environnement (2012). Rapport annuel de gestion 2011–12. Québec.

Bond, A, CV Viegas, CCSR Coelho and PM Selig (2010). Informal knowledge processes: the underpinning for sustainability outcomes in EIA? *Journal of Cleaner Production*, 18, 6–13.

Brendecke, C and L Ortolano (1981). Environmental considerations in Corps planning. *Water Resources Bulletin*, 17, 248–254.

CEN, European Committee for Standardization (2004). CWA 14924-5 *European Guide to Good Practice in Knowledge Management — Part 5: KM Terminology*. Brussels: CEN.

Chávez, BV and AS Bernal (2008). Planning hydroelectric power plants with the public: a case of organizational and social learning in Mexico. *Impact Assessment and Project Appraisal*, 26, 163–176.

Costanzo, BP and LE Sánchez (2014). Knowledge management in environmental consultancy firms. *Revista Produção* 24(4), 742–750 (in Portuguese).

Davenport, TH, DW De Long and MC Beers (1998). Successful knowledge management projects. *Sloan Management Review*, Winter, 43–57.

Fiol, CM and MA Lyles (1985). Organizational learning. *The Academy of Management Review*, 10, 803–813.

Fischer, TB, P Gazzola, U Jha-Thakur, S Kidd and D Peel (2009). Learning through EC directive based SEA in spatial planning? Evidence from the Brunswick Region in Germany. *Environmental Impact Assessment Review*, 29(6), 421–428.

Fitzpatrick, P (2006). In it together: organizational learning through participation in environmental assessment. *Journal of Environmental Assessment Policy and Management*, 8, 157–182.

Frappaolo, C (2006). *Knowledge Management*, 2nd ed. Chichester: Capstone.

Gibson, RB and S Hassan (2005). *Sustainability Assessment: Criteria and Processes.* London: Earthscan.

Handzic, M (2005). Managing knowledge with technology: mission possible. In *Knowledge Management Through the Technology Glass*, M Handzic (ed.), pp. 21–38. Singapore: World Scientific.

Hansen, MT, N Nohria and T Tierney (1999). What's your strategy for managing knowledge? *Harvard Business Review*, 77(2), 106–166.

Jha-Thakur, U, P Gazzola, TB Fischer, D Peel and S Kidd (2009). SEA effectiveness — the significance of learning. *Impact Assessment and Project Appraisal*, 27(2), 133–144.

Kim, DH (1993). The link between individual and organizational learning. *Sloan Management Review*, Fall, 37–50.

Leduc, R (2005). *Guide de la Modélisation de la Dispersion Atmosphérique.* Québec, Ministère du Développement Durable, de l'Environnement et des Parcs, Direction du suivi de l'état de l'environnement, envirodoq no ENV/2005/0072, rapport no QA/49, 38p. Available at http://www.mddep.gouv.qc.ca/air/atmosphere/guide-mod-dispersion.pdf.

Middle, G and I Middle (2010). The inefficiency of environmental impact assessment: reality or myth? *Impact Assessment and Project Appraisal*, 28(2), 159–168.

Morgan, RK (2012). Environmental impact assessment: the state of the art. *Impact Assessment and Project Appraisal*, 30(1), 5–14.

Québec (1988). L'évaluation environnementale: une pratique à généraliser, une procédure à parfaire. Rapport du Comité de révision de la procédure d'évaluation et d'examen des impacts environnementaux, Décembre 1988, Québec, 169 p.

Québec (1992). La procédure d'évaluation des impacts sur l'environnement. Commission de l'aménagement et des équipements, Avril 1992, 72 p.

Québec (2006). Guide d'analyse des projets d'intervention dans les écosystèmes aquatiques, humides et riverains assujettis à l'article 22 de la Loi sur la qualité de l'environnement. Available at http://www.mddep.gouv.qc.ca/eau/rives/delimitation.pdf.

Rueschemeyer, D (1983). Professional autonomy and the social control of expertise. In *The Sociology of Professions*, R Dingwall and PSC Lewis (eds.). London: Macmillan.

Saarikoski, H (2000). Environmental impact assessment (EIA) as collaborative learning process. *Environmental Impact Assessment Review*, 20, 681–700.

Sánchez, LE (2012). Information and knowledge management. In *Furthering Environmental Impact Assessment*, A Perdocoúlis, B Durning and L Palframan (eds.), *Furthering Environmental Impact Assessment: Towards a Seamless Connection between EIA and EMS*, pp. 19–38. Cheltenham: Edward Elgar.

Sánchez, LE and P Croal (2012). Environmental impact assessment from Rio-92 to Rio+20 and beyond. *Ambiente & Sociedade*, 15(3), 41–54. http://dx.doi.org/10.1590/S1414-753X2012000300004

Sánchez, LE and A Morrison-Saunders (2011). Learning about knowledge management for improving environmental impact assessment in a government agency: the Western Australian experience. *Journal of Environmental Management*, 92, 2260–2271.

Sinclair, AJ, A Diduck and P Fitzpatrick (2008). Conceptualizing learning for sustainability through environmental assessment: critical reflections on 15 years of research. *Environmental Impact Assessment Review*, 28, 415–428.

Standards Australia (2005). AS 5037–2005. *Knowledge Management — A Guide.*

Taminiau, Y, W Smit and A de Lange (2009). Innovation in management consulting firms through informal knowledge sharing. *Journal of Knowledge Management*, 13(1), 42–55.

Webler, T, H Kastenholz and O Renn (1995). Public participation in impact assessment: a social learning perspective. *Environmental Impact Assessment Review*, 15, 443–463.

CHAPTER 12

Environmental Assessment and Management Related Master Level Degree Programmes in the EU: Baseline, Trends, Challenges and Opportunities

Thomas B. Fischer* and Urmila Jha-Thakur†

School of Environmental Sciences
Department of Geography and Planning
University of Liverpool, 74 Bedford Street South
Liverpool L69 7ZQ
**fischer@liverpool.ac.uk*
†urmila1@liverpool.ac.uk

Environmental assessment and management (EAM) is an inherently interdisciplinary subject, drawing on science (e.g. ecology, biology, physical geography), social science (e.g. decision-making theory, political sciences, planning) and economics. Due to the absence of a clear disciplinary home, understanding of EAM-related higher education degree programmes has remained poor. Furthermore, teaching of EAM is influenced by how it is practised and therefore differs from country to country. The aim of this paper is to shed light on this still under-researched area by establishing a comprehensive overview of EAM-related master level degree programmes across the EU. In this context, current baseline and emerging trends as well as challenges and opportunities are presented and discussed. The paper is aimed specifically at those interested in EAM higher education, e.g. researchers, university teachers and (prospective) students.

Keywords: Environmental assessment; environmental management; higher education; Europe.

*Corresponding author.

Introduction

An upsurge in the interest of research on Environmental Assessment and Management (EAM) education has been observed over recent years (Stelmack *et al.*, 2005; Fischer *et al.*, 2008; Barker and Jones, 2008; Ramos *et al.*, 2008; Gazzola, 2008; Cherp, 2008; Gazzola and Jha-Thakur, 2009; Sanchez *et al.*, 2010; Fischer *et al.*, 2010; Sanchez and Morrison-Saunders, 2010; Fischer *et al.*, 2011; Fischer, 2012; Weiland, 2012). This is connected with a growing prominence of environmental issues in public and private policy, plan, programme and project decision-making processes, internationally, nationally, regionally and locally (Jordan, 2005). Connected with this, interest in EAM-related degree programmes has also steadily increased and associated programmes have gained in popularity (Fischer *et al.*, 2010). EAM is currently mainly taught in higher education institutions (Sanchez and Morrison-Saunders, 2010). However, to date, its understanding has remained poor and research on the subject has stayed embryonic (Sanchez, 2010). However, along with a continuously improving EAM practice, associated education endeavours also need to improve (Alderson, 1992; Sinclair and Diduck, 1995). In order to judge whether this is indeed happening, a better understanding of current EAM-related education is required. It is within this context that the paper establishes a comprehensive overview of EU-wide EAM master level degree programmes. It thus acts as a platform against which further EAM-related higher education research can be undertaken.

This paper builds on some of the results of the European T_{wo}EA-M (**E**nhancing **A**ttractiveness of **E**nvironmental **A**ssessment and **M**anagement Higher Education) project, which ran from 2008 to 2010. This project had the following five objectives (Fischer *et al.*, 2011):

(1) To establish a European and East Asian T_{wo}EA-M network on EAM education;
(2) To produce a guidebook of major types of current EAM programmes and their specialism in EU member states;
(3) To test the usefulness of European EAM books for East Asian higher education;
(4) To organise four EAM conferences with both, European and East Asian participants in China, South Korea, Malaysia and Austria; and
(5) To produce a report on specific East Asian EAM interests and education requirements.

The focus of this paper is on the outcomes of point two regarding current EAM programmes, looking at baseline, trends, challenges and opportunities of

EAM-related higher education in the EU. For this purpose, the remainder of the paper is divided into five sections. Firstly, the background of the paper is explained. Secondly, the methodology is introduced. Thirdly, EAM-related higher degree programmes across EU member states are presented and analysed. In the fourth section, emerging challenges and opportunities for EAM education are discussed. Finally, conclusions are drawn and the scope for further research is established.

Background

This paper provides for a comprehensive overview of EAM-related master programmes across the 27 EU member states in mid-2010, looking at academic home, implementation, requirements and scope. Subsequently, first the link between EA and EM is explained and the decision to look at both is justified. This is followed by an explanation as to why the focus here is on the EU.

Link between environmental assessment and management

To date, studies on environmental assessment and management-related higher education have mainly focused on EA (Fischer *et al.*, 2008; Ramos *et al.*, 2008; Gazzola, 2008; Sanchez, 2010; Weiland, 2012), with an emphasis on the disciplinary context within which it is taught. Such an emphasis has been useful in unveiling the multidisciplinary nature of EA and the dominance of specific disciplines teaching it, acting as "gatekeepers" for the subject area (Gazzola, 2011). As EA and EM are essentially intertwined concepts (Morrison-Saunders and Bailey, 1999), however, we suggest that these should be considered together.

Environmental Assessment (EA) is an administrative decision-making procedure, which comes with a number of methods that may be applied in different situations, ranging from individual projects (Environmental Impact Assessment — EIA) to policies, plans and programmes (PPPs; Strategic Environmental Assessment — Strategic Environmental Assessment (SEA)). Environmental Management (EM), on the other hand, is a set of activities and tools which are set out to achieve environmentally related objectives by managing impacts during production or construction processes (Fischer *et al.*, 2010). EA and EM complement each other. EA techniques help in assessing impacts in an *ex-ante* manner (i.e. during policy, plan, programme and project making), while EM techniques (e.g. environmental management systems — EMS) help in actually managing impacts (Cherp, 2008).

That EA and EM are two sides of the same coin is reflected in the scope of many academic journals (e.g. *Journal of Environmental Assessment Policy and*

Management (JEAPM); *Journal of Environmental Planning and Management* (JEPM); *African Journal of Environmental Assessment and Management* (AJEAM); *Integrated Environmental Assessment and Management* (IEAM)). Also, the world's biggest environmental professional organisation and accreditation body, the UK-based Institute of Environmental Management and Assessment (IEMA), focuses on both EA and EM.

To date, the relationship between EA and EM has remained underexplored. However, the combined delivery of *ex-ante* and *ex-post* management has been described as being critical for achieving effective environmental protection and management (Bailey, 1997; Morrison-Saunders *et al.*, 2004; Jha-Thakur, 2011). This is also one of the reasons why monitoring and other EIA follow-up activities are still seen as being weak (Glasson *et al.*, 1999; Jha-Thakur and Fischer, 2008).

Focus on EU degrees

The rationale for looking at practice in EU member states stems from the fact that these have played a crucial role in the development of EAM. For example, the EU introduced environmental standards, such as EMAS (the Eco-Management Auditing Scheme; Council Regulation 1836/93), and environmental assessment directives, including the (project) EIA Directive 337/85/EEC and the (plan and programme) SEA Directive 42/2001/EC that are considered to be leading pieces of legislation (Gazzola and Jha-Thakur, 2009). Furthermore, the EU has extensively supported associated research and has been involved in international capacity building and higher education support (Rennings *et al.*, 2006; Dalal-Clayton and Sadler, 2005; European Commission, 2010).

European EAM standards and directives have been reviewed and commented on in the international literature, as well as at many international conferences and workshops (Fischer, 2003, 2010; Risse, 2005; Aschemann, 2002; Arts *et al.*, 2001; Barker and Wood, 1999). This, together with some progressive policies on environmental issues, including e.g. on global climate change (Van Schaik, 2012) has led to the EU being seen as a global leader in the field of EAM. The EU's EAM legal framework is looked upon by non-European developed and developing countries, as well as development banks, as a milestone in the development of EAM (Gazzola and Jha-Thakur, 2009). Furthermore, a major share of EA and EM research activities and publications are of European origin. For strategic environmental assessment (SEA), for example, Fischer and Onyango (2012) recently established that European authors were listed in about two-thirds of all publications in refereed journals internationally. Based on a SCOPUS search, it can be concluded that for environmental impact assessment (EIA) of projects, there is also a heavy emphasis in the literature on European practices and authors.

European higher education institutions have achieved an excellent international reputation for their work on EAM. Furthermore, EU member states have been undergoing a process of standardisation and harmonisation in an effort to internationalise educational degrees across all EU member states (Jha-Thakur, 2008). The associated so-called Bologna process deserves special mentioning here, as this aims at streamlining the process of integrating the international dimension into different EU member states' national policies for higher education, achieving compatibility of degree programmes throughout the EU. This has been facilitated through the introduction of a wide range of mechanisms, such as the Two-Tiered Study Structure (TTSS; consisting of BA and MA levels, which has been common practice in the English-speaking world for many decades, but is new to many continental European countries), as well as the European Credit Transfer System (ECTS; with one ECTS meant to represent 25–30 hours of work), the Diploma Supplement, the Europass and the Tuning Project (Jha-Thakur, 2008).

Methodology

Both on-site as well as distance learning courses of higher education institutions were included in the analysis underlying this paper. Information on degree programmes was mainly collected through internet surveys, primarily because the internet is a preferred source of information for international students (Capper, 2001; Gordon, 2005). The main search engine used was Google. Searches made were made via international as well as national web pages. Furthermore, the use of other search engines was also explored. In this context, www.masterportal.eu, www.findmasters.com and www.masterstudies.com were found to be particularly useful. In collecting data for this work, four researchers were involved from three EU member states (Austria, UK, Slovakia), collecting information in several languages, at times with the help of academic colleagues from other member states (English, German, Spanish, Italian, French, Slovak, Polish, Danish, Swedish, Portuguese, Greek, Czech, Romanian, Hungarian, Dutch and Finnish). In an attempt to standardise search techniques, key phrases and words were chosen in different European languages, including "*Masters Environmental Assessment and Management*" (country name) and "*Masters Environmental Assessment*" (country name).

Data collection was standardised with the help of a template which was specifically designed for this purpose (see Table 1). Each completed template

Table 1. Categories and sub-categories for the templates.

No.	Country name	Template no.
	Headings	Sub-categories
1	Academic Home	Name of university
2		Weblink
3		Degree title
4		Emphasis (EA or EM)
5		Faculty/School/Department
6	Programme Implementation	Duration
7		Delivery language
8		Delivery method & technique
9		Assessment method
10		Focus (taught/research-based)
11		Programme structure
12	Programme Requirements & Scope	Geographical specialism
13		Entry requirements
14		Fees (Euro)
15		Credit structure
16		Career opportunities
17	Remarks	

obtained a title, based on the country in which the degree programme was offered and a code. Information was collected under three categories, including:

(a) Academic home;
(b) Programme implementation; and
(c) Programme requirements and scope.

Point (a) sheds light on "what" is taught, the degree title, the university name, the emphasis on either EA or EM and the faculty/school/department within which it is taught. An overall degree emphasis was identified, based on the choice of compulsory and optional modules offered. For example, if EM is a compulsory module and EA an optional one, the programme was said to have an EM emphasis. If the emphasis is not clear, for example in the case of a combined degree programme, this was noted, too. Point (b) mainly dealt with answering "how" the programmes were taught. Programme duration, language, methods and techniques used in delivering the programme, the structure of the programme and its focus (i.e. taught or research based) were identified. Point (c) finally established "what" and "where" questions, looking at programme requirements and scope.

Geographical specialism, entry requirements, fees, credit structure, and career opportunities were also identified. Finally, "who" is considered eligible to study a certain programme and "where" students are likely to find post-graduate job opportunities was established.

Some universities/faculties/schools/departments were found to offer more than one EAM-related programme. In these cases, different programmes were often found to have overlapping content. As a consequence, only one programme from any one institution was included. However, if this is the case, other programmes are then mentioned in the template. Whilst, as explained earlier, internet searches were mainly used for identifying degree programmes, other sources were used to supplement the information thus obtained, including in particular information provided by country-specific experts. In the initial search, no EAM-related master level programmes were found in six EU member states, including Romania, Portugal, Cyprus, Malta, Luxembourg and Greece. Subsequently, additional experts were contacted from these countries, leading to the identification of a couple of further programmes from Greece and Malta. Finally, two well-known international EAM professors were employed to check the adequacy of the findings.

There are four potential limitations of the survey underlying this paper. Firstly, as described above, the data collected was mainly based on information available on the internet. Therefore, programmes that were not advertised on the web would have not been found. However, this should only apply to a very small number of programmes, as the experts consulted on the results did not suggest that any other programmes should be added. Secondly, the boundaries with related subjects, like environmental engineering, were at times difficult to draw (and are probably somewhat artificial). Thirdly, EM is a term that comprises a very broad area of activities. This could be the main reason behind establishing a larger number of EM than EA programmes. Finally, EU member states are going through a transition initiated by the Bologna process. Hence, certain "old-style" programmes are not listed, which are still taught, though (e.g. 5-year *"Diplom"* degrees in Germany). It is important to note that the data collected and presented here was true at the time of the survey (summer 2010) and are subject to changes over time.

Findings of the Survey: Overview of EAM-Related Master Level Programmes

EAM-related master level degree programmes were found in 23 EU member states. No programmes were offered in Cyprus, Romania, Portugal and Luxembourg.

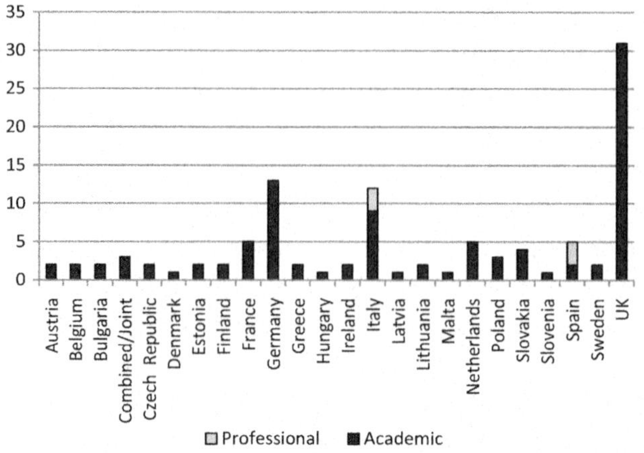

Figure 1. Country-wise distribution of EAM-related Master level programmes.

A total of 121 EAM-related master level programmes were identified, out of which a total of 106 programmes were analysed (only one programme from any one university was considered, as explained above).

Figure 1 presents an overview of the programmes. UK universities offered a total of 31 programmes. Furthermore, Italy and Germany accounted for 12 and 13 degree programmes, respectively. In Italy, three out of these were of a professional nature. Such programmes are usually shorter and, comparatively speaking, more expensive than their academic counterparts. Five programmes were offered in each of the Netherlands, France and Spain. In the case of Spain, three out of the five were professional programmes. Slovakia was identified to offer four programmes and Poland three. Ireland, Austria, Finland, Belgium, Bulgaria, Estonia, Lithuania, Greece, Sweden and the Czech Republic were found to be offering two programmes each. Finally, Denmark, Latvia, Hungary, Malta and Slovenia all offered one programme each. Some programmes were structured as combined degrees and not affiliated with any one particular country. These are represented in Figure 1 as "combined/joint". Out of these, two were European Commission Erasmus Mundus supported programmes (European Commission, 2010). The third programme was offered by the Baltic University, which is a co-operation of 225 universities across 14 EU member states. Other combined degrees are represented by the country of the leading university. Table 2 lists the 106 master level degree programmes. The following sections provide an overview of these programmes, based on the three categories introduced above (i.e. academic home, programme implementation, programme requirements and scope).

Table 2. EAM-related Master level degree programmes in the EU (summer, 2010).

Country	Name of university	Degree offered
1. Austria	1. University of Klagenfurt	1. MSc Management of Protected Areas
	2. Graz University	2. MSc Natural Sciences/MA Social and Economic Sciences.
2. Belgium	3. Free University of Brussels	3. MSc Human Ecology
	4. Ghent University	4. MSc Environmental Sanitation
3. Bulgaria	5. University of Architecture, Civic Engineering and Geodesy, Sofia	5. MSc Urbanism
	6. University of Sofia	6. MSc Environmental Protection
4. Combined	7. Baltic University	7. MSc Sustainable Community Development/ MSc Sustainable Water Management
	8. University of Girona (Spain); University of Southern Denmark (Denmark); University of Ljubljana (Slovenia)	8. European Master in Tourism Management (EMTM)
	9. Mining School of Nantes (EMNATES) (France); Politechnical University of Madrid (Spain); Royal University of Technology Sweden (KTH); Budapest University of Technology and Economics (BME) (Hungary); Queens University of Belfast (Northern Ireland)	9. The European joint MSc Management and Engineering of Environment and Energy (ME3)
5. Czech Republic	10. Charles University Prague	10. MSc Ecology and Environmental Protection
	11. Czech University of Life Sciences	11. MSc Engineering Ecology
6. Denmark	12. Aalborg University	12. MSc Environmental Management
7. Estonia	13. Estonian University of Life Sciences	13. MSc Natural Sciences
	14. Tallin University of Technology	14. MSc Engineering
8. Finland	15. University of Jyväskylä	15. MSc Corporate Environmental Management
	16. University of Turku	16. MSc Environmental Sciences
9. France	17. University of Technology of Troyes	17. Master of Engineering, Environmental Management and Sustainable Development (IMEDD)

Table 2. (*Continued*)

Country	Name of university	Degree offered
	18. University Paris Panthéon-Sorbonne	18. Professional Master of Sustainable Development, Environmental Management and Geomatics
	19. National Institute of Applied Sciences of Lyon	19. Master of Researched Environmental, Industrial & Urban Sciences (SEIU)
	20. University Michel des Montaigne Bordeaux	20. Professional Master of Landscape & Environmental Evaluation in Projects of Urbanism
	21. University of Paris	21. Masters in Sustainable Development, Environmental Management and Geomatics
10. Germany	22. University of Applied Sciences Oldenburg-Ostfriesland-Wilhelmshaven (FH Oldenburg) (Combined)	22. MSc Environmental Technology and Management
	23. University of Technology Berlin	23. Master of Environmental Planning (MEP)
	24. University of Technology Cottbus	24. MSc Environmental & Resource Management
	25. University of Technology Munich	25. MA Environmental Planning and Engineering Ecology
	26. Hohenheim University (Combined)	26. MSc Environmental Science-Soil, Water & Biodiversity
	27. University of Trier	27. MSc Environmental Assessment and Management
	28. University of Freiburg	28. MSc Environmental Governance
	29. Free University of Berlin	29. MA Public and Private Environmental Management
	30. University of Hannover	30. MSc Environmental Planning
	31. Berlin Humboldt University	31. MSc Integrated Natural Resource Management
	32. Hamburg University of Technology	32. MSc Joint European Master Programme in Environmental Studies (JEMES)
	33. University of Applied Sciences Anhalt	33. MSc Environmental Protection and Landscape Planning/MA Landscape Architecture
	34. University of Applied Sciences Wiesbaden-Rüsselsheim-Geisenheim	34. MSc Engineering in Environmental Management and Town Planning in Metropolitan Areas

Table 2. (*Continued*)

Country	Name of university	Degree offered
11. Greece	35. National Technical University of Athens, including University of Abertay Dundee and Mansoura University of Egypt as partners (combined)	35. MSc Environmental Management, Engineering and Technology (EnvMed)
	36. International Centre of Higher Agronomical Studies of the Mediterranean	36. MSc Geoinformation in Environmental Management
12. Hungary	37. Central European University Budapest	37. MSc Environmental Sciences, Policy and Management (MESPOM)
13. Ireland	38. University of Dublin, Trinity College	38. PgDip in Environmental Engineering
	39. Dublin Institute of Technology	39. MSc Sustainable Development
14. Italy (all programmes are called '2nd level degrees')	40. University of Parma	40. Sciences and Technologies for Environment and Natural Resources
	41. University of Modena and Reggio Emilia	41. Engineering for Environmental Sustainability
	42. University of Trento	42. Engineering
	43. University of Venice	43. Environmental Planning and Policies
	44. University of Siena	44. Environment Development & Economics
	45. Politechnical University of Turin	45. Territorial, Urban and Environmental Planning
	46. University of Milano-Bicocca	46. Sciences and Technologies for Environment and Territory
	47. University of Padua	47. Natural Sciences
	48. University of Salento	48. EIA & Auditing
	49. University of Camerino	49. Planning and Management of Protected Areas
	50. University of Ferrara	50. International Master Degree in Environmental and Regional Policy for Sustainability in Local Developm.

Table 2. (*Continued*)

Country	Name of university	Degree offered
	51. University of Rome	51. Economics of Environmental Governance and Territory (MEGAT)
15. Latvia	52. Riga Technical University	52. MSc Environmental Sciences
16. Lithuania	53. Kaunas University of Technology	53. MSc Environmental Engineering
	54. Vilnius Gediminas Technical University	54. MSc Environmental Engineering
17. Malta	55. University of Malta	55. Master in Environmental Management and Planning
18. Poland	56. Technical University of Lodz	56. MSc Nat. Sciences/Architecture & Urban Planning
	57. University of Wroclaw	57. MSc Geography & Environmental Management
	58. University of Silesia in Katowice	58. MSc Landscaping & Environmental Protection
19. Slovakia	59. University of Central Europe in Skalica	59. MSc Environmental Management
	60. Technical University of Zvolen	60. MSc Natural Science
	61. Slovak University of Technology	61. Master of Art-Dipl.Ing
	62. Comenius University in Bratislava	62. MSc-Dipl.Ing
20. Slovenia	63. The University of Ljubljana	63. MSc Urbanism
21. Spain	64. University Complutense de Madrid (UCM)	64. Master in Environment: Human and Socio Economic Dimensions
	65. University of València	65. Master Environmental Engineering
	66. University of Girona	66. Master in Environment
	67. Madrid EOI Business School	67. Master of Environment and Energy (Engineering & Environmental Management)
	68. Politechnical University of Madrid	68. Master in Management of Environmental Engineering Projects
22. Sweden	69. University of Gothenburg	69. MSc Environmental Management and Economics
	70. University of Lund	70. MSc Environmental Management & Policy (EMP)
23. The Netherlands	71. Free University Amsterdam	71. MSc Environment & Resource Management

Table 2. (*Continued*)

Country	Name of university	Degree offered
	72. Radboud University Nijmegen	72. MSc European Spatial & Environment Planning
	73. University of Twente	73. MSc Environmental and Energy Management
	74. Wageningen University	74. MSc Urban Environmental Management
	75. Utrecht University	75. MSc Sustainable Development
24. United Kingdom	76. University of East Anglia	76. MSc EA & Management
	77. Oxford Brookes University	77. MSc EA & Management
	78. University of Manchester	78. MA in EIA and Management
	79. University of Salford	79. MSc EA & Management
	80. University of Glamorgan	80. MSc Environmental Management
	81. University of Aberystwyth	81. MSc Managing the Environment
	82. Newcastle University	82. MSc Environmental Resource Assessment
	83. University of Brighton	83. MSc EA & Management
	84. University of Hull	84. MSc Environmental Technology
	85. University of Nottingham	85. MA/MSc Environmental Management
	86. University of Plymouth	86. MSc Sustainable Environmental Management
	87. University of Strathclyde	87. MSc Environmental Studies
	88. University of Gloucestershire	88. MA/MSc Environmental Policy and Management
	89. University of Edinburgh	89. MSc Environmental Sustainability
	90. London School of Economics and Political Science	90. MSc Environmental Policy & Regulation
	91. Heriot-Watt University	91. MSc Sustainability Engineering
	92. Royal Holloway University of London	92. MSc Environmental Diagnosis and Management
	93. Lancaster University	93. MA Environmental Management and Consultancy
	94. University of Liverpool	94. MA Environmental Management & Planning
	95. Coventry University	95. MSc Environmental Management
	96. Queens University Belfast	96. MSc Environmental Management

Table 2. (*Continued*)

Country	Name of university	Degree offered
	97. University of Southampton	97. MSc/MRes Environmental Management
	98. University of Glasgow	98. MSc Global Water Sustainability
	99. University of Stirling	99. MSc Environmental Management
	100. University of Wolverhampton	100. MSc Environmental Management
	101. University of Sheffield	101. MSc Env. Management of Urban Land & Water
	102. University of Hertfordshire	102. MSc Environmental Management
	103. University of Sunderland	103. MSc Environmental Management
	104. Kings College London (London University)	104. MSc Environmental Monitoring, Modelling and Management
	105. University of St. Andrews	105. MSc Sustainable Development
	106. University of Dundee	106. MSc Spatial Planning with EA

Academic Home

As is evident from Table 2, certain words appear particularly frequently in the degree titles of the identified programmes. The word "environment", for example, appeared in 86 degree programme titles. The next most frequently used word was "management", appearing in 50 titles. "Assessment" as such featured in nine titles only. Other popular words included "engineering" and "planning/policy". These were used in 12 and 15 titles, respectively. Finally, "sustainable", "urban", "development" and "technology" were also used five or more times.

As far as the EA/EM emphasis of the 106 programmes is concerned, half of them taught EA and EM complementarily, and amongst the remaining, the focus of 35 programmes was on EM, while 15 described their main focus to be EA. The host faculty/department/school was also identified in order to appreciate the context within which the programmes were delivered, and the basic orientation of them. Deciphering the concrete subject background was not always straightforward, as different universities are organised differently, resulting in information being provided in different styles. Figure 2 gives an overview of the degree programmes, based on the subject areas within which they were taught. For the purpose of this exercise, the terms "department" and "school" were used.

Figure 2 which is a sub-analysis of the right-hand column of Table 2 shows that 15 programmes were taught within Environment/Earth Science departments/schools. Engineering institutions offered 21, the largest number of EAM-related master level programmes of any one type of institution. Planning related institutions (including Architecture) offered 11 programmes and Management/Business institutions (including Economics) delivered 8 programmes. Somewhat unexpectedly, five EAM-related degree programmes were also offered in

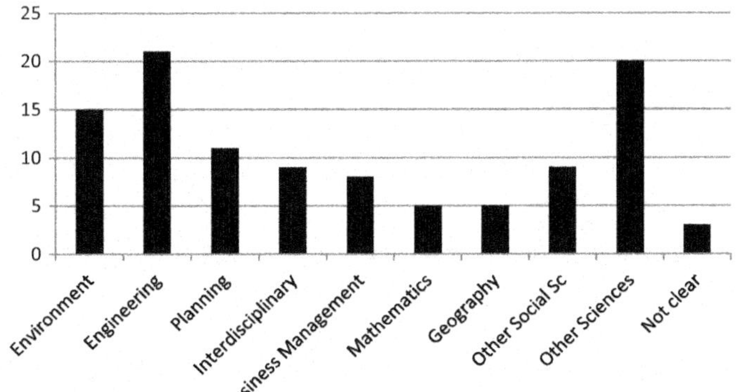

Figure 2. Subject areas within which EAM-related degree programmes were offered.

Mathematics-related institutions. This is, however, mainly down to some engineering/geosciences departments being part of a Mathematics faculty. Geography and Geosciences institutions offered five programmes while nine programmes were offered in an interdisciplinary manner, mostly as combined degree programmes. Two further categories of "other sciences" and "other social sciences" were also formed. A total of 29 programmes fell into these categories, 20 into the former, 9 into the latter. Apart from technical studies, natural sciences and life sciences, these also included some unusual institutions, such as "Medicine and Pharmacy" and "other social sciences" (e.g. Law, Policy and Social Studies and Tourism Studies).

In order to obtain an indication of whether there was a good spread of disciplines representing the main ingredients of sustainable development, programmes were mapped in terms of the basic orientation (i.e. the faculty home), which was divided into Science, Social Science and Economics (reflecting the main ingredients of sustainable development). A fourth category; "interdisciplinary" was also included. Whilst 66 programmes were Science-based, 20 were Social Science-based and three were Economics-based. Finally, 17 programmes were offered in an interdisciplinary manner.

Figure 3 provides for a breakdown of the different streams within which EAM-related programmes were offered by member states. It is important that despite a disciplinary home, programmes may still be taught in an interdisciplinary manner, e.g. by drafting in teaching staff from other disciplines. In most countries, EAM-related master level programmes fell into the (environmental) science area. In Germany, the UK and Spain, EAM was taught in three streams, while in Italy it was taught in all four streams. Slovakia offered two programmes in Environmental

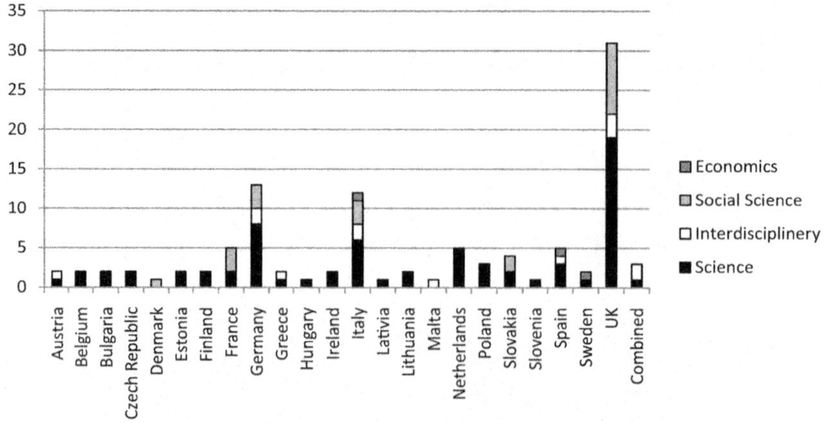

Figure 3. Country-wide distribution of streams within which EAM-related degrees are offered.

Science and two in Social Science. France was the only country where out of the five master level programmes offered, the majority (three) were from within the Social Sciences. EAM was taught within Economics-related departments/schools in three countries only, namely Spain, Italy and Sweden.

Programme implementation

The duration of the programmes was mainly either two years or one year. Seven programmes were identified, where the duration was less than one year or more than two years, with one programme lasting for one and a half years. Figure 4 shows that the UK, Ireland, the Netherlands and Bulgaria offered courses which were predominantly one year in duration. Two-year programmes were mainly offered in Italy, Germany, France, other central European countries, as well as the Nordic and Baltic countries. Out of the 12 programmes offered in Italy, three professional degrees were one-year long.

In Germany, one-year programmes were offered by the University of Applied Sciences in Oldenburg and the University of Technology Munich. Austria, Belgium and Greece both offered one- and two-year programmes. As far as the 'others' category is concerned, in Spain, two of the programmes offered were less than one year in duration, including the University of Girona and Madrid EOI Business School. In the UK, two degree programmes offered by the University of Strathclyde and the University of Hertfordshire, were ranging from one to three years and from one to two years, respectively. In Poland, the Technical University of Lodz offered its Master in Architecture and Urban Planning over a one and a half year period. The combined programme degree offered by the Baltic University was six months.

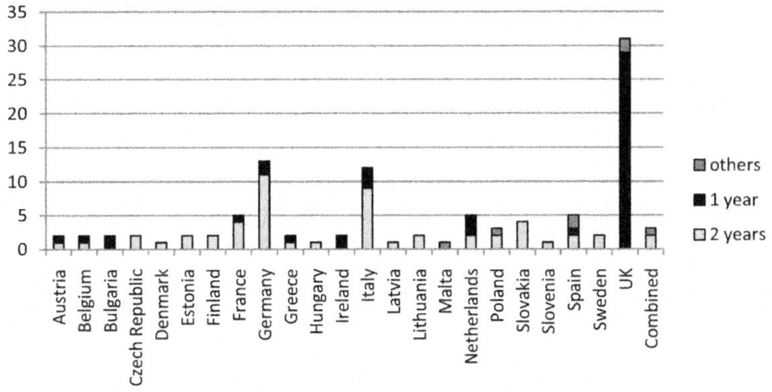

Figure 4. Duration-wise distribution of EAM-related programmes across Europe.

In terms of the delivery language of the programmes, English was dominant, with 71 of the 106 programmes being delivered in this language. This means that universities from non-native English-speaking countries offered such programmes, too. For example, in Germany, out of a total of 13 degree programmes offered, eight were found to be taught in English. Some programmes were offered in more than one language. For example, in Germany, four programmes were taught as a combination of German and English, while only one was taught in German alone. With regards to whether programmes were taught or research-based, nearly all fell into the first category (104 out of 106). The two research-based programmes were offered by the University of Wolverhampton and Kings College London in the UK.

Programmes did not always provide data with regards to delivery method and technique, assessment method and programme structure. One of the most distinct characteristics of master level programmes across the EU in terms of programme structure is the arrangement of modules into two categories: "compulsory" and "optional". As far as the delivery method is concerned, most programmes were restricted to on-site learning. There were, however, a few exceptions, for example, the University of Klagenfurt in Austria which offered an MSc in Management of Protected Areas (MPA), providing 190 out of 710 teaching hours as e-learning units. Furthermore, in certain cases, students were able to go overseas or to other places for their work placement. In Italy, all three professional degrees were two-third University on-site and one-third internship-based. Available data suggest that generally speaking, a wide variety of assessment techniques were in place across the EU, ranging from exams, assignments in the form of reports and essays, seminar papers, internships, projects in practice, dissertations and lab-work.

Programme requirements and scope

As far as a specific geographical focus of the 106 EAM-related master level programmes is concerned, whilst one-third of the programmes were not explicit about this, others made reference to specific geographical areas. For example, the Free University of Berlin was co-operating closely with Indian institutions, while the combined degree offered by the Baltic University had a Baltic Sea region focus. Whilst overall programme credits across the 106 programmes were mainly based on the European Credit Transfer System (ECTS; see Jha-Thakur, 2008), specific requirements varied from programme to programme. To some extent this can be explained by the different lengths of the programmes. Forty-nine programmes consisted of 120 ECTS, 21 of 60 ECTS, 18 of 90 ECTS and seven programmes of 180 ECTS. For 11 programmes no information was available.

Professional degrees were all based on 60 ECTS, which is the minimum requirement for a university-based MA according to the European Bologna process specifications.

In many EU member states, non-EU international students were charged different fees from EU students. This is in line with practice in other world regions, e.g. the USA, Canada and Australia. Universities of some countries, including Spain, Italy (most programmes), Germany and Slovenia charged the same fees (if there were any) to all students. In some countries, students were charged nothing or only very small registration fees. These included Austria, Finland, Germany, France and Sweden. Table 3 shows the fees charged to non-EU international students by the universities of the 23 countries. The highest fees payable by international students were found in the UK and Ireland. In some cases, these went beyond 15,000 Euros per year.

Table 3. Fee distribution (in Euro) of EAM-related Master programmes for international students.

Country name	No fees	up to 2500	2500 > up to 5000	5000 > up to 10,000	10,000 > up to 15,000	15,000 > 20,000	N.A
Austria	1	0	0	0	1	0	
Belgium	0	0	0	0	0	0	2
Bulgaria	0	0	1	1	0	0	
Czech Republic	0	0	1	1	0	0	
Denmark	0	0	0	0	1	0	
Estonia	0	1	1	0	0	0	
Finland	2	0	0	0	0	0	
France	3	0	0	0	0	0	2
Germany	8	3	2	0	0	0	
Greece	0	0	0	0	0	0	2
Hungary	0	0	0	1	0	0	
Ireland	0	0	0	0	1	1	
Italy	0	10	2	0	0	0	
Latvia	0	1	0	0	0	0	
Lithuania	0	0	2	0	0	0	
Malta	0	0	0	0	0	0	1
Netherlands	0	0	0	4	1	0	
Poland	0	0	3	0	0	0	
Slovakia	0	3	0	0	0	0	1
Slovenia	0	1	0	0	0	0	
Spain	0	1	2	1	1	0	
Sweden	0	0	0	0	1	0	
UK	0	0	0	2	22	5	2

Entry requirements were usually formulated in terms of:

(a) Nature and grade of first degree;
(b) Knowledge of language; and
(c) Work experience.

There were usually some language requirements. These either meant that prospective students should have a good knowledge of the language of study or that they had the duty to study this language before starting a programme. For studies in English, language requirements were standardised on the basis of the International English Language Testing System (IELTS) and Test of English as a Foreign language (TOEFL; see www.ielts.org and www.uk.toefl.eu). For master level students, there was usually the need for a "good first degree". However, the interpretation of "good" is ambiguous and appears to vary quite substantially. Some programme specifications considered work experiences as equivalent to a formal university degree. As far as career opportunities are concerned, the information available was wide ranging from private to public sectors, environmental consultancies, NGOs and organisations that deal with EM, as well as academia.

Emerging Trends, Opportunities and Challenges

EAM education degrees and EAM education research: Research on EAM-related education is still embryonic (Sanchez, 2010). Yet one emerging trend is that EAM-related master programmes are widely available all over Europe. This indicates a growing gap between EAM-related higher education degrees available and research on them. Furthermore, tertiary institutions play a critical role in environmental education due to their power to influence habits, thoughts and actions of future leaders (Ali-Khan, 1992; Thomas & Nicita, 2002). This is perhaps one of the reasons why previous research in the EU suggested that EA education is influenced by how EA is practised rather than 'EA practice and understanding being dependent on EA education' (Gazzola, 2008). Research on EAM-related higher education is thus important in order to be able to positively support the growing number of EAM-related degree programmes.

Disciplinary home of EAM-related degree programmes: Based on the data available, it was not always easy to decipher the disciplinary background within which EAM-related higher degree programmes are offered. Whilst programmes are often offered within a specific disciplinary context, they are not necessarily restricted to that discipline. As noted by Gazzola (2011), certain disciplines act as "gate-keepers" for EA-related education. However, many of these disciplines like planning, geography and management are themselves interdisciplinary in nature. Student eligibility requirements were quite flexible for most programmes, and generally speaking, students from a wide variety of disciplinary backgrounds were allowed to take such courses. These observations allow us to draw some broad conclusions with regards to European EAM-related higher degree programmes, namely that:

(a) They are not compartmentalised;
(b) They all contain interdisciplinary aspects; and
(c) They are accessible to students with a wide range of backgrounds.

These findings are not necessarily transferable to other countries outside the EU, like China, India or Pakistan, where EAM-related higher degrees are still overly circumscribed within science and engineering (Fischer, 2012; Gazzola and Jha-Thakur, 2009; Sinha, 2001; Wang *et al.*, 2003). EU EAM-related degree programmes are flexible and allow for possible career changes, thereby enhancing the attractiveness of these programmes for international students.

Transferability of skills for international students: Following on from the discussion on the disciplinary context, data collected further reveal that the discipline within which EAM-related higher degrees are offered depends on how EAM is practised and understood in different countries. This is in line with the findings from previous research (see the various documents produced in the TwoEA-M project on www.twoeam-eu.net). This raises questions of transferability of skills for international students. For example, in the UK, EA education is often spatial/land use planning-based. This is perhaps owing to the fact that the EA-related regulations in the UK were introduced through the Town and Country Planning Regulations under which 70% of the EIAs are undertaken here (Arts *et al.*, 2012; Glasson and Bellanger, 2003). However, in other countries, this may not be the case and hence if the EA practice in an international student's home country is mainly 'engineering'-based, transferability of skills may be problematic.

Disparity in credit structure: Despite the Bologna process, there are still some considerable differences between master level EAM programmes in different EU member states in terms of the credits required to complete the degrees. In current UK practice, as a general rule, one-year master degrees are said to require 90 ECTS. However, there is a discrepancy with regards to how the ECTS is translated into practice in terms of effort hours. According to the Bologna requirements, one ECTS credit denotes 25–30 hours of work. In the case of the UK, however, this is understood to be only 20 hours (Jha-Thakur, 2008). This implies that workload requirements in the UK are lower compared with most other countries, and 90 credits can actually be translated into 75 credits (Roberts, 2007). This is mainly explained by the fact that UK degrees are mostly one-year programmes only.

English dominates other national languages: The prominence of English as the delivery language is evident amongst EAM-related higher degrees, where it was the delivery language for 71 of the 106 programmes. This appears to be an exceptionally large number for a specific degree topic. A search on the http://www.mastersportal.eu/ website, which lists over 20,000 master level degree programmes in the EU (of which nearly half are from the UK), shows that only about 3,500 of those were offered in English (March 2013). This choice of delivery language makes the degree programmes more attractive for the vast majority of aspiring international students. However, this trend may have long-term implications for domestic EAM-related higher education and related practice. To start with, in terms of international recruitment, this clearly disadvantages countries not using English as their delivery language. Therefore, case studies and practices of these countries are more likely to be ignored in international curricula of EAM. Also, the delivery of degree programmes in a language different from the national language has led to a "fear of loss of cultural heritage" (Huisman and Wende, 2004, 355).

Conclusions

In this paper, we have presented the results of an analysis of environmental assessment and management (EAM)-related higher education master level degree programmes throughout the EU. Twenty-three of the 27 member states were found to offer related programmes in 106 institutions, with UK universities offering nearly a third of them. Whilst other countries on average offer considerably fewer programmes, this does not mean environmental assessment and management is necessarily given little attention or is ignored. In the cases of e.g. Germany, Italy and France, lower numbers can be partly explained by the only recent introduction of two-tier degree programmes (i.e. BAs and post-graduate MAs), following the Bologna process on standardisation of European higher education.

The EAM-related master levels programmes were analysed in terms of their academic home, their emphasis on EA or EM, as well as the Faculty/School/Department within which they were taught, programme implementation and duration, delivery language, delivery method and technique, assessment method, taught or research-based, and programme structure, as well as programme requirements and scope, geographical specialism, entry requirements, fees, credit structure and career opportunities. Overall, about two-thirds of the programmes were taught at a science institution and one-third at a social science institution.

Whilst about half of the programmes were taught as a combination of both EA and EM, of the other half, over 65% had an environmental management focus. Whilst in the UK, nearly all full-time programmes take one year to complete, in most other countries, the duration is two years. Most programmes that specified a regional focus of study said this was international, about one-third said they were mainly national. Nearly all programmes were run on-site, with very few e-learning options on offer. Furthermore, nearly all programmes were based on a combination of "compulsory" and "optional" modules. Assessment was done mostly in a rather traditional way, ranging from exams, assignments, seminar papers and reports over lab-work to dissertations.

There is plenty of scope for further research on EAM-related higher education. To start with, similar projects could be conducted in other countries or for other degree programmes, such as bachelor level degrees. Furthermore, more in-depth questions could be asked on each of the programmes covered here. This could happen through e.g. interviews with representatives of the programmes. It would also be beneficial to establish student numbers and their origins. Finally, a survey with students of the programmes covered would allow making judgements on e.g. programme quality.

Acknowledgements

The paper is an outcome of the T_{wo}EA-M project, a two-year project financed by the European Commission. The authors wish to acknowledge the other members of the T_{wo}EA-M team: Ingrid Belcakova (Faculty of Architecture, Slovak University of Technology) and Ralf Aschemann (University of Graz, Austria). Information about the T_{wo}EA-M project can be found on http://www.twoeam-eu.net/
.

References

Alderson, G (1992). Focus on current purposes of public legal education and information in Canada. National Public Legal Education and Information (PLEI), Policy Consultation Package, Ottawa.

Ali-Khan, S (1992). *Colleges Going Green: A Guide to Environmental Action in Further Education Colleges*. London: Further Education Unit.

Arts, J, P Caldwell and A Morrison-Saunders (2001). Environmental impact assessment follow-up: good practice and future directions-findings from a workshop at the IAIA 2000 conference. *Impact Assessment and Project Appraisal*, 19(3), 175–185.

Arts, J, H Runhaar, TB Fischer, U Jha-Thakur, F Laerhoven, P Driessen and V Onyango (2012). The effectiveness of EIA as an instrument for environmental governance – a comparison of 25 years of EIA practice in the Netherlands and the UK. *Journal of Environmental Assessment Policy and Management*, 14(4), 1250025.

Aschemann, R (2002). Pros and cons of the EU SEA directive. *Paper presented to the Annual Meeting of the International Association of Impact Assessment, The Hague*. http://www.anidea.at.

Bailey, J (1997). Environmental impact assessment and management: an underexplored relationship. *Environmental Management*, 21(3), 317–327.

Barker, A and C Jones (2008). Reassessing the direction of postgraduate environmental assessment education: the Manchester experience 1996–present. In *Environmental Assessment Lecturers' Handbook*, ed. TB Fischer *et al.*, pp. 18–23; www.twoeam-eu.net.

Barker, A and C Wood (1999). An evaluation of EIA system performance in eight EU countries. *Environmental Impact Assessment Review*, 19(4), 387–404.

Capper, J (2001). The emerging market for on-line learning: insights from the corporate sector. *European Journal of Education*, 36(2), 237–245.

Cherp, A (2008). Teaching environmental assessment in the context of postgraduate environmental courses-challenges for environmental postgraduate education and five mindsets for sustainability. In *Environmental Assessment Lecturers' Handbook*, ed. TB Fischer *et al.*, pp. 24–31; www.twoeam-eu.net.

Dalal-Clayton, B and B Sadler (2005). *Strategic Environmental Assessment (SEA): A Sourcebook and Reference Guide to International Experience.* London: Earthscan.

European Commission (2010). Action 1: Joint Programmes including Scholarships. http://eacea.ec.europa.eu/Erasmus_Mundus/programme/action1_en.php.

Fischer, TB (2012). EIA teaching at tertiary level academic institutions in Pakistan — current baseline, development needs and further action. NIAP, IUCN Pakistan. http://niap.pk/docs/Knowledge%20Repository/Reports/Report%20EIA%20education%20Pakistan%20Fischer.pdf.

Fischer, TB and V Onyango (2012). Strategic environmental assessment-related research projects and journal articles: an overview of the past 20 years. *Journal of Impact Assessment and Project Appraisal*, 30(4), 253–263.

Fischer, TB, U Jha-Thakur, I Belcakova and R Aschemann (2011). *A Report on East and South East Asian Higher Education Postgraduate Environmental Assessment and Management Requirements and Interests*. Liverpool: Liverpool University Press. www.twoeam-eu.net.

Fischer, TB (2010). Reviewing the quality of strategic environmental assessment reports for English spatial plan core strategies. *EIA Review*, 30(1), 62–69.

Fischer, TB, U Jha-Thakur, I Belcakova and R Aschemann (2010). *A Guidebook to Environmental Assessment and Management Related Master Level Higher Education in European Union Member States*, www.twoeam-eu.net.

Fischer, TB, P Gazzola, U Jha-Thakur, I Belcakova and R Aschemann (2008). *Environmental Assessment Lecturers' Handbook*. Bratislava: Road Bratislava Press. www.twoeam-eu.net.

Fischer, TB (2003). Strategic environmental assessment in post-modern times. *EIA Review*, 23(2), 155–170.

Gazzola, P (2011). Can environmental appraisal be truly interdisciplinary? *Journal of Environmental Planning and Management*, 54(9), 1189–1208.

Gazzola, P and U Jha-Thakur (2009). Internationalisation and standardisation of European environmental assessment—relevance to India. *Environmental Education Research*, 15(6), 625–641.

Gazzola, P (2008). Existing EA-related master degree programmes in the EU: an analysis—concepts, principles and key modules. In *Environmental Assessment Lecturers' Handbook*, eds. Fischer, TB, P Gazzola, U Jha-Thakur, I Belcakova, R Aschemann. Bratislava: Road Bratislava Press.

Glasson, J, R Therivel and A Chadwick (1999). *Introduction to Environmental Impact Assessment*. 2nd ed, London: UCL Press.

Glasson, J and C Bellanger (2003). Divergent practice in a converging system? The case of EIA in France and the UK. *Environmental Impact Assessment Review*, 23(5), 605–624.

Gordon, J (2005). Editorial. *European Journal of Education*, 40(4), 353–360.

Huisman, J and MVD Wende (2004). The EU and Bologna: are supra-and international initiatives threatening domestic agendas? *European Journal of Education*, 39(3), 349–357.

Jha-Thakur, U (2011). Exploring regional variation in EIA follow-up. *Journal of Environmental Assessment Policy and Management*, 13(3), 435–458.

Jha-Thakur, U (2008). Internationalisation of master degree programmes—reflecting on the European Bologna process. In *EA Lecturers' Handbook and Curriculum for EA-Related Master Programmes*, ed. TB Fischer *et al.*, Bratislava: Road Bratislava Press. pp. 10–17; www.twoeam-eu.net.

Jha-Thakur, U and TB Fischer (2008). Are open cast coal mines in India casting a shadow on the Indian environment? *International Planning Development Review*, 30(4), 441–458.

Jordan, A (2005). *Environmental Policy in the EU*, 2nd ed. London: Earthscan.

Morrison-Saunders, A and J Bailey (1999). Exploring the EIA/Environmental management relationship. *Environmental Management*, 24(3), 281–295.

Morrison-Saunders, A, B Jenkins and J Bailey (2004). EIA follow-up and adaptive management. In *Assessing Impact, Handbook of EIA and SEA Follow-up*, ed. A Morrison-Saunders and J Arts. pp. 154–157. London: Earthscan.

Ramos, T, T Cecílio and JJ Melo (2008). Environmental impact assessment in higher education and training in Portugal. *Journal of Cleaner Production*, 16(5), 639–645.

Rennings, K, A Ziegler, K Ankele and E Hoffmann (2006). The influence of different characteristics of the EU environmental management and auditing scheme on technical environmental innovations and economic performance. *Ecological Economics*, 57(1), 45–59.

Risse, N (2005). *Application of Strategic Environmental Assessment in the Member States of the European Union: Synthesis of the Questionnaire's Answers*. http://plannet.difu.de/2005/proceedings/2005_plannet-proceedings.pdf.

Roberts, G (2007). The Bologna process: an update. *The Higher Education Academy, Centre for Education in the Built Environment*, 16.

Sanchez, LE (2010). Environmental impact assessment teaching at the university of Sao Paulo: evolving approaches to different needs, *Journal of Environmental Assessment Policy and Management*, 12(3), 245–262.

Sanchez, LE and A Morrison-Saunders (2010). Teaching impact assessment: results of an international survey. *Impact Assessment and Project Appraisal*, 28(3), 245–250.

Sinclair, J and A Diduck (1995). Public education—an undervalued component of the environmental assessment public involvement process. *Environmental Impact Assessment Review*, 15(3), 219–40.

Sinha, IN (2001). *A Framework of EIA for Environmental Sustainability*, ENVIS Monographs No. 8, September. Dhanbad, India: ENVIS Centre on Mining Environment, Indian School of Mines.

Stelmack, CM, AJ Sinclair and P Fitzpatrick (2005). An overview of the state of environmental assessment education at Canadian universities. *International Journal of Sustainability in Higher Education*, 6(1), 36–53.

Thomas, I and J Nicita (2002). Sustainability education and Australian universities. *Environmental Education Research*, 8(4), 475–492.

Van Schaik, L (2012). The EU and the progressive alliance negotiating in Durban: saving the climate? Overseas Development Institute and Climate and Development Knowledge Network 2012. http://www.odi.org.uk/sites/odi.org.uk/files/odi-assets/publications-opinion-files/7845.pdf.

Wang, Y, R Morgan and M Cashmore (2003). Environmental impact assessment of projects in the People's Republic of China: new law, old problems. *Environmental Impact Assessment Review*, 23(5), 543–579.

Weiland, U (2012). Teachings on environmental impact assessments in geography reflecting demands of the working practice: a German case study. *Journal of Environmental Assessment Policy and Management*, 14(3), 1250019.

CHAPTER 13

Current State of the SEA System in Brazil: A Comparative Study

Marcelo Montaño*, Priscila Oppermann,
Anne Caroline Malvestio and Marcelo Pereira Souza

Sao Carlos School of Engineering – University of Sao Paulo
Sao Carlos, SP Brazil
**minduim@sc.usp.br*

After the first applications in the early 1990s, the practice of Strategic Environmental Assessment (SEA) in Latin America has been going through a slow, but important evolution. In this paper we elaborate on the main aspects of this evolution, based on the findings of a comparative study that put the practice of SEA in different contexts into perspective. SEA systems considered included consolidated (UK, Spain and Portugal) and unconsolidated (South Africa, Mexico, Angola and Mozambique) systems. The Brazilian system is characterised by a disperse structure and a low level of self-organisation, which makes integration with plan making difficult while, at the same time, stimulating "flexible" and "adaptive" SEA processes. We suggest that the lack of both institutional drivers and a clear procedural framework are hindering further development of the Brazilian system.

Keywords: SEA systems; comparative studies; Brazil.

Introduction

Since its early development in the 1980s as an EIA-based decision support instrument aiming to anticipate environmental impacts at the early decision levels of policies, plans and programmes, Strategic Environmental Assessment (SEA) has been recognised as an important instrument to foster sustainability in decision-making. Whilst according to Tetlow and Hanusch (2012, p. 16) SEA

*Corresponding author.

has evolved "from a largely EIA-based and responsive mechanism to a far more proactive process of developing sustainable solutions as an integral part of strategic planning activities", Fischer (2006) argued that it is the combination of strategic proactive approaches, in particular at policy levels, together with more EIA-based responsive mechanisms in particular at plan and programme levels in an overall SEA/EIA framework which is essential for overall effectiveness of the instrument. In the context of the "evolution" of SEA, as reported in the professional literature, enforcement instruments such as the European Directive on SEA and the UNECE SEA Protocol to the Espoo Convention about transboundary effects of policies, plans and programmes have played an important role in making this a routine tool, which is applied thousands of times every year throughout the world (Fischer and Onyango, 2012).

Another important trigger for the spread of SEA is the role played by donor agencies in the development co-operation context. These have established SEA as a tool for capacity building and mainstreaming of environmental issues in developing countries and countries in transition (Loayza, 2012; Dalal-Clayton, 2013), enforcing the role of SEA as a safeguard tool.

There is an increasing interest in comparative studies focused on the development and the performance of SEA systems, which have put into perspective their strengths and weaknesses (as, for example, Therivel, 1993; Chaker *et al.*, 2006; Gazzola, 2008), reinforcing the importance of the context where SEA is applied (Marsden, 1998; Fischer, 2006; Runhaar and Driessen, 2007; Zhang *et al.*, 2013), how context influences SEA effectiveness (Gazzola, 2008) and also the importance of flexible and adaptive SEA systems (in a European context, discussed by e.g. Fischer and Gazzola, 2006).

The number of developing countries currently applying SEA in different situations in increasing (e.g. Dalal-Clayton and Sadler, 2005; Lemos *et al.*, 2012) although it doesn't necessarily mean it is used in routine procedures of planning. In SEA in Latin America, for example, Chile, Peru and Bolivia have national legislation for the use of SEA in certain strategic decisions, but only Chile has adopted formal requirements to integrate this instrument in plan making. Unsurprisingly, the number of SEAs prepared in Chile (211 SEAs from 2011 to 2014, according to the Environmental Ministry[1]) is much larger than in Peru (only 3 SEAs reported so far[2]) and in Bolivia (13 SEAs from 2002 to 2012[3]).

[1] http://eae.mma.gob.cl/index.php/ficha.

[2] http://dar.org.pe/eae/eae_antecedentes2.html.

[3] http://www.eia.nl/en/countries/sa/bolivia/sea.

Given the fact that the literature is scarce with regards to comparative studies focused on these countries, in this paper we look at the main aspects of the Brazilian SEA system, contrasting it with SEA systems of developed (UK, Spain and Portugal) and developing (South Africa, Mexico, Angola and Mozambique) countries.

The paper unfolds in the following way; the first section describes the context and sets for the relevance of the adopted objectives; in the second section, the main aspects of a SEA system are provided as a background, together with a brief description of the context of SEA in Latin America; in the third section we present the methodological framework adopted for the comparative study, presenting the categories and criteria used; and in the last section we discuss the outcomes and conclusions of the paper.

Background

Different approaches have been used to study the organisation and performance of SEA systems. Therivel (1993) defined an SEA system as being organised around a set of main components: the regulatory basis for SEA application; the type of policies, plans and programmes SEA has been applied to; the guidelines defining the procedures to its application; and the process of review, approval and follow-up; among others. Jones *et al.* (2005) related the organisation of different SEA systems to the institutional framework supporting SEA: legal basis; level of integration; guidance; coverage and tiering; and Chaker *et al.* (2006) compared the SEA systems in 12 countries, taking into account the existing legal, institutional and procedural frameworks. Fischer (2007) presented a comparative study of 11 SEA systems based on the context of application, the procedures and methodological aspects applied to SEA, along with the strategicness of application in terms of the decision tiers' policies, plans and programmes. Others, like Nilsson *et al.* (2009), relied on the growing recognition of the need for a SEA system to be "decision-oriented", which means running in parallel with actual planning rather than formally anticipated decision-making processes, and formulating conditions for coordinating *ex-ante* and *ex-post* stages of the process.

Specifically concerning the design and introduction of an instrument like SEA within the political system, Wirutskulshai *et al.* (2011) argued that progress on implementation depends on institutional factors such as legal-structural environment factors and the political and administrative culture of a system. They stressed that the professional literature is recurrent in pointing out two factors that particularly affect the momentum and progress on SEA implementation — legislative support and political culture of participation. "In a given institutional culture of limited collaborative planning and domination of powerful vested interests, the legal provision becomes very important to facilitate SEA initiation" (*ibid.*, p. 353). This statement was corroborated by Madrid *et al.* (2011) who analysed four SEA cases related to a large investment programme in infrastructure projects in South America. The authors found that the lack of clear objectives and definitions of SEA and the lack of a legal basis hinders the full appropriation of the principles established by funding agencies when an SEA is required.

In Brazil, the experiences with SEA are based on few applications over the last 20 years (around 40 SEAs by early 2014). Here, SEAs have been said to have had no significant influence on strategic decision-making (Malvestio and Montaño, 2013). Still, there has been a lively and ongoing debate about the best format and practices of SEA in the country (Kirchhoff, 2006; Sánchez and Silva-Sánchez, 2008; Gallardo and Bond, 2011; Pizella and Souza, 2012; Malvestio and Montaño, 2013). SEA has been prepared without the definition of procedural guidelines, methodological aspects or screening/scoping criteria (Malvestio and Montaño, 2013). Sánchez and Silva-Sánchez (2008) argued that in particular the lack of tiering at strategic levels of decision-making are making SEA just another loop to go through in an already long pathway to project approval, thus transferring significant conflict to project-EIA decision-making. Moreover, Pellin *et al.* (2011) argued that the practice of SEA in Latin America as required for multilateral funding agencies through their safeguard policies cannot be considered effective due to the lack of a consistent endogenous framework of environmental planning.

Taking into account the initiatives to build a regulatory basis for the instrument in Brazil, Montaño *et al.* (2013) recognised some patterns of institutional learning through the practice of SEA down the years. According to these authors, the first initiatives had used SEA as a voluntary instrument focused on the assessment of cumulative effects in EIA of large projects, budget plans and other governmental decisions. In a second period (2003–2010) the vision of SEA changed to a mandatory instrument focused on the effects of specific plans and programmes, revealing the influence of concepts such as flexibility and adaptability of SEA systems. The last period (since 2010) has been characterised by an understanding of SEA as an instrument to be mandatorily applied in policy, plan and programme (PPP making), focusing on tiering with the project-EIA level as well. SEA had been evolving slowly, but firmly, with perceptions of progress as new legislative breakthroughs and mandatory implementation are expected in the short term (ibid). Montaño *et al.* (2013) concluded that there were still important gaps to be filled, related to the definition of guidance on the procedural aspects to be followed and the timing of SEA application.

Methodological Aspects

The comparative study underlying this paper was designed, considering the approach provided by Schneider and Schmitt (1996), which consists of:

(i) a selection of two or more comparable phenomena with clear definition in time and space, thus allowing to recognise the empirical universe that is to be investigated;
(ii) the definition of characteristics to be compared, which means the descriptive elements to support the comparative study; and
(iii) generalisation, usually based on the common aspects identified in the study.

Subsequently, systems of SEA and their development are considered from Brazil and seven other countries, including Angola, Mozambique, South Africa, England, Portugal and Spain, and reflecting the situation between 2011 and 2013. The sample comprises countries with different levels of development, planning structure and political/institutional contexts, all taking SEA as an instrument to support strategic decisions/actions (Table 1).

SEA systems were reflected on in terms of:

- Institutional framework — defined by the normative elements (rules and other criteria) set for the organisations/agents responsible for SEA coordination and operation (upstream and downstream) along side strategic decision-making; and
- Procedural framework — comprising of objectives, methodological approach and procedures for use in SEA.

The following third element was also included:

- Community of research and practice — to further the development of the system, promoting institutional and social learning (see e.g. Cashmore *et al.*, 2004 and Jha-Thakur *et al.*, 2009); and to increase expertise.

Figure 1 illustrates the major elements of an SEA system as considered in this paper.

Seven descriptive characteristics were chosen for the comparative review of the different systems, as follows:

- clarity of objectives and procedures;
- existence of coordination framework;
- clearly allocated responsibility for the preparation of SEA report;
- existence of strategic level of application;
- existence and use of guidelines/guidance;

Table 1. General aspects of SEA systems in selected countries.

Tradition	Development stage	Country	General aspects of SEA system
British	developed	UK (England)	consolidated, applied to plans and programmes (EU Directive 42/2001); influence of Sustainability Appraisal
	emergent	South Africa	non-mandatory SEA, but with guidelines provided by public authorities; influence of the National Environment Management Act (basis for IA legislation)
Spanish	developed	Spain	consolidating system, from the transposition of EU Directive; difficulties to coordinate/integrate national level with autonomous departments
	developing	Mexico	non-mandatory SEA; experience in sectoral plans (mainly Tourism and Energy); influence of National Development Plans; influence of funding agencies
Portuguese	developed	Portugal	consolidated (Law-decree 232/2007, EU Directive); influence of land-use planning
	emergent	Brazil	non-mandatory SEA; experience in sectoral plans (mainly Tourism and Energy); strong influence of the "environmental licensing culture"; influence of funding agencies; different initiatives to implement SEA in legislation (federal and state levels)
	developing	Angola	non-mandatory SEA; influence of the National Programme of Environmental Management; influence of funding agencies
	developing	Mozambique	non-mandatory SEA; recognition of the relevance in development plans; pilot-programme conducted by Netherlands and the World Bank to the definition of strategies and guidelines to SEA application

- level of experience with SEA; and
- research expertise and level of communication to a broader audience in the professional field.

Data were collected on the basis of existing documents, including SEA reports, the professional literature (including e.g. academic papers and government agencies reports), normative documentation (e.g. legislation, directives, guidelines), and from

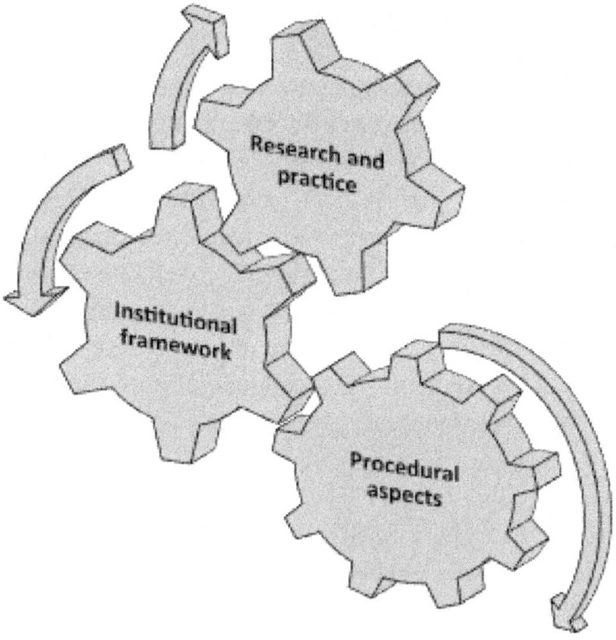

Figure 1. Elements of an SEA system.

personal interviews/questionnaires with representatives of the different countries included in the comparative study. These interviews had involved 29 personal meetings between October 2011 and July 2013 with managers and planners from government agencies, consultants and academics. In a complementary way, five open-question questionnaires were sent out and responded to by e-mail.

Results and Discussion

In this section, results of the review are presented for the eight countries in alphabetic order.

Angola

Project-EIA procedures and environmental licensing had been formalised since 2004 through federal legislation. There is no regulation for SEA and the experience with this instrument is difficult to quantify due to the lack of information. Interviewees reported SEA as almost exclusively related to the requirements of funding agencies such as, for example, the African Development Bank. Although no SEA report was available for consultation, data gathered provided evidence for six SEAs, prepared from 2010 to 2013 in the context of territorial planning, industry complexes and for the National Programme of Water Resources 2010–2013.

Usually, the process of SEA is spontaneously coordinated by the proponent of the plan or programme since there are no formal requirements or procedures to be followed. The quality of SEA reports is not reviewed.

SEA is applied mainly to large projects subject to the safeguard instruments of funding agencies. Some of the interviewees mentioned the application of SEA to territorial planning, but there wasn't any publicly available information to confirm this. Academic communities in Angola are not yet focusing on Strategic Environmental Assessment (SEA), although there are some undergraduate programmes in the field of Impact Assessment. No academic communications in regular scientific databases could be identified.

Brazil

It has been 20 years since the first SEA was conducted in Brazil. To date, there have only been limited experiences with this instrument with almost 40 SEAs having been conducted so far. SEA is predominantly applied to assess the effects of large development projects, to comply with the requirements of funding agencies or to facilitate project-EIA approval. Therefore, it is normally coordinated by the proponent, which might be the government body conducting a loan agreement or the environmental agency conducting a project-EIA.

Over the years, initiatives to formalise SEA and to make it mandatory as an instrument of the National Environment Policy have lacked political support and there are currently no concrete advances at the national level. According to Montaño *et al.* (2013), the main triggers to apply SEA at the federal level include simple recommendations of governmental agencies or ministries. At the state level there is currently more support for initiatives to formalise SEA, and three states (Bahia, Minas Gerais and Sao Paulo) have already included SEA amongst their environmental policy instruments. Currently, there is no clear definition of the timing and procedures to be applied.

Academic research is becoming more consolidated and the practice of SEA in Latin America has been systematically investigated in recent years, with a small but significant presence in relevant publications in the field. Also, over the last decade, a growing proximity between scholars and government (mainly at the federal level) in terms of capacity building has been observed.

England

SEA has been a mandatory instrument since 2004, after the European Directive on SEA was transposed into national law. There is a clear definition of competences and responsibilities to the application of SEA, a clear designation of institutions and government bodies to be consulted, prescribed administrative procedures, quality review and follow-up, and so on (Therivel and Walsh, 2006). The proponent authority is responsible for the preparation of SEA reports, often involving private consultancies in this process.

The country has a wide range of experiences with the application of SEA to plans and programmes, with a prevalence of territorial and urban planning, where SEA is applied within the overall regime of Sustainability Appraisal (see Therivel and Fischer, 2012), transportation, electricity and waste planning (Fischer, 2012), supported by a diverse number of guidelines specifically designed to their respective sectors of application. In the first year after the transposition of the EU Directive there were about 400/450 SEAs under preparation in the UK (Therivel and Walsh, 2006). In 2014, well over 1,000 SEAs had been conducted in England (Therivel and Fischer, 2012).

Academic expertise is one of the main characters of the English system, contributing with critical reflexions concerning the use of the instrument. According to Fischer and Gazzola (2006), academics from UK were in fact responsible for a majority of communications in the international literature, too. To the researchers of this paper, a significant parcel of the improvements in the English system must be credited to the contributions of the academics to the field.

According to the data collected, the main topics of discussion concerning the development and further improvement of the SEA system in England involves how to ensure adequate public participation in SEA processes, the better use of baseline data, a better definition of options, effectiveness of follow-up measures, and on how to increase the actual influence of SEA in decision-making.

Mexico

Besides having no legal provisions for the application of SEA in the country, there is also no procedural guidance. Therefore, when required, SEA is coordinated by the proponent of the strategic action and supervised by the national environmental agency. Some of the interviewees reported the use of Spanish guidelines when an SEA is required. Also, the information collected suggests regular contact of practitioners with international organisations such as the International Association for Impact Assessment (IAIA), as well as development agencies, such as the World Bank, and to a smaller extent, academics.

The number of SEAs prepared so far in Mexico is very low. Data collected indicates that between three and five SEAs have been prepared by government bodies, as well as another eight SEAs in the context of World Bank-funded projects. From an academic perspective, there haven't been any major inputs into the SEA system. There are virtually no academics that have been routinely developing research in the field.

Mozambique

Although EIA became part of federal legislation in 2006, to date there have been no formal regulations for SEA and no national guidelines on the use of SEA in PPP making. Generally speaking, Mozambique has only recently started to develop an interest in SEA and the country has been using the instrument in the context of some loan agreements, following guidance provided by funding agencies. There were also voluntary initiatives by national government. Up to the beginning of 2012 this means 3 SEAs and 10 other strategic assessments required by the World Bank had been conducted.

According to the data collected, the proponent of the strategic action (normally the Ministry involved in the loan agreement) and the Ministry of Environment have been responsible for the coordination and integration of SEA into PPP making, and for commissioning consultants to prepare the SEA reports. Finally, in Mozambique, there are several undergraduate programmes in the field of Impact Assessment, but there were no postgraduate level activities and academic research in the field.

Portugal

SEA has been mandatory to plans and programmes since 2007, after the transposition of the EU Directive on SEA (which came three years late). Portuguese regulations comprise the definition of the strategic level of application, responsibilities related to coordination, preparation of SEA reports, consultation and review.

At the time of the interviews in 2012, three main documents existed for providing guidance on SEA, and focusing on territorial and municipal planning, and on the integration of best practices within SEA procedures. The national environmental agency was responsible for the coordination of the information flow and conformity review.

The proponent of the strategic action was responsible in preparing the SEA report and ensure its quality, supported by other institutions eventually consulted during the process. This is a characteristic of the SEA system which was pointed out by the interviewees as one of the weaknesses of SEA in the country. One of the interviewees (an employee of the environmental agency) reported that the Portuguese experience comprised 537 SEA reports up to October 2012, most of them prepared for municipal development plans.

The academic expertise was seen by interviewees as a positive aspect to the development of SEA in Latin America. There is a reasonable number of groups that embrace SEA as a field of research and contribute to the development of procedures, guidelines, and methodological approaches (e.g. at the universities of Algarve, Aveiro, Lisbon and Porto). The contribution of various academics to the relevant professional literature is also consolidated.

South Africa

Despite the widespread awareness of the concept of environmental assessment of policies, plans and programmes, introduced through the national environmental management law, there are currently no formal requirements for the use of SEA, which is spontaneously applied in the country. Territorial planning at the local (municipal) level and port developments are supported by specific guidelines and regulations.

The environmental authority at the federal level recommends the use of SEA, which is supported by the general objectives of the national environmental management law. Without a formal institutional framework, however, there is no definition of the responsibilities concerning coordination, preparation of SEA reports and quality review. According to Retief *et al.* (2007), virtually all SEAs are prepared by the private sector.

Whilst there is a lack of consolidated information concerning the number of SEAs prepared, this can be taken as rather high when compared with other developing countries. The information gathered reported about 50 SEAs prepared up to 2003 and then the preparation of about 15 SEAs per year in South Africa by 2005, mainly focusing on development requirements imposed by the installation of large projects as industrial clusters and port development. The Department of Environmental Affairs recently declared its intention to develop regulations to legislate the development of SEAs and is currently "working towards the development of a Strategic Environmental Management Framework for energy applications, specifically renewable energy applications" (DEA, 2013: p. 15).

The contributions of South African academics to the development of SEA within the country are significant in terms of education, professional training and research communication, including a consolidated presence in the relevant professional literature. The main topics discussed in the professional literature are related to the concepts, application and regulation to the use of the instrument, as well as its effectiveness in PPP making (Retief, 2007; Retief *et al.*, 2007; Rossouw and Wiseman, 2004).

Spain

The European Directive on SEA was transposed into national legislation in 2006, and SEA has been a mandatory instrument since then. Given the specificities of the administrative system in the country, SEA legislation is comprehended much more as a bridge between regional and national levels of planning than an applied decision support instrument. It is expected that regional governments (from autonomous communities) complement national legislation, thus adapting SEA to each context of application. Despite the existence of a national legislation there is not any formal guidance to apply SEA at the national level. Only a few documents have been published so far, providing guidance to SEA applied at regional levels.

The coordination of SEA processes is clearly defined by the European Directive according to which SEA must be applied to every plan or programme for which preparation and approval is required by legislation and which is likely to cause significant effects on the environment. SEA is the responsibility of the Ministry of Environment in cases of national plans or programmes. Furthermore, regional and local environmental authorities can recommend the use of SEA, and it is their responsibility to coordinate the integration of SEA in decision-making considering the information provided by a Sustainability Report. The proponent of the plan or programme is responsible for the preparation of the Environmental Report.

SEA has been applied to water resources plans and programmes, territorial planning, transportation, and to energy and solid waste management at a lower scale (Gómez Orea, 2007). Currently the number of SEAs produced in Spain is estimated to be in the range of hundreds per year (only in the Autonomous Community of Cataluña, one of the most populous of the country, the environmental authority reports[4] more than 400 SEAs prepared over the last 8 years). Interviewees mentioned that procedural requirements are of particular importance, which is not surprising considering the nature of the SEA Directive.

Academics are actively involved in SEA, although so far academic outputs have not occupied much space in mainstream journals. The main topics discussed by academics are related to the conceptual and procedural framework of the use of SEA (Gómez Orea, 2007; Jiliberto, 2007).

---

[4] Information available on the website of the environmental agency of Cataluña — http://www20.gencat.cat/portal/site/mediambient/menuitem.685af0bd03466a424e9cac3bb0c0e1a0/?vgnextoid=84d3dc1871db8310VgnVCM2000009b0c1e0aRCRD&vgnextchannel=84d3dc1871db8310VgnVCM2000009b0c1e0aRCRD&vgnextfmt=default.

Outcomes from the Comparative Study

Table 2 provides for a comparison of the different characteristics of the eight SEA systems.

Broadly speaking, systems differ with regards to their stage of evolution. This was previously described by other authors, including e.g. Bina (2007), who interpreted SEA under an "evolutionary" perspective, comprising of three stages: (1) an instrument similar to EIA but able to embrace the "strategic" dimension of PPP making in the first stage; (2) the search for improvements of technical (methods and tools) and procedural aspects of SEA (to ensure a more effective integration with planning and decision-making); and (3) the introduction of sustainability issues in the third stage.

Common characteristics of consolidated SEA systems can be described as follows: SEA is a mandatory instrument and there are clearly defined procedures and objectives for SEA in place. Practice tends to adhere to international best practice principles, largely influenced by academic expertise, stimulating institutional and social learning. This was the case particularly in the English and Portuguese systems. Furthermore, unconsolidated systems were found to use non-mandatory SEA, where objectives and procedures were unclear and where practice and resulting learning is limited and rather disperse. Despite certain differences of their SEA systems, this was the case for South Africa, Brazil, Mexico, Mozambique and Angola. The South African system is more advanced than the others, presenting a good level of guidance and academic expertise/communication, and a reasonable number of SEAs produced every year. On the other hand, Mozambique and Angola were only at the initial stages with no advances in terms of legal provisions, objectives for assessment and procedures, guidance and academic expertise.

The Brazilian system occupies an intermediate position between consolidated and unconsolidated systems of SEA, showing some advances in terms of its coordination framework, the responsibility for the preparation of SEA reports and also in terms of the strategic level of application (although the prevailing use of SEA is to large projects). Also, there is a well-established community of academic research, which actively contributes to further developments in the field. In this context, Montaño *et al.* (2013) provided for some evidence of institutional learning through the practice of SEA. However, individual and social learning has only been marginal to date, suggesting a significant shortcoming in the current practice of SEA in the country.

Table 2. Comparison matrix.

Comparison criteria	England	Portugal	Spain	South Africa	Brazil	Mexico	Mozambique	Angola
Legal provisions	✓ Environmental Assessment of Plans and Programmes Regulations 2004, transposition of the EU Directive	✓ Law 232/2007, transposition of the EU Directive	✓ Law N° 9/2006, transposition of the EU Directive	×	× Sao Paulo State Policy on Climate Change	Ø	Ø	Ø
Coordination framework	✓ National, regional or local authorities	✓ Promoting authority, supervised by Portuguese Environmental Agency	✓ At the discretion of Autonomous Communities	⇔ Recommended and supervised by the national environmental authority and regional/local promoter	⇔ Partially defined to the states of Sao Paulo and Minas Gerais	× Without legal prescription; national environmental authority and regional/ local promoter	× Without legal provisions; Ministry of Environmental Coordination and proponent authorities	Ø
Guidelines	✓✓ Large number of guidelines	✓ 3 national guidelines	✓ To the regional level	✓ Guidelines prepared by the environmental agency	Ø Draft to federal government, without approval since 2010	Ø	Ø	Ø
Preparation SEA report	✓ Public authorities, academic groups or private consultants	✓ Public authorities, academic groups or private consultants	✓ Public authorities, or private consultants; prevalence of public authorities	⇔ Without legal provisions; normally private consultants	⇔ Without legal provisions; prepared by private consultants	⇔ Without legal provisions; prepared by public authorities and private consultants	⇔ Without legal provisions; normally private consultants	⇔ Without legal provisions; normally private consultants

Table 2. (*Continued*)

Comparison criteria	England	Portugal	Spain	South Africa	Brazil	Mexico	Mozambique	Angola
Strategic level	✓ Plans and programmes	✓ Plans and programmes	✓ Plans and programmes	⇔ Land-use plans; large projects	⇔ Mainly sectoral programmes; large projects	⇔ Mainly sectoral programmes	× Mainly large projects	× Mainly large projects
Experience	✓✓ Large experience with all types of plans and programmes	✓ Large experience with sectoral and territorial plans and programmes	✓ Large experience in water resources, transport and territorial planning	⇔ Some experience with municipal planning; industrial and port development	× Few cases, mainly tourism and energy programmes and infrastructure	× Few cases, mainly tourism and energy programmes	× Few cases, mainly territorial planning and infrastructure	× Few cases, mainly territorial planning and infrastructure
Research expertise	✓✓ Consolidated research groups with frequent relevant publications	✓ Consolidated research groups with some relevant publications	⇔ Unconsolidated research groups and small number of relevant publications	✓ Consolidated research groups and small number of relevant publications	✓ Consolidated research groups and small number of relevant publications	× Unconsolidated research groups and no relevant publications	Ø	Ø

Note: Ø no advances × very initial stage ⇔ some advanced ✓ advanced ✓✓ far advanced.

Conclusions

Eight SEA systems have been compared, including three consolidated systems from the EU with clear definitions about SEA objectives and procedures. Furthermore, five developing countries were explored that are applying SEA as a non-mandatory instrument and without definitions of a suitable institutional frameworks, objectives and procedures. Our findings reinforce the suggestions made by Malvestio and Montaño (2013) with regards to the connection of identified shortcomings in the Brazilian system and the lack of formal requirements and guidance. The main characteristics of the Brazilian SEA system are connected with an instrument perceived to be suitable for supporting a wide range of decisions. This perception contributes to an overall acceptance of SEA here as a suitable instrument for application at higher levels of decision-making. Although the Brazilian SEA system is currently "flexible" and "adaptive", it is quite clear that the lack of both institutional drivers and a defined procedural framework are hindering substantive improvements on the performance of SEA in Latin America.

Acknowledgement

The authors would like to acknowledge FAPESP (Sao Paulo State's Research Support Foundation) for the financial support provided to this research.

References

Bina, O (2007). A critical review of the dominant lines of argumentation on the need for SEA. *Environmental Impact Assessment Review*, 27(7), 585–606.

Cashmore, M, R Gwilliam, R Morgan, D Cobb and A Bond (2004). The interminable issue of effectiveness: substantive purposes, outcomes and research challenges in the advancement of environmental impact assessment theory. *Impact Assessment and Project Appraisal*, 22(4), 295–310.

Chaker, A, K El-Fadl, L Chamas and B Hatjian (2006). A review of strategic environmental assessment in 12 selected countries. *Environmental Impact Assessment Review*, 26, 15–56.

Dalal-Clayton, B (2013). Turning green the strategic way — the role and potential of strategic environmental assessment in securing a green economy. IIED, http://pubs.iied.org/pdfs/16537IIED.pdf.

Dalal-Clayton, B and B Sadler (2005). *Strategic Environmental Assessment: A Sourcebook and Reference Guide to International Experience*. London: Earthscan.

DEA (2013). *Annual report 2012/13*. Department of Environmental Affairs, Republic of South Africa.

Fischer, TB (2006). SEA and transport planning: towards a generic framework for evaluating practice and developing guidance. *Impact Assessment and Project Appraisal*, 24 (3), 183–197.

Fischer, TB (2007). *Theory and Practice of Strategic Environmental Assessment: Towards a More Systematic Approach*. UK; USA: Earthscan.

Fischer, TB (2012). The quality of local transport plan (LTP) 3 SEA reports in North West England – a comparative review with English spatial and waste management planning related SEA practice. *Town and Country Planning*, 81(6), 281–286.

Fischer, TB and P Gazzola (2006). SEA effectiveness criteria — equally valid in all countries? The case of Italy. *Environmental Impact Assessment Review*, 26, 396–409.

Fischer, TB and V Onyango (2012). SEA related research projects and journal articles: an overview of the past 20 years. *Impact Assessment and Project Appraisal*, 30(4), 253–263.

Gallardo, ALCF and A Bond (2011). Capturing the implications of land use changes in Brazil through environmental assessment: time for a strategic approach? *Environmental Impact Assessment Review*, 31(3), 261–270.

Gazzola, P (2008). What appears to make SEA effective in different planning systems? *Journal of Environmental Assessment Policy and Management*, 10(1), 1–24.

Gómez Orea, D (2007). *Evaluación ambiental estratégica: un instrumento para integrar el medio en la elaboración de planes y programas*. [Strategic environmental assessment: an instrument to the integration of environment in plans and programmes]. Madrid: Ediciones Mundi-Prensa.

Jha-Thakur, U, P Gazzola, TB Fischer, D Peel and S Kidd (2009). SEA effectiveness — the significance of learning. *Impact Assessment and Project Appraisal*, 27(2), 133–144.

Jiliberto, RH (2007). Strategic environmental assessment: the need to transform the environmental assessment paradigms. *Journal of Environmental Assessment Policy and Management*, 9, 1–24.

Jones, C, M Baker, J Carter, S Jay, M Short and C Wood (eds.) (2005). *Strategic Environmental Assessment and Land use Planning: An International Evaluation*. London: Earthscan.

Kirchhoff, D (2006). Capacity building for EIA in Brazil: preliminary considerations and problems to be overcome. *Journal of Environmental Assessment Policy and Management*, 8(1), 1–18.

Lemos, C, TB Fischer and MP Souza (2012). Strategic environmental assessment for tourism planning — extent of application and quality of documentation. *Environmental Impact Assessment Review*, 35, 1–10.

Loayza, F (ed.) (2012). *Strategic Environmental Assessment in the World Bank — Learning from Recent Experiences and Challenges*. Washington DC: The World Bank SEA Community of Practice.

Madrid, CK, GM Hickey and MA Bouchard (2011). Strategic environmental assessment effectiveness and the initiative for the integration of regional infrastructure in South

America (IIRSA): a multiple case review. *Journal of Environmental Assessment Policy and Management*, 13(4), 515–540.

Malvestio, AC and M Montaño (2013). Effectiveness of SEA applied to renewable energy in Brazil. *Journal of Environmental Assessment Policy and Management*, 15(2), 1340007/1–1340007/21.

Marsden, S (1998). Importance of context in measuring the effectiveness of strategic environmental assessment. *Impact Assessment and Project Appraisal*, 16(4), 255–266.

Montaño, M, AC Malvestio and PA Oppermann (2013). Institutional learning by SEA practice in Brazil. *UVP Report*, 27(4+5), 201–206.

Nilsson M, H Wiklund, G Finnveden, DK Jonsson, K Lundberg, S Tyskeng and O Wallgren (2009). Analytical framework and tool kit for SEA follow up. *Environmental Impact Assessment Review*, 29(3), 186–199.

Pellin, A, CC de Lemos, A Tachard, ISD de Oliveira and MP de Souza (2011). Avaliação Ambiental Estratégica no Brasil: considerações a respeito do papel das agências multilaterais de desenvolvimento. [SEA in Brazil: considerations about the role of multilateral development agencies]. *Engenharia Sanitária e Ambiental*, 16(1), 1–10.

Pizella, DG and MP Souza (2012). Brazilian GMO regulation: does it have an environmental approach? *Journal of Environmental Assessment Policy and Management*, 14(2), 1–16.

Retief, F (2007). Effectiveness of strategic environmental assessment (SEA) in South Africa. *Journal of Environmental Assessment, Policy and Management*, 9(1), 83–101.

Retief, F, C Jones and S Jay (2007). The emperor's new clothes — reflections on strategic environmental assessment (SEA) practice in South Africa. *Environmental Impact Assessment Review*, 28, 504–514.

Rossouw, N and K Wiseman (2004). Learning from the implementation of environmental public policy instruments after the first ten years of democracy in South Africa. *Impact Assessment and Project Appraisal*, 22, 131–140.

Runhaar, H and PPJ Driessen (2007). What makes strategic environmental assessment successful environmental assessment? The role of context in the contribution of SEA to decision-making. *Impact Assessment and Project Appraisal*, 25(1), 2–14.

Sánchez, LE and SS Silva-Sánchez (2008). Tiering strategic environmental assessment and project environmental impact assessment in highway planning in São Paulo, Brazil. *Environmental Impact Assessment Review*, 28(7), 515–522.

Schneider, S and CJ Schmitt (1998). O uso do método comparativo nas Ciências Sociais [The use of comparative method in social sciences]. *Cadernos de Sociologia*, 9, 49–87.

Tetlow, MF and M Hanusch (2012). Strategic environmental assessment: the state of the art. *Impact Assessment and Project Appraisal*, 30(1), 15–24.

Therivel, R (1993). Systems of strategic environmental assessment. *Environmental Impact Assessment Review*, 13, 145–168.

Therivel, R and F Walsh (2006). The strategic environmental assessment directive in the UK: 1 year onwards. *Environmental Impact Assessment Review*, 26(7), 663–675.

Therivel, R and TB Fischer (2012). Sustainability appraisal in England. *UVP Report*, 26(1), 16–21.

Wirutskulshai, U, E Sajor and N Coowanitwong (2011). Importance of context in adoption and progress in application of strategic environmental assessment: experience of Thailand. *Environmental Impact Assessment Review*, 31, 352–359.

Zhang, J, P Christensen and L Kørnøv (2013). Review of critical factors for SEA implementation. *Environmental Impact Assessment Review*, 38(7), 515–522.

CHAPTER 14

Environmental Impacts of Civil Emergency Plans and their Exemption from SEA

Steve Swain* and Riki Therivel

*steveeswain@gmail.com

Emergency plans are exempt from undergoing Strategic Environmental Assessment (SEA). However, the measures they include to minimise the effects of disasters can have environmental impacts. The study underlying this paper assessed UK emergency plans for their possible environmental impacts, identified which types of plan would not fulfil the criteria necessary to require SEA and which ones would, but are then exempt. Further consideration of the likely practical involvement of environmental authorities in the plan-making process of regulated processes and in emergency responses provided a view on whether the SEA exemption results in environmental impacts not being minimised. This work has highlighted that the implementation of emergency plans is as important as the plans themselves in terms of providing scope for the protection of the environment for the management of spatially generic disasters. Site-specific disaster management tends to include detailed environmental protection measures through regulatory processes.

Keywords: Strategic environmental assessment; civil emergency; emergency planning; environmental impacts; regulated sites; disaster events.

Introduction

Local, national and global emergencies form part of our everyday lives, if not directly then indirectly in the media. The Pitt Review (Pitt, 2008) into the 2007 UK floods described the floods as the country's largest peacetime emergency since World War II and stated that the "impact of climate change means that the

*Corresponding author.

probability of events of a similar scale happening in future is increasing". Since then, in 2010, the UK Government's Adaptation Sub-Committee included emergency planning as one of its five priority areas for immediate action in preparing for climate change. Furthermore, the UK Government's National Adaptation Programme 2013 also includes an objective to promote and strengthen emergency response and recovery to help build climate change-resilient communities. The possible increase in climatic disasters places even greater importance on the emergency plans that are designed to minimise their effects on society and the environment because they are likely to be implemented more frequently. An aspect of their implementation, which forms the focus of this study, is the degree to which the implementation of these plans can have impacts on the environment and, should this be the case, the mechanisms in place to reduce these impacts, without affecting the efficacy of the emergency response. An example of an emergency response that resulted in a negative environmental impact occurred in 2005 at the Buncefield oil storage and transfer depot, in south-east England. It was largely destroyed by a series of explosions and fire. Overfilling of a tank and the resultant loss of primary containment was followed by losses of both secondary and tertiary containment when a bund-retaining wall around the tank and a system of drains and catchment area also failed. This led to fuel and fire-fighting foam and water contaminating nearby soil and water (Health and Safety Executive, 2011).

Examples of environmental impacts of emergency plans, combined with the increased likelihood of climate change-related disaster events, led to a literature survey to investigate the current state of knowledge about the mechanisms in place to ensure any possible negative effects of these plans are minimised. There appears to be a dearth of available information, which then prompted this study.

The main method by which the environmental impacts of plans are limited in the UK is by them undergoing SEA during their production. However, the SEA Directive (2001/42/EC) exempts plans and programmes "the sole purpose of which is to serve national defence and civil emergency" from undergoing SEA. A similar exemption exists for projects solely serving civil emergency within the Environmental Impact Assessment (EIA) Directive (85/337/EEC as amended). However, the value of applying both SEA and EIA has been shown by various authors, including e.g. Benson (2007), Gore and Fischer (2014) and in this special JEAPM issue Tajima *et al.* (2014) and Azechi and Nishikizawa (2014). So does the SEA exemption mean that risks of environmental damage caused by mitigation measures within emergency plans are not being minimised to the greatest degree possible? This is separate to the consideration of the environmental impacts of the disaster they respond to.

In the UK, the creation of emergency plans to help protect people and the environment from such events is vital and necessary, primarily through the requirements of the Civil Contingencies Act 2004 (CCA) and through site-specific legislation, like the Control of Major Accident Hazards (COMAH) Regulations 1999 (as amended 2005 and 2008).

The aim of the study underlying this paper was to analyse a variety of emergency plans to identify and categorise the mitigation measures included within them and then determine whether these could cause environmental impacts. The various plans were then assessed to see whether they would be subject to SEA if the exemption did not exist. This helped identify which plans are subject to the exemption and which are not because they do not fulfil all the criteria necessary to be subject to SEA.

This article firstly introduces the types of emergencies and emergency plans that are considered relevant to the UK, then discusses the mitigation measures they employ and the possible environmental impacts these could have. The results of the assessment of whether they would be subject to SEA, with and without the exemption are then provided, including a summary of any possible environmental impacts. However, any possible impacts drawn from this desk study would only be theoretically possible. To identify any actual negative effects, they would have to be considered alongside the procedures required to implement them, which can involve real-time decision-making to protect the environment. The relative merits of the Strategic Environmental Assessment (SEA) process before and during emergency plan implementation are, therefore, discussed prior to the conclusion.

Types of Emergency and Emergency Plans

Emergency plans outline what should be done in the short period prior to anticipated events, to prevent their occurrence; during an incident in the response phase, to reduce, control or mitigate the incident's effects and to help communities deal with the immediate effects; and afterwards, in the recovery phase, to rebuild, restore and rehabilitate affected communities (Cabinet Office, 2012). The types of events for which plans should be produced are set out in UK legislation. The primary UK legislation relating to strategic emergency planning is the CCA. The CCA does not override other legislation for site-specific emergencies, which can also require emergency plans to be made. The CCA defines an emergency as:

> "an event or situation that threatens damage to human welfare only if it involves, causes or may cause;

- loss of life,
- human illness or injury,
- homelessness,
- damage to property,
- disruption of a supply of money, food, water, energy or fuel,
- disruption of a system of communication,
- disruption of facilities for transport, or
- disruption of services relating to health, an event or situation that threatens damage to the environment only if it involves, causes or may cause;
- contamination of land, water or air with biological, chemical or radio-active matter, or
- disruption or destruction of plant life or animal life."

Figure 1, taken from the UK Government's National Risk Register of Civil Emergencies from 2010, gives an overview of the relative impacts and likelihoods of civil emergencies in the UK. The risks include natural hazards such as human and animal disease, and the results of severe weather; major accidents; and malicious attacks on people, infrastructure, and computer systems. In 2012, the UK Government produced an updated National Risk Register, which contains two

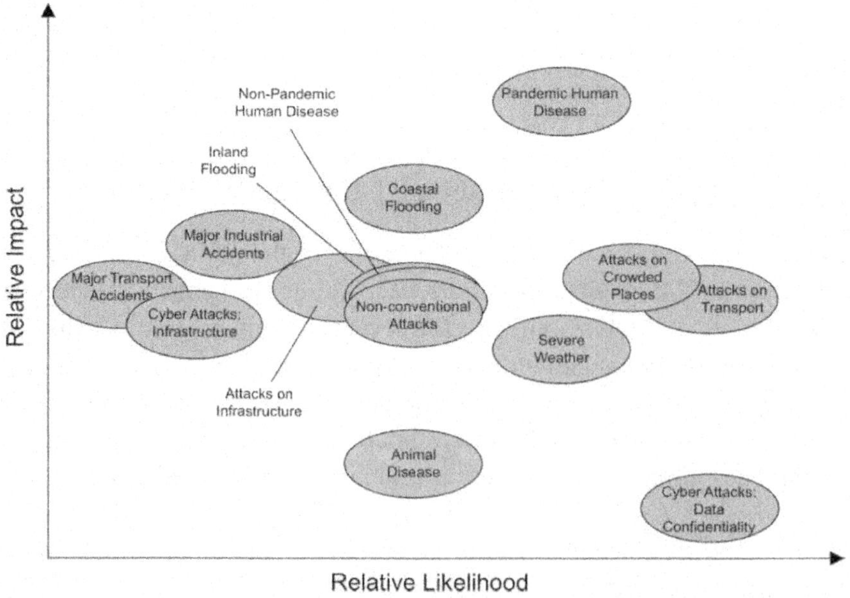

Figure 1. An illustration of the high consequence risks facing the United Kingdom.

matrices that represent these risks in greater detail. There are small changes to the 2010 version identified within the document, caused by updated evidence concerning: the 2009 influenza pandemic; severe space weather; volcanic ash hazard; planning for high impact terrorism; and social disruption and industrial action. However, the 2010 version's diagram is more suitable in providing a general overview of the types of risks identified and quantified.

The CCA places duties on various organisations and agencies involved in emergency planning and responses by identifying them as Category 1 or Category 2 responders. Category 1 responders tend to be the 'blue light' services; the police, fire and rescue, health and maritime and coastguard organisations; and local authorities. However, in Northern Ireland only the police and maritime and coastguard agencies are equivalent to Category 1 responders. From an environmental perspective, the Environment Agency (EA) in England, Natural Resources Wales (NRW) and the Scottish Environmental Protection Agency (SEPA) are also Category 1 responders. Due to the different set up in Northern Ireland the Northern Ireland Environment Agency (NIEA) is not a Category 1 responder, However, they are still expected to act in accordance with the Northern Ireland Contingency Framework, which ties together all relevant organisations.

Category 2 responders provide support to the Category 1 responders and include organisations that transmit and distribute electricity, water and sewerage undertakers, transport companies and the Health and Safety Executive. In Northern Ireland telecommunications providers are the only Category 2 responders.

All responders have responsibilities to assess the risks relating to their functions and plan for, advise on and respond to these risks. Table 1 gives an indication of which plans are required by what legislation and which organisations must develop and implement them.

Other legislation also requires the production of emergency plans but these are not covered in this study. They include the Genetically Modified Organisms (Contained Use) Regulations 2000, Ionising Radiation Regulations 1999, Safety of Sports Grounds Act 1975 and the Fire Safety and Safety of Places of Sport Act 1987, the Merchant Shipping (Oil Pollution Preparedness, Response and Co-operation Convention) Regulations 1998 and Flood and Water Management Act 2010.

Emergency plans specify the roles and responsibilities of the people and organisations involved before, during and after the emergency, highlighting the triggers for various actions to be initiated, including marking of different phases of the response/recovery and references to other relevant plans. They are all also affected by controls on the public accessibility of specific information due to security and/or commercial confidentiality considerations. However, variations exist in the degree to which; they are single or multiple authority focused; they target single or multiple

Table 1. Types of emergency plans, their producers and legislative background.

Legislation	Producer of plan	Plan type
Civil Contingencies Act 2004 (CCA)	Local, regional, national authorities	Area-wide EP
	Category 1 Responders	Business Continuity Plan
Control of Major Accident Hazards (COMAH) Regulations 1999	Local authorities Site operators	"Off-site" EP "On-site" EP
Radiation Emergency Preparedness Public information (REPPIR) Regulations 2001	Local authorities Site operators and companies transporting radioactive substances	"Off-site" EP "On-site" EP
Pipeline Safety Regulations (PSR) 1996	Local authorities through which pipelines pass	"Off-site" EP

EP = Emergency Plan.

hazards; they are general or only consider one element of the operational response; and whether they are specific to the response or recovery phase.

There are also large variations in the information included within them, depending on the legislation requiring them and the resultant purpose of the plan. For example, plans produced for specific known sites, like those required by COMAH and REPPIR, contain detailed analyses of possible hazards and similarly detailed mitigation measures to combat them. Whereas plans produced by authorities in accordance with the CCA tend to be less detailed in providing flexibility for responders given the lesser certainty surrounding the exact locations of the hazards involved.

All of these plans should connect to each other in an emergency planning framework, which includes business continuity plans, and which is regularly tested and reviewed. The mitigation measures contained within them and possible environmental impacts are covered in the following section.

Emergency Mitigation Measures and Their Environmental and Health Impacts

Emergency plans include the following categories of measures to mitigate for the various emergency scenarios:

- Managing the behaviour of professionals and the public during emergencies, for instance through guidance on congregation, evacuation, quarantine and animal culls, with full consideration paid to human rights issues;

- Providing guidance for emergency responders on tangible (structural and non-structural) elements of the response phase, including on organisational logistics, providing temporary shelters and waste transfer measures; and
- Helping the various elements of the recovery phase, for instance by restoring housing following a flood, liaising with affected communities and debriefing to learn lessons for future events.

The following possible environmental impacts of these mitigation measures include hypothetical scenarios and actual impacts that have occurred but do not include measures that are not outlined in emergency plans, like those that are implemented under emergency powers or duties, like flood defence works. Over 40 emergency plans were studied to conduct this assessment.

Population and human health

Civil emergency plans minimise impacts on populations and human health through the physical treatment of people during an event, including the provision of shelter, health care, food, and, if necessary, measures to help the bereaved. Negative environmental impacts could arise through the siting of temporary structures, like rest centres, humanitarian assistance centres, demountable facilities large enough to house 600 fatalities and sites for the temporary storage of medical waste in environmentally sensitive areas. The likelihood of temporary centres having impacts on the environment is reduced by the wide geographical spread, and therefore use, of existing schools, community halls, council buildings (Mid Sussex District Council, 2011; Essex County Council, 2007) and developed areas close to hospitals (Department of Health, 2003).

Material assets

Infrastructure provided in response to a civil emergency could include the repair, replacement, and/or temporary provision of:

- Waste management facilities or processing areas: large quantities of waste could result, either from an event, like an oil spill, or by the response, through the use of fire-fighting water and foam (Health and Safety Executive, 2011; IMPEL, 2011) and the culling of animals after foot-and-mouth or rabies infections.
- Temporary bunded sites might be needed to store waste until it can be dealt with (Southampton City Council, 2008), for example, two sites were set up after the flooding in Carlisle in 2005 (Carlisle City Council, 2005). These could lie within environmentally sensitive areas, as occurred in the response to the *Sea Empress* oil spill in 1996 (Colcomb *et al.*, 1997).

- Accommodation centres would be needed to house and provide logistical support to volunteers involved in the clean-up of shoreline oil pollution.
- Health care facilities and emergency mortuaries: see the previous section on population and human health.
- Dams, bunds and riverbed barriers: these can be pollution prevention measures that can halt the flow of pollution in to groundwater or surface water receptors, as undertaken to reduce the spread of contaminants following the failure of a containment bund at Kolontar in Hungary in 2010 (IMPEL, 2011). However, this could mean the pollution affects other areas, which could also be environmentally sensitive.
- Recovery infrastructure: given that almost any structure or community could be affected by one or more of the events listed in the National Risk Register and that recovery includes rebuilding and regeneration, it is reasonable to consider that almost all infrastructure could need to be replaced or upgraded as part of the recovery from a certain event. This includes transport, energy and communications infrastructure, housing, educational facilities and commercial and industrial buildings. Other elements of the national infrastructure would support affected areas until recovery has restored local capacity for these services. For example, gas storage capacity has increased to compensate for a disruption in supply, both for domestic and industrial electricity generation (Parliamentary Office of Science and Technology, 2004). However, these developments are not specified in pre-existing recovery plans. Instead, they set the platform to enable decisions to be made.

Air and climatic factors

Impacts from civil emergency plans on air quality, like those that involve pyres for culled cattle, are unlikely to be significant. However, the use of in-situ burning to remove oil from the sea would be more significant, depending on the size of the oil release. This is not considered a viable option in UK waters and is not government policy, due to the atmospheric pollution it would cause. Large-scale transport, for instance of waste materials or for the construction of replacement infrastructure, would also increase carbon dioxide and local air pollution emissions. Releases from industrial sites can cause atmospheric pollution. For example, a loss of containment occurred at Heilbronn in Germany in 2010, when an accidental build-up of a cloud of hydrochloric acid within a chemical plant building was released into the atmosphere when the building ventilation was turned on (IMPEL, 2011). However, the release of gaseous pollutants into the atmosphere is advised against if they would be harmful to people or the environment. Instead, the suppression of any explosion or construction of plant strong enough to withstand the pressures is advised to prevent the uncontrolled loss of containment.

Soil and water

Civil emergency plans generally aim to minimise contamination of soil and water through spills or accidents. For instance, some plans to tackle marine oil spills involve the use of booms and absorbents to recover oil from the water, when the weather allows. Dispersants have been seen to have potentially negative effects on marine corals (Gupta, 2010), although they are not necessarily more toxic than the oil they are dispersing (Fuller *et al.*, 2004) and any negative effects would have to be balanced against the improvement caused by oil not reaching the shoreline. Plans for industrial accidents could lead to the construction of bunds, dams or riverbed barriers to contain pollution from accidents, as mentioned in the material assets section previously, or indeed from the foam or fire-fighting water used to combat them. The fire-fighting foam and water can have negative environmental impacts if not contained, as occurred at Buncefield (Health and Safety Executive, 2011) and following fires at the Universal Freight Warehouse in Yorkshire in 1982 (Health and Safety Executive, 2012), at Allied Colloids Ltd in Bradford in 1992 (Health and Safety Executive, 2012) and at a wood recycling plant in Saint-Cyprien in France in 2008 (IMPEL, 2011). Recovered oil, pollution absorbent and affected sand and soil must be treated as hazardous waste. This must go to a licensed waste facility, although it can initially be stored on a bunded or contained paved area. If available waste facilities are overwhelmed, temporary sites would be needed. The initial storage area, transport and temporary sites all have the potential to pollute nearby soils, surface water and groundwater, if the sites are not contained and vehicles not adequately cleaned.

Biodiversity, fauna, flora

These could be affected by several types of civil emergency plans, for instance those that involve temporary site emergency centres, pyres for culled animals or waste storage sites in sensitive areas. Water diversion could affect riverine biodiversity. Coasts and estuaries host a large number of environmentally sensitive areas like Special Protection Areas and Ramsar sites. In the case of an oil spill or shipping accident, emergency facilities, like accommodation centres, or equipment may need to be provided on these sites, or waste and materials may need to be moved through them. The use of heavy vehicles and aggressive clean-up procedures could also damage the existing habitats (Norfolk County Council, 2009). However, the possible environmental impacts of any mitigation measures would need to be weighed up against the effects of not using them. For example, dispersants, as covered in the previous section, are often used in oil spills to avoid the oil reaching sensitive locations.

Cultural heritage and landscape

Effects on cultural heritage sites or landscapes would only really occur if sensitive areas were affected by temporary structures. Churches could be impacted as they are community-focused buildings that could be used in times of need to provide temporary shelter, although most settlements are likely to also have schools or community halls nearby. Otherwise, the only effects that the emergency mitigation measures could have would tend to result from the construction of emergency infrastructure, like dams, pits and bunds that would need to be undertaken without the archaeological assessment usually required for planning permission. Therefore, the development could either unwittingly impact on an unknown heritage site or would need to occur before the normal mitigation measures to preserve the site or artefacts have occurred or before they have been documented.

The possible worse-case scenarios described above are generally likely to be far less significant or undesirable than the overall effects of the emergencies they are designed to mitigate for.

Emergency Plans and the SEA Selection Criteria and Exemption

The SEA Directive (2001/42/EC) exempts plans and programmes "the sole purpose of which is to serve national defence and civil emergency" from undergoing SEA. A similar exemption exists for projects solely serving civil emergency within the Environmental Impact Assessment (EIA) Directive (85/337/EEC as amended).

This exemption does not include plans for the proactive reduction of risks, for example through the construction of infrastructure, like the Thames Barrier, to avoid flooding in London from tidal surges. The European Commission's 2003 guidance on the implementation of SEA is clear that only reactive plans should be thought of as serving civil emergency.

The emergency plans investigated previously for their possible environmental impacts, were assessed to see if they would be subject to the SEA Directive's exemption. This assessment categorised the plans into those that would not be subject to SEA because they do not meet the required criteria and those that meet the criteria but are exempt.

If any of the answers to the following questions were 'no' then the emergency plan would not be subject to SEA. If the answers were all 'yes', they would be subject to SEA, if it were not for the exemption.

(1) Is the plan subject to preparation and/or adoption by a national, regional or local authority OR prepared by an authority for adoption through legislative procedure by Parliament or Government?
(2) Is the plan required by legislative, regulatory or administrative provisions?
(3) Is the plan prepared for agriculture, forestry, fisheries, energy, industry, transport, waste management, water management, telecommunications, tourism, town and country planning or land use AND does it set a framework for future development consent of projects in Annexes I and II to the EIA Directive? OR will it, in view of its likely effect on sites, require an assessment under the Habitats Directive?

If so:

Does it determine more than the use of small areas at a local level and is it not just a minor modification of a plan already subject to SEA? OR is it likely to have a significant effect on the environment?

If not:

Does it set the future framework for future development consent of projects that are not within Annexes I and II to the EIA Directive AND is it likely to have a significant effect on the environment?

The only emergency plans considered to meet the SEA criteria and thus be exempt from SEA were those plans produced by local authorities that tackle oil spill emergencies as required through CCA, "off-site" industrial and pipeline emergencies as required through COMAH and PSR, respectively, and CCA generic recovery plans.

The potential environmental impacts of the mitigation measures in the exempt plans include:

- The potential pollution of the marine environment through the use of dispersants;
- Ecological impacts from the use of heavy vehicles to transport oil-spill waste and from the siting of waste storage and transfer centres close to sensitive environmental areas;
- The pollution of surface water and groundwater through the use of fire-fighting water and foam, should containment be unsuccessful;
- Localised habitat and potentially historical or cultural damage due to the construction of bunds, dams or riverbed barriers, and possible effects of altered flows on sensitive local areas;

- Resource implications of the replacement and regeneration of affected areas and communities in the recovery process, including energy and carbon costs; and
- Ecological damage as a result of having to re-site affected infrastructure to non-brownfield sites.

These impacts would only be equivalent to the effects of the exemption if SEA would definitely be effective in reducing or nullifying them and if no current processes exist to counter them, for example, the involvement of environmental organisations with statutory duties in the emergency response process.

To benefit the environmental performance of emergency responses, environmental assessment would need to either improve the plans or the ways they are implemented. As mentioned earlier the degree of information included in different types of plans varies. The less detailed non-site-specific plans do not have such a definitive impact on the emergency response, instead providing flexibility and scope for real-time decision-making. With this in mind, possible benefits of environmental assessment are discussed below.

Possible Benefits of Environmental Assessment Prior to an Emergency Occurring

The use of environmental assessment procedures during emergency plan production to benefit the environmental performance of emergency responses would be more applicable to site-specific than non-site-specific hazard plans because the hazards and resultant mitigation measures and any subsequent environmental impacts can be assessed more thoroughly and they define, more closely, the likely emergency response.

Site-specific hazards are already strictly regulated through COMAH, REPPIR and PSR, requiring detailed information on hazard characteristics to inform the use of preventative measures to make the risks as low as reasonably practicable (ALARP). COMAH Regulations require that risks to the environment are considered and COMAH sites are regulated by the Competent Authority, which consists of the various health and safety executives and environmental agencies specific to England (EA), Wales (NRW), Scotland (SEPA) and Northern Ireland (NIEA). Changes in primary legislation would be needed before the REPPIR and PSR emergency plans would need to consider environmental impacts.

Considering the possible environmental impacts of the SEA exemption identified previously, the degree to which off-site flow paths for fire water or foam or hazardous substances held on-site are assessed for environmental impacts, including those of the measures used to direct flow down them, would depend on the degree of modelling and information required by the regulator to define the risks and mitigate for them on-site and the degree to which these are considered by the local authority within their contingency plans. If this does not already routinely happen then environmental assessment could provide a mechanism to analyse the effects and options for mitigation.

Off-site plans are required to dovetail with CCA plans for the area concerned. These consider the environmental impact, largely through the involvement of the same local authority emergency planners who would have been informed of environmental sensitivities during the plan creation process and through real-time contact with environmental agencies as Category One responders during any emergency response and recovery. Environmental data would also be accessible to the emergency planner through the local authority town planning department that will have received the data from the environmental agencies as statutory consultees in the Town Planning process. This should ensure that strategic decisions concerning the siting of temporary shelters are well informed as regards environmental impacts.

Two possible advantages of incorporating environmental assessment into the production of emergency plans could be the formalisation of the need for the environment to be considered and certain aspects of the procedures required. From initial contact had with officers regulating these sites, it is likely that no formalisation would be needed for COMAH sites and improvement upon the mitigation measures in the emergency plans is not likely to be gained from the provision of alternatives, because of the necessity to keep risks ALARP and use of risk hierarchies. Any improvements through increased public consultation could well be outweighed by the current reasons for restrictions on information, namely: security; commercial confidentiality; and personal privacy.

Further study is suggested into the processes of site-specific emergency plan creation and further liaison with practitioners about the potential efficacy of these benefits.

Possible Benefits of Environmental Assessment During an Emergency

Any benefits, during an emergency, would only result if a concise form of environmental assessment could be developed that would not significantly slow down the emergency response. The environmental consequences of plan implementation should already be included in the real-time decision-making via the statutory involvement of the environmental agencies as Category One responders under the CCA, with alternative measures in place in Northern Ireland. Well defined roles for the staff of environmental agencies for different types of emergencies exist that enable information on the environmental sensitivities of areas affected or surrounding emergency sites to be provided and technical expertise pertinent to various hazard types, like those associated with regulated industrial sites, to be included. The effectiveness of this mechanism is dependent on the environmental agencies' own business continuity plans but regular exercises help to embed behaviour and thinking, and highlight potential areas to work on.

In terms of the specific possible impacts of the SEA exemption identified previously in this article, the use of oil dispersants is controlled by environmental agencies that license the deposition of substances at sea and can exempt the need for a licence if the purpose is to treat oil. These agencies are the Marine Management Organisation in England and Wales, Marine Scotland and the NIEA. There are conditions to this exemption and approval is still required, ensuring any possible effects on the marine environment are considered and balanced with the possible major benefits to the shoreline of their usage.

Similar real-time decision-making is needed to consider how to collect, transport and treat any waste resulting from an oil slick reaching the shore. Given the unknowable locations of affected shoreline prior to an incident, it may not be practical to have identified the lowest impact environmental sites or routes within an area that could provide for effective waste infrastructure for all possible scenarios. Real-time environmental agency involvement in decision-making concerning siting and routes would be expected to result in any environmental impacts being as minimal as possible.

Similarly, given the unknown nature of disasters it is not possible to have specific recovery plans in place prior to their occurrence. Instead, recovery plans and decisions made for recovery resulting from the frameworks they set out will include environmental issues, such as limiting the impact of waste movement and treatment, and through the siting of new structures and buildings by the involvement of the environmental agencies within this process. Adopted present day frameworks for future development could also provide a guideline for recovery, which will already have been assessed under SEA, for example within local plans.

Conclusions and Future Developments

Emergency plans contain various mitigation measures, of which some could have environmental impacts, but all are exempt from SEA. While the majority of emergency plans do not meet the SEA criteria anyway some do but are exempt from SEA. Their possible impacts could affect the marine environment, surface and ground water quality, localised habitat, historical or cultural features and have resource use and wider carbon and energy use implications, in the absence of other environmental protection mechanisms. However, there are other environmental protection mechanisms, especially the involvement of environmental authorities in the plan creation and implementation processes. Further study is advised to assess the possible benefits to site-specific emergency management of aspects of SEA, within the plan production stage. A real-time form of environmental assessment would be needed to be applied during the implementation of CCA plans, to potentially improve their environmental impact.

References

Azechi, K and S Nishikizawa (2014). The EIA system and hazard management of sediment-related disaster in Japan — a case study in wind farm projects. *Journal of Environmental Assessment Policy and Management*, 16(3), 1450023.

Benson, C (2007). Environmental assessment. In C Benson and J Twigg (eds), *Tools for Mainstreaming Disaster Risk Reduction*. Geneva: Provention Consortium, pp. 79–89.

Cabinet Office (2012). Cabinet Office: National Risk Register of Civil Emergencies, 2012 edition. http://www.gov.uk/government/uploads/system/uploads/attachment_data/file/61929/CO_NationalRiskRegister_2012_acc.pdf [Accessed 28 May 2013].

Carlisle City Council (2005). Carlisle City Council: Report to Executive, Ref TC01/05.

Colcomb *et al.* (1997). Shoreline cleanup and waste disposal issues during the Sea Empress incident, *1997 International Oil Spill Conference Proceedings*, 1917(1), 195–198. http://www.iosc.org/papers_posters/02005.pdf.

Department of Health (2003). Guidelines for smallpox response and management in the post-eradication era (smallpox plan), Appendix 5 specifications for Smallpox Care Centres. http://webarchive.nationalarchives.gov.uk/20110907164128/http://www.dh.gov.uk/prod_consum_dh/groups/dh_digitalassets/@dh/@en/documents/digitalasset/dh_4071515.pdf [Accessed 28 May 2013].

Essex County Council (2007). Essex County Council: Bradwell Off Site Emergency Plan (June 2007) http://www.essex.gov.uk/Your-Council/Local-Government-Essex/Documents/Your_Safety_Bradwell_doc.pdf [Accessed 28 May 2013].

Fuller, C, J Bonner, C Page, A Ernest, T McDonald and S McDonald (2004). Comparative toxicity of oil, dispersant, and oil plus dispersant to several marine species. *Environmental Toxic Chemical*, 23(12), 2941–2949.

Gore, T and T B Fischer (2014). Uncovering the factors that can support and impede post-disaster EIA practice in developing countries: the case of Aceh Province, Indonesia. *Environmental Impact Assessment Review*, 44, 67–75.

Gupta, S (2010). Oil spill dispersant could damage coral populations. *NewScientist*, 3 August 2010. www.newscientist.com/article/dn19260-oil-spill-dispersant-could-damage-coral-populations.html [Accessed 28 May 2013].

Health and Safety Executive (2011). COMAH, Buncefield: why did it happen? http://www.hse.gov.uk/comah/buncefield/buncefield-report.pdf [Accessed 28 May 2013].

Health and Safety Executive (2012). Fire at Universal Freight warehouse, Yorkshire, 13 February 1982 – Accident Summary. http://www.hse.gov.uk/comah/sragtech/caseuniversal82.htm [Accessed 28 May 2013].

Health and Safety Executive (2012). The Fire at Allied Colloids Limited, Low Moor; Bradford, 21 July 1992 – Accident Summary. http://www.hse.gov.uk/comah/sragtech/casealliedcol92.htm [Accessed 28 May 2013].

IMPEL (2011). European Union Network for the Implementation and Enforcement of Environmental Law (IMPEL): IMPEL Seminar in Aix-en-Provence 16 and 17 November 2011, lessons learnt from industrial accidents. http://impel.eu/wp-content/uploads/2012/03/brochure_gb_impel2011.pdf [Accessed 28 May 2013].

Mid Sussex District Council (2011). Mid Sussex District Council Emergency Plan. http://www.midsussex.gov.uk/media/MSDC_MIPBTEdit2012.pdf [Accessed 28 May 2013].

Norfolk County Council (2009). Norfolk Coastal Pollution Emergency Response Plan 2009. http://www.norfolkprepared.gov.uk/download/NCC112544 [Accessed 28 May 2013].

Parliamentary Office of Science and Technology (2004). October 2004, No. 230. The Future of UK Gas Supplies. http://www.parliament.uk/documents/post/postpn230.pdf [Accessed 28 May 2013].

Pitt, M (2008). The Pitt review: learning lessons from the 2007 floods. http://webarchive.nationalarchives.gov.uk/20100807034701/http://archive.cabinetoffice.gov.uk/pittreview/_/media/assets/www.cabinetoffice.gov.uk/flooding_review/pitt_review_full%20pdf.pdf [Accessed 28 May 2013].

Southampton City Council (2008). Southampton City Council — Oil and Chemical Pollution Plan. http://www.southampton.gov.uk/Images/OilChem%20Pollution%20Plan%20V1.2%20June%202013UNRESTRICTED_tcm46-336070.doc [Accessed 28 May 2013].

Tajima, R, T Gore and TB Fischer (2014). Policy integration between environmental assessment and disaster management. *Journal of Environmental Assessment Policy and Management*, 16(3), 1450028.

INDEX

CPSIA information can be obtained at www.ICGtesting.com
Printed in the USA
BVOW09*1238120216

436507BV00002B/5/P

9 781783 268375